Springer Oceanography

The Springer Oceanography series seeks to publish a broad portfolio of scientific books, aiming at researchers, students, and everyone interested in marine sciences. The series includes peer-reviewed monographs, edited volumes, textbooks, and conference proceedings. It covers the entire area of oceanography including, but not limited to, Coastal Sciences, Biological/Chemical/Geological/Physical Oceanography, Paleoceanography, and related subjects.

More information about this series at http://www.springer.com/series/10175

Yonggang Jia · Xiaolei Liu ·
Shaotong Zhang · Hongxian Shan ·
Jiewen Zheng

Wave-Forced Sediment Erosion and Resuspension in the Yellow River Delta

上海交通大學出版社
SHANGHAI JIAO TONG UNIVERSITY PRESS

Springer

Yonggang Jia
Shandong Provincial Key Laboratory of Marine Environment and Geological Engineering, College of Environmental Science and Engineering
Ocean University of China
Qingdao, Shandong, China

Xiaolei Liu
Shandong Provincial Key Laboratory of Marine Environment and Geological Engineering, College of Environmental Science and Engineering
Ocean University of China
Qingdao, Shandong, China

Shaotong Zhang
College of Marine Geosciences
Ocean University of China
Qingdao, Shandong, China

Hongxian Shan
Shandong Provincial Key Laboratory of Marine Environment and Geological Engineering, College of Environmental Science and Engineering
Ocean University of China
Qingdao, Shandong, China

Jiewen Zheng
Pilot National Laboratory for Marine Science and Technology (Qingdao)
Qingdao, Shandong, China

ISSN 2365-7677 ISSN 2365-7685 (electronic)
Springer Oceanography
ISBN 978-981-13-7031-1 ISBN 978-981-13-7032-8 (eBook)
https://doi.org/10.1007/978-981-13-7032-8

Jointly published with Shanghai Jiao Tong University Press, Shanghai, China
The print edition is not for sale in Mainland China. Customers from Mainland China please order the print book from: Shanghai Jiao Tong University Press.

Library of Congress Control Number: 2019934364

This Springer imprint is published by the registered company Springer Nature Singapore Pte Ltd.
The registered company address is: 152 Beach Road, #21-01/04 Gateway East, Singapore 189721, Singapore

Preface

Sediment erosion and resuspension is a process in which soil particles on the surface of the seabed are activated by hydrodynamic force and eventually enter the overlying water body. It is the root cause of beach evolution as well as channel erosion and siltation, an important cause of instability of offshore engineering structures, and an important path of releasing and transporting seabed buried pollutants to the water body. Accurate prediction of sediment erosion and resuspension process has been a challenge for coastal engineers and scholars.

The traditional view is that sediment erosion and resuspension is induced by wave-current combined shear stress, and the quantitative evaluation method is mainly based on the balance between erosion force and erosion resistance. However, in a wave-dominant hydrodynamic environment, the fine-grained seabed is prone to accumulation of pore water pressure and even seabed liquefaction under the action of wave loading. Such characteristic of wave–seabed interaction causes constant changes in the composition, structure, physical, and mechanical properties of the seabed sediments, and thus further affects the erosion and resuspension process of sediments. Due to the complexity of the process and the interdisciplinary nature of the problem, there is not only a lack of systematic understanding of the physical mechanism but also a lack of a highly universal computational model for quantitative prediction of the sediment erosion and resuspension under such wave–seabed interaction, which seriously constrains the development of offshore engineering calculations and numerical simulations.

In this monograph, the Yellow River Delta was selected as the research area. With the funding of multiple projects, we combined in situ long-term observations, field investigations, laboratory simulation experiments, and theoretical calculations to systematically study the physical mechanism by which wave-induced liquefaction of fine-grained seabed affects sediment erosion and resuspension. Based on the understanding of the mechanism, we modified the traditional shear erosion model and proposed a calculation model of liquefaction erosion. Finally, we verified the applicability of the model by comparison with measured data and used the new model to calculate and predict the sediment erosion and resuspension, source ratios, and their influence on long-term beach evolution in the Chengdao sea area of the Yellow River Delta under different sea conditions. The research results of this monograph are of great scientific values to understanding the dynamic changes of the fine-grained seabed sediments in response to waves, analysis and evaluation of the engineering geological environment conditions of the seabed, prediction, prevention, and control of geological disasters in the estuary delta, and understanding the resuspension, long-distance transport, and fate of sediments from rivers into the sea.

This monograph consists of eight chapters. Chapter 1 mainly introduces the research progress in related fields. Chapter 2 mainly introduces the engineering geological environment and sediment properties of the Yellow River Delta as a representative research area, including the formation and evolution, topography and geomorphology, and marine dynamic environment of the modern Yellow River Delta, and the sediment types, distribution, geological strata, grain size composition characteristics, mineral composition characteristics, microstructure characteristics, and physical and mechanical properties. Chapter 3 introduces the current erosion status in the Yellow River Delta, including the current status of erosion in the intertidal zone and on the underwater delta. Chapter 4 introduces the erodibility characteristic of seabed sediments in the typical study area of delta lobes formed in different sedimentary ages in the Yellow River Delta. Chapter 5 introduces the occurrence process of sediment resuspension in the Yellow River Delta. Chapter 6 introduces the wave-induced pore pressure response in relation to sediment erosion and resuspension in the Yellow River Delta. Chapter 7 introduces the physical mechanisms of sediment erosion and resuspension in the Yellow River Delta under the action of waves. Chapter 8 introduces the theoretical prediction of wave-induced sediment erosion and resuspension in the Yellow River Delta.

The contents covered in this book include major research outcomes of numerous research projects sponsored by National Natural Science Foundation of China (40876042; 41072215; 41072316; 41402253; 41427803), Qingdao National Laboratory for Marine Science and Technology (QNLM2016ORP0110). Several postgraduate students from the Shandong Provincial Key Laboratory of Marine Environmental and Geological Engineering participated in relevant research work, including Xiangmei Meng, Zhongnian Yang, Lei Guo, Liping Zhang, Chaoqi Zhu, Mingzheng Wen, Hong Zhang, etc. The editors in charge of this book Dan Li and Xiaofei Li also contributed great effort for the smooth publication of this book. I am grateful to Prof. Dong-Sheng Jeng (School of Engineering, Griffith University, Australia) for his kind support and help during the completion of the monograph. We hereby express our sincere gratitude to them.

The study on wave-induced sediment erosion and resuspension on fine-grained seabed represented by the Yellow River Delta is the intersection of estuarine sediment dynamics, marine soil mechanics, and marine geology. Although we have tried hard to do some tentative research, as limited by the authors' academic realm, level of knowledge, and the complexity of the academic problems studied, there are inevitably some inadequacies in the book. We sincerely request readers to criticize and correct.

Qingdao, China Yonggang Jia
January 2019

Contents

Chapter 1
Introduction

1.1 Sediment Dynamics in the World's Major Estuaries

Estuaries are the transitional zone from terrestrial to marine environment, becoming the most complex area of physical, chemistry, and biological processes on the earth. Therefore, nearshore and estuarial sediments dynamics are of common interests to the oceanographers, harbor and coastal engineers, environmental and fluid mechanics scientists.

To study the clay sediments transport, deposition, resuspension processes in estuary and near coastal areas, US Navy Research Center started the STRATAFORM (Strata Formation on Margins) plan at Eel River in 1994, and employed on-site monitoring and numerical modeling to investigate the clayey sediments transport process. This study is in-depth study of short-term- and long-term geological strata transformation of the estuary area (Charles 1999). European Union in its 4th Development Frame carried out coastal sediments, coastal environment, and engineering study under a project named "MAST III". A group of research institutions including Oxford University, Delft Technology University of Netherland, British HR Wallingford Centre, Wales University, Belgium Hydrology Research Institute, Holland Delft Hydrology Research Centre jointly conducted a research project about the sediment bed dynamics called "COSINUS" (Prediction of Cohesive Sediment transport and bed morphology dynamics in estuaries and coastal zones with integrated Numerical Simulation Models). This project involved the establishment of a sediment exchange equation, on-site sediment bed strength test, indoor deposition column test, sediment bed consolidation model, a sediment bed dynamic model based on Biot consolidation theory, as well as sediments erosion and transport indoor experiments. The project is accomplished with important founding (Dearnaley et al. 2002).

Y. Jia et al., *Wave-Forced Sediment Erosion and Resuspension in the Yellow River Delta*, Springer Oceanography, https://doi.org/10.1007/978-981-13-7032-8_1

British HR Wallingford Hydrology Centre conducted clayey sediments deposition characteristics investigation and modeling. In September 1998, they performed on-site monitoring at Calstock at Tamar River Estuary. Meanwhile, indoor deposition experiment was carried out at Delft University and Oxford University. The deposition column experiment was used to examine the relationship between deposition type and sediment bed density and strength. Sills (1997) discussed the clayey sediments deposition process based on the Oxford University deposition column test, measured sediments density and stress at bottom, middle and surface and their change over time; Ariathurai and Arulanandan (1986) employed on-site electric method tested the sediment density. The Netherlands is well known for its land reclamation from filling out the coastal sea area. They conducted detailed studies on marine land reclamation engineering including sediment consolidation process. Verbeek et al. (1993) investigated the sediment consolidation, transformation trend of the natural deposited silty sediments in the Netherlands; Mimura (1993) observed the erosion and deposition rates of the clayey sediments under the wave action; Merckelbach et al. (2001) tested the consolidation degree and strength of the bottom sediments using indoor experiments; Winterwerp et al. (2001) explored fast settlement of the saturated silt suspension; Kesteren and Kessel (2002) from WLP Delft Hydraulics studied the integration and extension of air traps in clayey sediments; Van (2002) from Rijakswaterstaat Limburg Directorate studied the suspended clay layer at Ems Estuary. Their results indicated that (1) the suspended clayey layer includes fluid and consolidated parts; (2) the key to investigate the sediments density along the vertical column including the bottom suspended clayey layer is the dissipation and mixing of the deposition and non-Laminar current.

1.2 Sediment Erosion and Resuspension

Sediment erosion and resuspension is the initial process of marine sediment dynamics, which is of great engineering, scientific, environmental, or economic significance, especially in the coastal zones, where distribute most of the world's population and cities. The global material cycle, maintenance of coastal construction, release of buried pollutants and the aquaculture which is strongly influenced by the turbidity and nutrition of seawater are all closely related to this process. Therefore, better understanding the magnitude and mechanism of sediment resuspension in the coastal area is quite important.

Sediment erosion and resuspension generally refers to the process of wearing away coastal materials due to the imbalance in the supply and export of material by natural forces and human activities, such as the high winds, waves, currents, tides, trawling, and dredging (Mohan et al. 2011). In coastal areas, this process generates an important redistribution of sediments and has particular indications for regional particulate matter budgets and export to deeper marine environment. Specifically speaking, erosion and resuspension can be divided into different items. Sediment resuspension and seabed erosion are two aspects of the same physical process, "sediment resuspension" is often discussed in the area of sediment dynamics, whereas "seabed erosion" is more mentioned in marine engineering. Sedimentologists are more concerned with how sediments move in the water, while marine engineers are more concerned with the evolution of seabed after the sediment leaves and suspended. Erosion refers to the response of seabed, as the seabed surface is lowed by hydrodynamics. However, lowing of seabed surface is now necessary during the resuspension of sediments in fluffy layers. Moreover, when sediments are resuspended in the interior of the seabed which is an important topic in this book, terminology "resuspension" is more appropriate than "erosion".

The physical mechanism of sediment erosion/resuspension has been mainly attributed to the tidal currents or the waves (Van Raaphorst et al. 1998). Tidal currents erode the bottom sediments by the friction between flow speed and the seabed surface, while waves are assumed to cause resuspension through the wave orbital velocity and resultant wave orbital shear stress. For the conditions with both presence of waves and currents, coupled wave-current shear stress was frequently argued to control sediment resuspension (Brand et al. 2010).

Historically, sediment resuspension mechanisms are either studied using controlled laboratory experiments or field observations and both the approaches have advantages. For example, in situ data is closer to the real law of the nature while the controlled indoor tests are more reliable for establishing quantitative relationship between detected parameters. Hence numerous investigations regarding sediment resuspension have been conducted around the globe, including site-specific instrumented tripod observations and benthic flume experiments, trying to characterize either the site-specific erosive property or spatial and temporal (seasonal) patterns of sediment resuspension within a scope of study area.

The subaqueous modern Yellow River Delta (YRD) is one of the world's most turbid sea, not only at the present river estuary due to the massive sediment discharge, fast deposition and disperse of plume front (Li et al. 2000), but also in the northern abandoned lobes, where distributes large numbers of offshore platforms of the Shengli Oilfield. In fact, the abandoned lobe has exposed to serious coastal erosion since 1976 when the river channel moved southward to the present estuary (Chu et al. 2006). Although parts of the coast are successfully protected from recession due to breakwaters, submarine seafloor are still experiencing severe erosion and an offshore water zone with high turbidity always exist in this area (Fig. 1.1).

With the objective of finding the physical mechanism for the serious erosion and massive resuspension in the modern Yellow River Delta, to finally improve the modeling effect of silty sediment erosion and resuspension, the research works of this book is conducted in the past decades.

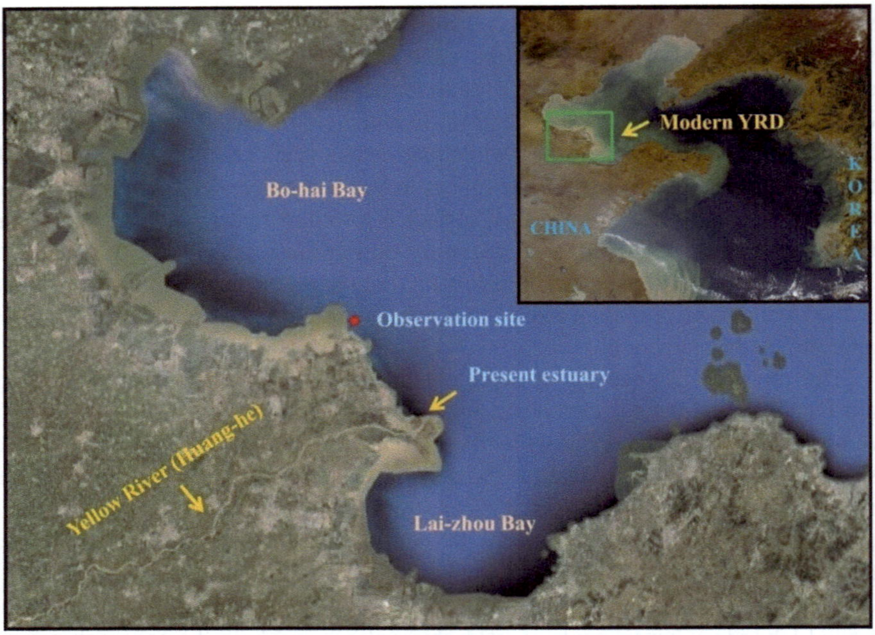

Fig. 1.1 The modern Yellow River Delta (YRD), Chengdao sub-sea, Bo-hai Bay, China

1.3 Research Advance

1.3.1 Research Advance on Sediment Erosion and Resuspension

(1) ***Erosion Power—Bottom Shear Stress of Hydrodynamics***

Sediment resuspension refers to the progress of particles or agglomerates of seabed moving away from the seabed into the overlying water by hydrodynamic forces (Henry and Minier 2014). Its occurrence mechanism is often different in different types (e.g., shape or hydrodynamic conditions) of the gulf or estuary delta. The coast/delta can be divided into tidal-controlled and wave-controlled coast/delta based on different hydrodynamic conditions. Erosion and resuspension of sediment in tidal-controlled coast is mainly controlled by the friction between tidal friction velocity (U_*) and seabed sediment particles (Van Raaphorst et al. 1998):

$$\tau_c = \rho U_*^2 \tag{1.1}$$

When the near-bottom shear stress τ_c is greater than the critical shear stress (τ_{cr}) of the seabed surface sediments, the equilibrium state of the sediments is broken. Sediment will be suspended into the overlying water, and becomes resuspended materials. In wave-controlled coast/delta, many studies have found that the contribution of wave to sediment erosion/resuspension was much greater than that of tidal current (Brand et al. 2010). It is generally considered that the orbital velocity (U_w) is the main driving force.

$$\tau_w = \frac{1}{2}\rho f_w U_w^2 \tag{1.2}$$

where ρ is the density of seawater, f_w is friction coefficient of wave, and U_w is the horizontal component of the maximum wave orbital velocity in a wave period (T). When the turbulence is well developed in the wave boundary layer, the friction coefficient of wave and the roughness of bed(k_b)are related to the radius of major axis of the near-bottom wave orbit (A_b).

Nielsen (1992) suggested that the friction coefficient of wave was estimated by the following relationship:

$$f_w = \exp\left[5.213(k_b/A_b)^{0.194} - 5.977\right] \tag{1.3a}$$

$$A_b = U_w T/2\pi \tag{1.3b}$$

Soulsby (1997) suggested that the friction coefficient of wave was estimated by the following relationship:

$$f_w = 1.39(k_b/A_b)^{-0.52} \tag{1.4a}$$

$$A_b = U_w T \tag{1.4b}$$

Based on the linear wave theory, the near-bottom wave orbital velocity (U_w) is related to wave height (H), wave period (T), and water depth (h).

$$U_w = \frac{\pi H}{T \sinh(kh)} \tag{1.5}$$

$$\omega^2 = gk \tanh(kh) \tag{1.6}$$

where $\omega = 2\pi/T$ is the angular frequency, $k = 2\pi/L$ is the wave number, $L = g\mathrm{T}^2/\pi \tanh(kh)$ is the wave length.

As sediment erosion and resuspension in many sea areas controlled by coupling effect of wave and tidal current, a series of coupled wave-current shear stress models are developed (Soulsby 1997).

$$\tau_{wc} = \sqrt{(\tau_{c,wc} + \tau_w \cos\psi)^2 + \tau_w \sin\psi^2} \tag{1.7}$$

$$\tau_{c,wc} = \tau_c\left\{1 + \left[1.2(\frac{\tau_w}{\tau_w + \tau_c})^{3.2}\right]\right\} \tag{1.8}$$

where τ_{wc} is the coupled near-bottom shear stress under the interaction of waves and currents, $\tau_{c,wc}$ is the flow-induced shear stress enhanced on the basis of pure flow shear stress (τ_c) under the interaction of waves and currents, and ψ is the angle between waves and currents.

(2) ***Erosion Resistance—Critical Entrainment Threshold Stress of Sediments***

Critical shear stress for sediment entrainment (τ_{cr}) is an important parameter to estimate erosion rate or resuspension flux. For sandy sediments, τ_{cr} can be characterized by a dimensionless Shields parameter θ (Fig. 1.2).

For cohesive sediments, many laboratory experiments and field studies have found that τ_{cr} is significantly affected by a series of seabed properties (Aberle et al. 2004) such as density, water content, particle size and biological indicators, and varies with time and space. Making the estimation of the critical erosion state of cohesive sediments quite complicated, even some scholars have questioned whether the critical erosion shear stress really exist? Therefore, most of the critical erosion problems of cohesive sediments depend on laboratory or in situ measurements. The mainstream methods/tools are annular flumes, Cohesive Strength Meter (CSM) (Tolhurst et al. 1999), and remote sensing (GOCI) (Ge et al. 2015), etc.

Even though, there are still some empirical methods trying to calculate the critical shear stress of cohesive sediments (Taki 2000):

$$\tau_{ce} = 0.05 + \beta \left\{ \frac{1}{\pi/6(1 + sW)^{1/3} - 1} \right\}^2 \tag{1.9}$$

where W is the water content, $s = \rho_s/\rho_w - 1$, β is the empirical coefficient.

Considering the cohesion and particle weight of cohesive sediments, Nouwakpo and Huang (2010) proposed that

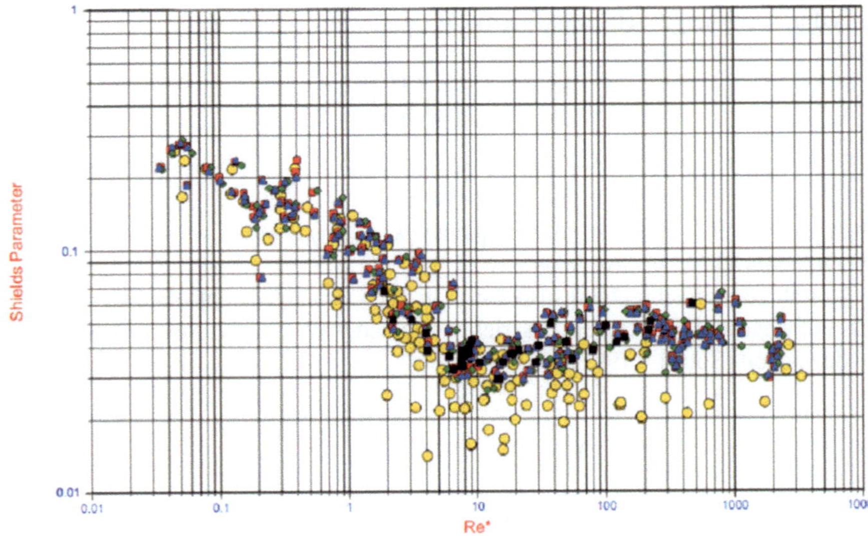

Fig. 1.2 Shield parameter curve

$$\tau_{cr} = \sqrt{\left(\frac{\frac{C_o}{L}D_{psm}}{2}\right)^2 + \left(\frac{\frac{4}{3}\pi\left(\frac{D_{psm}}{2}\right)^3(\rho_s - \rho_w)g}{\frac{1}{2}\pi D_{psm}^2}\right)^2} \quad (1.10)$$

where C_o is the cohesion, D_{psm} is the Sauter average particle size which characterizes the uneven degree of the sediment particles.

(3) ***Erosion and Resuspension Parameter I—Suspended Sediment Concentration***

Suspended sediment concentration (SSC) can be estimated based on the Shear Erosion Theory above. In early stage, some scholars found that there was a good relationship between SSC and wave orbital velocity or wave orbital shear stress (Clarke et al. 1982):

$$C_{ref} = C_a + \beta(u_w - u_{cr}) \quad (1.11)$$

where C_a is the SSC near the observation point, u_w is the wave orbital velocity, u_{cr} is the critical wave orbital velocity, β is the empirical coefficient.

Wright et al. (1988) proposed an parameterization equation based on Shields parameters

$$C_{ref} = A\rho_s\theta_{sf}^{\prime 3} \quad (1.12)$$

where θ'_{sf} is the surface friction Shields parameter defined by Nielsen (1986)

$$\theta'_{sf} = \frac{0.5\rho f_w U_w^2}{(\rho_s - \rho)gD} \quad (1.13)$$

Glenn and Grant (1987) proposed

$$\overline{C_{ref}} = \frac{\rho_s C_{bed}}{T}\int_{t=0}^{t=T}\frac{\gamma_0\psi'(t)}{1+\gamma_0\psi'(t)}dt \; \psi'(t) > 0 \quad (1.14a)$$

$$\psi'(t) = \frac{\tau' - \tau_{cr}}{\tau_{cr}} \quad (1.14b)$$

where γ_0 is the resuspension coefficient, C_{bed} is the volume density of bed, T is the wave period.

Lee et al. (2004) introduced the effect of settling velocity of particles and proposed a prediction relationship applicable for sandy sediments

$$C_{ref} = A\left[\theta_{sf}\frac{u_{*sf}}{\omega_s}\right]^B \quad (1.15)$$

where ω_s is the settling rate, u_{*sf} is the friction velocity, A, B are all empirical coefficients.

(4) ***Erosion and Resuspension Parameter II—Erosion Rate (Resuspension Flux)***

Resuspension flux and erosion rate of sediments mentioned in this paper can be considered as the same concept, but when we compare our results with experts from different fields, we consider the choice of pronouns.

As early as the 1960s–1970s, erosion of riverbed had become a hot issue in the field of hydraulic engineering and was extensively studied. It was generally believed that erosion rate can be expressed as a function of flow-induced near-bottom shear stress and critical shear stress (Dyer 1986)

$$E_{\mathrm{r}} = M_{\mathrm{e}} \left\{ (\tau/\tau_{\mathrm{cr}})^{\Phi} - 1 \right\} \tag{1.16}$$

where E_r is the erosion rate, M_e is the erosion constant, Φ is the bed parameter.

Later on, research results of hydraulic riverbed sediment transport in open channel were adopted by marine engineers to study the transport of seabed sediments. Considering the significant influence of wave action in coastal ocean environment, the coupled wave-current bottom shear stress parameters are introduced to form the calculation method for erosion and resuspension of seabed sediment in wave-current coexistence environment. Lavelle et al. (1984) derived an empirical relationship $E_r = \alpha\tau^{\beta}$ based on field data. Later on, calculation form has been unified (Sanford and Maa 2001)

$$E_{\mathrm{r}} = M_{\mathrm{e}} \left(\tau - \tau_{\mathrm{cr}} \right)^{\Phi} \tag{1.17}$$

Until now, the numerical simulation of sediment transport in the field of sedimentary dynamics still mainly uses this form of erosion model. The calculation method of erosion rate above attributed the sediment resuspension process to the stability of sediment particles on the seabed surface. When the continuous action of horizontal current-induced shear stress (Fig. 1.3a) or reciprocating wave-induced orbital shear stresses (Fig. 1.3b), exceeds the critical shear stress of sediments, the steady state of sediments is broken, erosion, and resuspension occur. This can be defined as waves lift up sediments and then currents transport them (Chen et al. 2004).

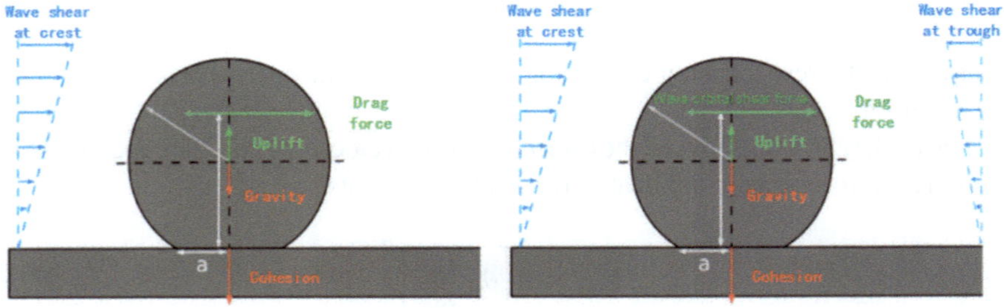

Fig. 1.3 Schematic diagram of mechanism of sediment erosion and resuspension

1.3.2 Research Advance on Wave-Induced Seabed Response

(1) ***Wave-Induced Excess Pore Pressure***

According to the principle of effective stress of saturated soil, the deformation and strength of soil are closely related to the effective stress. Only the stress transmitted through the contact point of the particle can cause the deformation of the soil and affect the strength of the soil. The pore water pressure (u), the effective stress (σ') and the total stress (σ) of the normally consolidated seabed are in equilibrium $\sigma = \sigma' + u$ (Terzaghi 1924). When an external load acts on the soil bed, a part of the additional stress is borne by the pore water, so that the pore pressure rises and the excess pore water pressure (P_{exc}) is generated. Seismic load can induce excess pore water pressure in the soil, and the rapid release of the excess pore pressure will lead to liquefaction, sand boiling, and then cause geological disasters such as earthquakes. In the marine environment, waves act as an additional cyclical dynamic load (Fig. 1.5a), which can also induce excess pore water pressure in seafloor sediments (Ishihara and Towhata 1983). The resulting pore pressure response mode is related to factors such as sediment type, density, initial stress state, and reciprocating stress intensity and frequency. The pore pressure response mechanism of the silty soil seabed under wave action is mainly divided into two types (Jeng 2013): one is the oscillating excess pore water pressure (P_{osc}), also known as the transient excess pore water pressure, the other is the residual excess pore water pressure (P_{res}), also known as the cumulative excess pore water pressure (Fig. 1.4b).

a. Transient (Oscillating) Excess Pore Water Pressure

The transient excess pore water pressure is periodically cyclically reciprocated around the equilibrium position (Fig. 1.4c). This excess pore water pressure is directly related to the transmission of sea surface wave pressure along the depth of the seabed and is therefore unique in the marine environment (Wang 2014). When the crest passes, the oscillating pore water pressure is at the peak, and when the trough passes, the oscillating pore water pressure is at the bottom. Zen and Yamazaki (1990) pointed out that the oscillation amplitude of transient pore water pressure decreases with the increase of seabed depth, which is related to the attenuation of the wave infiltration pressure with depth (Fig. 1.4d). Recent studies have shown that the distribution of transient pore water pressure is not absolute. The experimental results of Wang et al. (2014) show that the amplitude of the oscillating pore water pressure below the antinode of the standing wave decreases with depth, which is in accordance with the above rules. In the certain depth range below the standing wave node, there is no oscillating pore water pressure, but after exceeding a certain depth, the oscillating pore water pressure appears. The experimental results of Zhang (2016) also show that the maximum value of transient pore water pressure appears at a certain depth below the seabed surface, not the surface of the seabed. In essence, the transient pore water pressure response corresponds to the elastic deformation of the seabed soil (Wang et al. 2014), and the residual pore water pressure response corresponds to the plastic deformation of the soil (Sekiguchi et al. 1995). The generation of excess

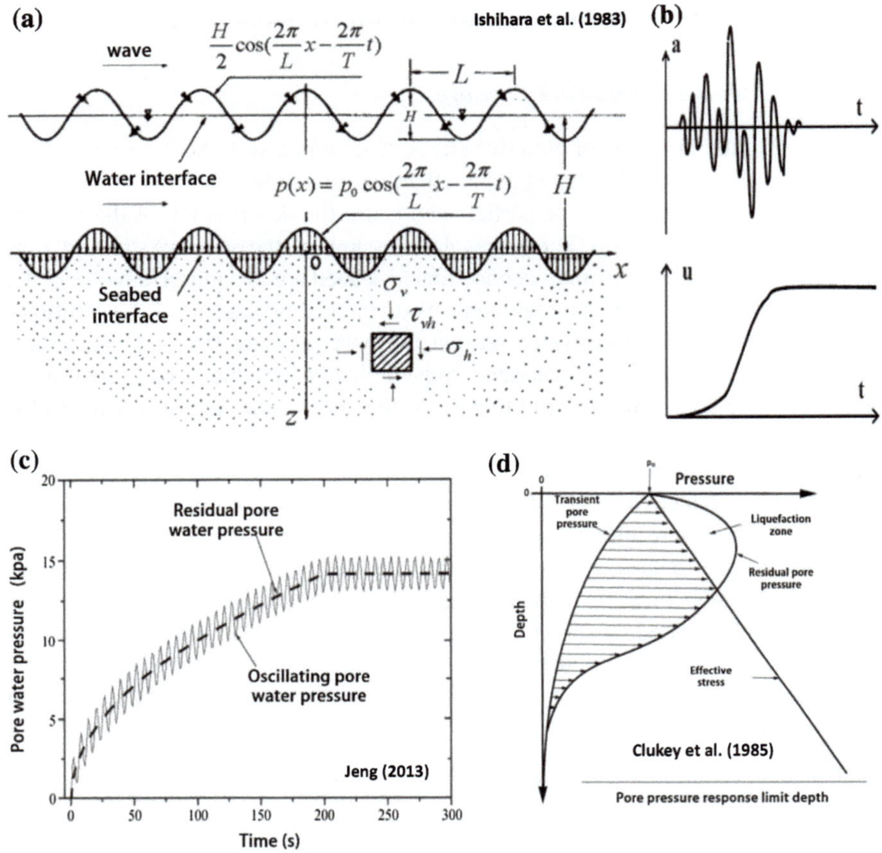

Fig. 1.4 Stress and pore pressure response in the seabed under wave action **a** Wave loading mode **b** Two pore pressure response modes **c** Pore pressure response time history curve **d** Pore pressure response depth profile

pore water pressure originates from the elastic and plastic deformation of the seabed under external loads.

$$\varepsilon = \varepsilon^e + \varepsilon^p \tag{1.18}$$

In the formula, ε^e and ε^p are elastic and plastic body changes, respectively. The elastic body can be recovered instantaneously, corresponding to the transient excess pore water pressure response; the unrecoverable plastic body deformation accumulates under the action of wave load cycle, and the incompressible pore fluid does not reach the drainage, which means that the cumulative pore water pressure increases, corresponding to the plastic body deformation of the soil (Sekiguchi et al. 1995). The centrifuge experiments of Sassa and Sekiguchi (1999) and the flume experiments of

Kirca et al. (2013) and the flume experiments of Wang et al. (2014) show that: both the antinode and the node of progressive wave and standing wave can generate transient and residual two kinds of excess pore water pressure can generate both transient and residual excess pore water pressure. It indicates that the cyclic normal stress or shear stress can simultaneously cause two kinds of pore water pressure responses. This recognition has led to an advancement in the view that initial residual pore water pressure originates from seabed shear stress and shear strain (Yamamoto et al. 1978).

The theoretical calculation of transient pore pressure response usually uses an elastic constitutive model (Yamamoto et al. 1978). The wave-induced pore pressure response of the sandy seabed only shows the characteristics of transient oscillations (Tzang 1992); and in the silty soil seabed with relatively poor permeability, the wave-induced pore pressure response also shows a cumulative rise, that is, residual excess pore water pressure response.

b. Residual (Cumulative) Excess Pore Water Pressure

Residual excess pore water pressure refers to the cumulative rise based on the initial pore water pressure, which tends to increase faster and dissipate relatively slowly. This excess pore water pressure is closely related to wave parameters (such as frequency, wave height, etc.) and the sediment properties. There are two necessary conditions for the generation of residual pore water pressure: (1) the soil skeleton has compressibility, some additional stress will be borne by the incompressible pore water, (2) the sediment permeability is poor, and the pore water can not freely leave the force zone in time under the cyclic extrusion. Silty sediments have a higher compressibility than clay because of their lower permeability than sand, and are relatively prone to residual excess pore water pressure (Clukey et al. 1985).

The generation and development of excess pore water pressure is related to the consolidation properties of the seabed. The Biot equation in the one-dimensional case has the same form as the Terzaghi equation:

$$\frac{\partial p}{\partial t} - c_v \frac{\partial^2 p}{\partial z^2} = f \tag{1.19}$$

where p is the excess pore water pressure, c_v is the consolidation coefficient of soil, z is the seabed depth, and $f = \partial u_g / \partial t$ is the source term, that is, the pore water pressure development mode. Based on the consolidation equation, the pore water pressure development model is added as the source term to calculate the cumulative pore water pressure. The pore water pressure development model is based on the empirical relationship between the number of dynamic load cycles and pore water pressure growth based on the indoor soil unit dynamic test, such as the anti-sinusoidal mode of sand (Seed and Rahman 1978):

$$u_g = \sigma_0' \frac{2}{\pi} \arcsin \frac{N}{N_l}^{1/2\theta} \tag{1.20}$$

In the formula, u_g is the excess pore water pressure, σ_0' is the initial vertical effective stress, MN is the number of dynamic load cycles, and N_l is the number of dynamic load cycles when the liquefaction or pore water pressure stability is no longer rising, and θ is the empirical coefficient.

$$N = \frac{\mathrm{t}}{\mathrm{T}} \quad N_l = \left(\frac{1}{\alpha}\frac{\tau}{\sigma_0'}\right)^{-1/\beta} \tag{1.21}$$

where t is time and T is the wave period, and α and β are empirical parameters related to soil type and relative density. τ is the amplitude of wave-induced seabed shear stress, which can be obtained by the Biot elastic model. When the seabed depth is greater than L/2

$$\tau = p_0 \lambda z \exp(-\lambda z) \tag{1.22}$$

$$P = \frac{\gamma_w H}{\cosh(\lambda h)} \cos(\lambda x - \omega t) = P_0 \cos(\lambda x - \omega t) \tag{1.23}$$

In the formula, γ_w is the seawater bulk density, H is the wave height, the wave number is $\lambda = 2\pi/L$, h is the water depth, and the angular velocity is $\omega = 2\pi/T$.

The linear development model (McDougal et al. 1989) has the following form:

$$u_g = \sigma_0' \frac{N}{N_l} \tag{1.24}$$

In recent years, the newly developed hyperbolic development model using elasto-plastic constitutive relations can simultaneously calculate transient and residual excess pore water pressure (Dunn et al. 2006). The development mode of silty soil is hyperbolic model (Chen et al. 2004). Among them, the exponential hyperbolic development model has the following form:

$$u_g = \frac{\sigma_0' \left(\frac{N}{N_l}\right)^{\mathrm{a}}}{\mathrm{b}\left(\frac{N}{N_l}\right)^{\mathrm{a}} + \mathrm{c}} \tag{1.25}$$

In the formula, a, b, and c are the empirical coefficients. When a = 1, it is the conventional hyperbolic mode.

c. Seabed Liquefaction Discrimination Method

There are two types of liquefaction criteria: the first is the effective stress criterion. The soil liquefies when the wave-induced average effective stress is equal to the initial average vertical effective stress. It also includes one dimension (Zen and Yamazaki, 1990) and three-dimensional effective stress criteria (Tsai and Lee 1995):

$$(\gamma_s - \gamma_w)\mathrm{z} \le \sigma_z' \tag{1.26a}$$

$$\frac{1}{3}\left[(\gamma_s - \gamma_w)(1 + 2K_0)z\right] \leq \left(\sigma'_x + \sigma'_y + \sigma'_z\right) \tag{1.26b}$$

In the formula, the first term represents the initial average vertical effective stress of the soil at the depth z, γ_s is the dry weight of the soil, K_0 is the static pressure coefficient of the soil, and the second term represents the average effective stress produced by wave.

Jeng (2013) pointed out that the effective stress standard does not apply to the liquefaction discrimination of a limited thick seabed. At present, the second criterion based on pore water pressure is adopted, that is, when the residual residual pore water pressure is equal to the initial average vertical effective stress, the soil is liquefied. Also includes one-dimensional and three-dimensional super-pore pressure criteria

$$(\gamma_s - \gamma_w)z \leq P_{res} \tag{1.27a}$$

$$\frac{1}{3}\left[(\gamma_s - \gamma_w)(1 + 2K_0)z\right] \leq P_{res} \tag{1.27b}$$

Tzang and Ou (2006) proposed that when the pore water pressure reached the static pressure of the overlying soil, the soil liquefaction would be

$$(1 - n)(\rho_s - \rho_w)gz \leq P_{res} \tag{1.28}$$

In the formula, n is the porosity of soil, ρ_s is dry density of soil, and ρ_w is density of seawater.

When Wang (2014) proposed liquefaction discrimination, $P_{exc} = P_{res} + P_{osc}$ is used on the right side of the equation because the oscillation pore water pressure and residual pore water pressure tend to exist at the same time. When the two pore water pressures are superposed, the pore pressure is the largest, and the liquefaction is most likely to occur at this time.

In addition, the hydraulics field has also proposed methods for determining soil bed liquefaction (seepage failure) based on hydraulic gradients, such as the Ergun formula (Yang 2003a, b), for non-cohesive soils

$$\frac{\Delta P}{L} = 150\frac{(1-\varepsilon)^2\mu V_s g}{\varepsilon^3 D_p^2} + 1.75\frac{(1-\varepsilon)\rho_w V_s^2 g}{\varepsilon^3 D_p} \tag{1.29}$$

In the formula, ε is porosity, L is bed thickness, μ is fluid dynamic viscosity coefficient, V_s is seepage velocity, D_p is particle size.

For cohesive seabed soils, the cohesion source term needs to be added to the Ergun formula

$$\frac{\Delta P}{L} = 150\frac{(1-\varepsilon)^2\mu V_s}{\varepsilon^3 D_{psm}^2} + 1.75\frac{(1-\varepsilon)\rho_w V_s^2}{\varepsilon^3 D_{psm}} = (\rho_p - \rho_w)g + \frac{C_0}{L} \tag{1.30}$$

In the formula, C_0 is the cohesive force and D_{psm} is the Sauter mean particle size, representing the inhomogeneity of the particles.

1.3.3 Research Advance of Sediment E&R Considering Wave-Seabed Response

(1) ***Physical Mechanism***

The applicability of sediment erosion/resuspension theory, which only considers horizontal shear stress, has been dubious in the wave-dominated hydrodynamic environment. Wave effect on the seabed sediment is not simply provide the horizontal reciprocating shear stress on the surface of the seabed produced by the orbital movement of water particles. Sea level fluctuations due to the passing-by of wave crest and wave trough also exerts vertical cyclic loading on the seabed, causing instantaneous and residual pore pressure responses, which in turn triggers oscillating and cumulative seepage flows in the seabed (Fig. 1.5). Transient seepage flows at the sediment–water interface can be subdivided into instantaneous infiltration flows and upwelling flows (Fig. 1.5a).

Infiltration flows in the surf zone has received widespread attention. Nielsen (1997) found that on the one hand, infiltration current would reduce the thickness of the bottom boundary layer and thus enhance the near-bottom shear stress; on the other hand, the downward drag effect of infiltration flows on the surface sediment would inhibit the sediment erosion process. Which effect dominates the entrainment determined by the balance of the sediment particle specific gravity and the seabed permeability. Based on the clear understanding of mechanism, Nielsen (1997) modified

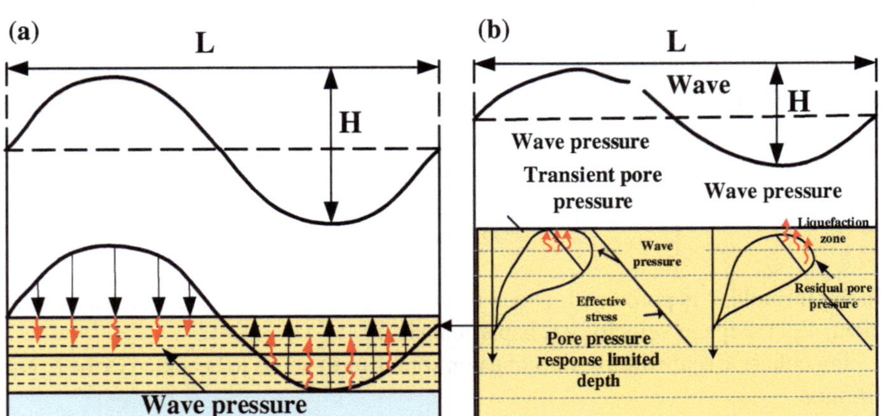

Fig. 1.5 Wave-induced seabed instantaneous (oscillation) and cumulative (one-way) seepage diagram

the traditional Shields parameters and successfully quantified the effect of infiltration flows on erosion and resuspension

Traditional Shields parameter

$$\theta = \frac{u_{*0}^2}{gd_{50}(s-1)} \tag{1.31}$$

Nielsen (1997) modified Shields parameter for considering the effects of downward seepage

$$\theta' = \frac{u_{*0}^2(1-\alpha\omega/\mathrm{u}_{*0})}{gd_{50}(\mathrm{s}-1-\beta\omega/K)} \tag{1.32}$$

where S is the grain specific gravity, K is the permeability coefficient, ω is the angular velocity, α, β are empirical coefficients.

The experimental results of Obhrai et al. (2002) supported the modification of Nielsen (1997) and found that infiltration flows can reduce sediment erosion amount up to 50%. Similarly, Myrhaug et al. (2014) also considered the wave-induced seepage from the perspective of Shield parameters and would not repeat them here.

More researches have been done on the effect of upwelling flows on erosion and resuspension, because not only the upwelling component of transient seepage flows, but also the nonuniform vertical distribution of wave-induced residual excess pore water pressure (P_{res}) in the seabed can generate vertical seepage (Fig. 1.5b). Since the seabed surface is the free drained boundary, seepage direction be vertically upward, as long as residual excess pore water pressure is accumulated large enough in the peak area of the P_{res}, to completely overcome the self-gravity of particles.

For sandy sediments, many studies have shown that the effect of vertical seepage is unimportant (Baldock and Holmes, 1999). Because the coarser and larger weight of sandy sediment particles make it not significantly affected by the slow seepage flows. However, some experiments also found that when the vertical seepage gradient is large enough, it does have a vertical injection effect on the surface sediments and further promotes the sediment erosion and resuspension (Cao and Chiew 2014). That is to say, the effect of the upward seepage flows is not absolute, there is a balance between seepage strength and the grain size of sediment or seabed permeability. It is noteworthy that anthropogenic vertical seepage flows in laboratory experiments are mostly used to simulate groundwater seepage (Smith et al. 2009), of which the intensity is often larger than that induced by the accumulation of pore water pressure, until it has liquefied the seabed. When the vertical seepage gradient makes the degree of liquefaction of the sediment reach 80%, the critical erosion flow speed decays by only 10% (Carstens et al. 1976).

Baldock and Holmes (1999) found that the effect of vertical seepage on the starting of cohesive sediments is not obvious, because of its low permeability. On the one hand, the seepage velocity of pore water is relatively small. On the other hand, the water head pressure applied to the bottom of the bed takes a long time to reach the surface of the seabed. However, some studies have also found that pore water pressure gradient can cause overall slump of riverbed blocks (Fox et al. 2007) under the action of groundwater seepage. Simon and Collison (2001) pointed out that the occurrence of this process compliance with Moore's Kulun guidelines

$$S_r = c' + (\delta - u)\tan\varphi' \tag{1.33}$$

where S_r is the shear strength of the bed, c' is the effective cohesion, $\delta = W\cos\beta$ is the total normal stress, W is the sliding block weight, β is the sliding surface angle, u is the pore water pressure, φ is the effective internal friction angle.

When the vertical seepage is in the limit case, that is, the vertical seepage gradient reaches the overlying effective stress (σ_v) and causes seabed liquefaction (Sumer 2014), its promotion effect on erosion and resuspension will become more significant. As early as the 1960s, some scholars pointed out that the critical erosion shear stress changed with time and was affected by the magnitude and duration of wave loadings (Alishahi and Krone 1964). Mehta et al. (1989) pointed out that the critical shear stress of sediments under unidirectional flow was one order of magnitude larger than the critical value under wave action. When the accumulation of wave-induced pore pressure equals the effective stress of overburden, sediments liquefy and are easily mixed into the overlying water in vertical direction by tidal currents. Indoor flume experiment of Tzang et al. (2009) more directly proved that seabed liquefaction can lead to a 10–20 times increase in suspended sediment concentration, but it has a certain lag effect, and its effect on erosion/resuspension is not significant in sandy sediments, because pore pressure accumulation is not obvious as silty ones.

(2) ***Numerical Simulation***

Considering the remarkable seepage effect in seabed when liquefaction occurs, some scholars have tried to take the influence of seepage effect into consideration in the traditional erosion model from different perspectives. Modify the formula of critical shear stress for erosion is a mainstream idea. By introducing seepage force into the force analysis of the sediments in the seabed boundary layer, Wang et al. (2014) derived the critical shear stress equation of surface sediment under seepage flow, and verified its rationality using an actual calculation example.

$$\tau_e = \frac{\left\{\frac{\pi}{6}d^3\left[(\rho_s - \rho)g - \frac{\Delta P}{\Delta L}\right]\right\}\tan\varphi}{\frac{\pi}{8}d^2(C_d + C_L\tan\varphi)} \tag{1.34}$$

where ΔL is the distance between two depth in seabed, ΔP is the excess pore pressure difference between the two points, τ_e is the critical shear stress, φ is the saturated soil static internal friction angle, d is the sand size (for the median diameter of d_{50}), ρ_s is the sediment particle density, ρ is the density of water, resistance coefficient $C_D = 0.4$, uplift coefficient $C_L = 0.1$.

Cheng et al. (2004) also carried out similar studies and will not be described here. Another representative modification method is Fox et al. (2007) who introduced the seepage velocity parameters into the traditional critical erosion shear stress formula

$$\tau_e = \frac{C_2 q}{(s-1)\varepsilon K} \tag{1.35}$$

where C_2 is the empirical coefficient, s is the ratio of particle to fluid density, q is the percolation rate, ε is the porosity, K is the permeability coefficient.

To consider the influence of wave-induced seabed liquefaction on erosion/resuspension in the Yellow River Delta, we evaluated the decay law of critical entrainment flow velocity (u_{cr}) and critical shear stress (τ_{cr}) of liquefied sediments under waves of different recurrence period through laboratory experiments. It is reported that u_{cr} decays 6–32% and τ_{cr} decays 12–53% under the waves of 5-year recurrence period. Under the waves of 50-year recurrence period, the maximum u_{cr} attenuation is about 46% and the τ_{cr} attenuation is up to 72%. Zhang et al. (2017a) also attempted to parameterize the erosion resistance of liquefied sediments in Hangzhou Bay and found that the erosion rate of liquefied sediments was significantly affected by its yield stress

$$E_r = 0.00027\left(\frac{\tau_b}{\tau_c} - 1\right)e^{-0.00076\tau_y} \tag{1.36}$$

where τ_y is the yield stress, τ_b is the bottom shear stress, τ_c is the critical shear stress.

It is generally accepted that the effect of wave-induced seabed liquefaction on erosion and resuspension is to reduce the erosion resistance of surface sediments. Moreover, some scholars also argued that the vertical seepage flows caused by residual pore pressure will also cause vertical internal transport of fine particles. Clarke et al. (1982) was the first to propose that waves would cause the movement of pore water in surface sediments and carry fine sediments into the water to suspend. The effect of wave-induced pore water movement on sediment resuspension was initially proposed. Maa et al. (1998) proposed that the surface of cohesive seabed would form floating mud under the action of waves and the thickness of the floating mud layer would also be affected by the transient water level fluctuation. The floating mud can easily be suspended in the case of unidirectional flow, which has a significant impact on cohesive sediment transport process. Nichols et al. (1994) found that significant upwelling and overflow of pore fluid and sediment flow occurred after liquefaction. However, due to its simulation method of artificial hydraulic gradient applied in a tank, the true effect of wave-induced pore pressure accumulation cannot be completely simulated effectively (Clukey et al. 1985). Tzang (1998) suggested that wave-induced pore pressure response would promote the movement of pore water, resulting in "internal sediment suspension" in the seabed. Sterpi (2003) designed an flume experiment to study the grain size of the internal eroded sediments that carried by the vertical seepage erosion, and established a rough estimation of the erosion amount. Fine-grained sediments on the intertidal seabed of the Yellow River Delta after the storm surges and speculated that the source was the fine-grained material (5–8 φ) which was "pumped" from the interior of the tidal flat to the surface of the seabed during the dissipation of wave-induced residual excess pore pressure. According to the results of flume experiment, we found that the seabed liquefaction would further increase the amount of resuspension after wave orbital shear had suspended all it could suspend, and the source was the “pumping” transport of fine particulate material vertically by the seepage within the seabed. We found that the “pumping” phenomenon in the original (normally consolidated) seabed was not significant but was evident in the rapid consolidation phase of the newly deposited seabed, and proposed the “pumping” mechanism of fine-grained sediments inside the seabed after liquefaction.

Moreover, some scholars carried out parameterization works of liquefaction affecting erosion and resuspension through in situ observations. Due to the complexity of the site environment, these works do not make a detailed distinction between what mechanism of seabed liquefaction affects the erosion and resuspension. Instead, the overall effect of liquefaction is taken into account in traditional erosion models, to more accurately simulate sediment erosion/resuspension processes and thus better reproduce suspended sediment concentrations. At present, this is the forefront research thinking of this scientific problem in the field of mathematical simulation. Now the typical work of several international counterparts will be introduced.

Rodriguez and Mehta (2000) found that the erosion constant (resuspension coefficient) is the function of wave height

$$\begin{aligned} M_{\mathrm{e}} &\propto (\mathrm{H}-\mathrm{H}_0)^3 \quad \text{when}\,\mathrm{H} > \mathrm{H}_0 \\ &= 0 \qquad\qquad\quad \text{when}\,\mathrm{H} < \mathrm{H}_0 \end{aligned} \tag{1.37}$$

Foda and Huang (2001) put forward a similar empirical relationship

$$M_{\mathrm{e}} \propto \mathrm{H}^3 \tag{1.38}$$

where H is the wave height, H_0 is the critical wave height. As erosion constants (M_{e}) can reflect the erosion resistance of the seabed. The parameterization equations of Foda-Huang and Rodriguez-Mehta were actually attempting to reflect the influence of waves on the erosion resistance of sediments into the erosion constants.

However, the physical meaning of using wave height parameters to reflect the influence of waves on the sediment erosion is not clear. More directly, Lambrechts et al. (2010) attempted to demonstrate the effect of liquefaction on erosion and resuspension using wave-induced pore pressure parameters and successfully introduced it into traditional erosion models:

$$E_1 = M_1\ (\tau - \tau_{cr})^{\Phi} \tag{1.39}$$

where $E_2 = M_2 H_{\mathrm{s}}^3 \mathrm{F}(\omega, \mathrm{H})$ is the traditional erosion model, $E_2 = M_2 H_{\mathrm{s}}^3 \mathrm{F}(\omega, \mathrm{H})$ is the newly added liquefied resuspension sources, M_2 is the resuspension coefficient due to seabed liquefaction, H_{s} is the wave height, power index 3 is obtained by the regression of a large number of field data of effective wave height and corresponding SSC.

1.3.4 Research Advance of Sediment E&R in the Modern Yellow River Delta

As one of the most active hydrodynamic forces in the Yellow River Delta, the wave can remodel coasts. Further, wave-induced shear stress is also the significant factor that influences sediment resuspension in the inner continental shelf; besides, the suspended sediments are sources of longshore and offshore transportation (Wright et al. 1988). The field observations and satellite images indicate that the nearshore wave action is the major driving force for the deltaic erosion since the distribution of wave-induced bottom shear stress is spatially coinciding well with the deltaic erosion off the abandoned Shenxiangou-Diaokou delta lobe. Both the wave energy releasing and induced high bottom shear stress in the shallow area result in significant local sediment resuspension, which produces a highly turbid zone off the abandoned delta lobe (Wang et al. 2006). Waves vary significantly with seasons in the Yellow River mouth. Southern winds are prevailing in summer and have a very small impact on sediments and coasts. However, strong northerly winds in winter can suspend sediments and seafloor stability (Wang et al. 2006). The strong waves and wind-driven currents in the winter seasons are highly effective in resuspending the previously deposited sediments (Wright et al. 1988), and abundant evidence of several different types of post-depositional slope failure and subaqueous mass-movement phenomena was acquired (Prior et al. 1986).

The freshwater and sediments are transported into sea via estuary areas; and therefore marine dynamic conditions and sediment transport are affected by river runoff and sediment discharge. The waves in this area are mostly driven by the winds in the Bohai Sea and thus wave characteristics have distinctly seasonal variability associated closely with the activities of monsoon. In the summer season, accompanied by flood discharge from river, waves are induced by the southern winds from the land with short fetches, and thus have little impact on the river-laden sediment transport and coastal geomorphology (Wang et al. 2006). In the winter season, the cyclones from Siberia usually produce strong northeastern winds with long fetches in the Bohai Sea, causing strong waves and wind-driven currents in the coastal areas of the Yellow River delta, and having significant impacts on resuspension of the previously deposited sediments, as well as on the seasonal variation of the subaqueous slope. Results from Wang et al. (2007) show shear front in nearshore region stops sediments from transported and make them stay in shallow waters, therefore resuspension is very important in transport sediments in winter.

Simulation, in which the effect of resuspension is considered, shows sediments from Yellow River have a migration trend to the offshore area in Bohai Sea. However, the parameters in this simulation are not from observation. With hydrological and sedimentological investigations carried out off the Yellow River mouth and the nearby Bohai Sea during August–September 2007, Qiao et al. (2010) concluded that resuspension of surface sediments by the tidal current is very common in estuaries and coastal environments. However, tidally induced resuspension may play a less important role in suspended sediment distribution due to low tidal range (<1.5 m) and weak tidal currents (<1.0 m/s) near the Yellow River mouth area. Based on the data on the current velocity, water temperature, salinity, turbidity and concentration of suspended sediment collected in November 2006, Yang et al. (2011) found the abandoned Diaokou Yellow River mouth and the present Yellow River mouth, and the third one in the bottom layer around the abandoned Qingshuigou Yellow River mouth have highest concentration of suspended sediment. Besides, the intensity of sediment transport in winter is much stronger than in summer due to the powerful effect of winter storms although the river water and sediment discharges to the sea were much greater in summer. Without consideration of sediment resuspension and its effect, it is difficult to understand sediment transport in the present Yellow River delta.

Many questions are waiting to be answered about sediment resuspension in the Yellow River delta, such as determining criterion, process, and its role in remodeling of seafloor and sediment properties. The long-term dynamic change of environment and its influence on subsequent resuspension process is also involved. All aforementioned questions touch upon dynamic deposition process (e.g., transport and accumulation of sediments), sediment response (e.g., liquefaction, thixotropy, resuspension, transport, and deposition) and changes in composition, structure, physical, and mechanical properties. However, little is known about aforementioned questions and wave-induced sediment resuspension.

1.4 Outline of the Book

In summary, many scholars have proposed different physical mechanisms for seabed liquefaction influencing sediment erosion and resuspension. In the field of mathematical prediction methods, some scholars modify the critical state of sediment erosion, such as correction of Shields parameters or critical shear stress (τ_{cr}); some scholars also improve the model from the perspective of modifying erosion rate (E_r) or erosion coefficient (M_e). Due to the complexity and interdisciplinary of the problem, no universal liquefaction erosion model has been developed, to be popular as the traditional power law erosion shear model. Therefore, the research of this book attempts to discuss the physical mechanism of the erosion and resuspension caused by liquefied seabed. Based on the innovative understanding of the mechanism, the liquefaction index (pore pressure parameters) is incorporated into the traditional shear erosion calculation model, and the liquefaction model is built to optimize the reproducing of silty sediment erosion process, and then finally promotes the prediction model of beach evolution in the Yellow River Delta.

References

Aberle J, Nikora V, Walters R (2004) Effects of bed material properties on cohesive sediment erosion. Mar Geol 207(1):83–93
Alishahi MR, Krone RB (1964) Suspension of cohesive sediment by wind-generated waves. University of California, Hydraulic Engineering Laboratory, Wave Research Projects, Institute of Engineering Research
Ariathurai R, Arulanandan K (1986) An electrical method to measure in-situ sediment densities. In: Estuarine cohesive sediment dynamics. Springer, New York, NY, pp 206–218
Baldock TE, Holmes P (1999) Seepage effects on sediment transport by waves and currents. In Coastal Engineering pp 3601–3614
Brand A, Lacy JR, Hsu K et al (2010) Wind-enhanced resuspension in the shallow waters of South San Francisco Bay: mechanisms and potential implications for cohesive sediment transport. JGR: Oceans 115(C11)
Cao D, Chiew YM (2014) Suction effects on sediment transport in closed-conduit flows. J Hydraul Eng 140(5):04014008
Carstens T, Brebner A, Kamphuis JW (1976) Seabed mobility under vertical pressure gradients. Proceeding of behaviour of offshore structures (Boss' 76) 423–438
Charles A (1999) Overview of its design and synthesis of its results. Mar Geol 15(4):3–12
Chen GX, Liu XZ (2004) Study on dynamic pore water pressure in silty clay interbedded with fine sand of Nanjing. Chinese J Geotech Eng 01:79–82
Chen SL, Zhang GA, Yang SL et al (2004) Temporal and spatial changes of suspended sediment concentration and resuspension in the Yangtze River Estuary and its adjacent waters. Acta Geogr Sin 59(2):260–266
Cheng YZ, Jiang CB, Pan Y et al (2004) Effect of wave-induced seepage force on incipient sediment motion. Advanc Water Sci 59(2):260–266
Chu ZX, Sun XG, Zhai SK et al (2006) Changing pattern of accretion/erosion of the modern Yellow River (Huanghe) subaerial delta, China: based on remote sensing images. Mar Geol 227(1):13–30
Clarke TL, Lesht B, Young RA et al (1982) Sediment resuspension by surface-wave action: an examination of possible mechanisms. Mar Geol 49(1–2):43–59
Clukey EC, Kulhawy FH, Liu PLF et al (1985) The impact of wave loads and pore-water pressure generation on initiation of sediment transport. Geo-Mar Lett 5(3):177–183
Dearnaley MP, Roberts W, Jones S et al (2002) Measurement and modeling of the properties of cohesive sediment deposits. In: Winterwerp JC, Kranenburg C (eds) Fine sediment dynamics in the marine environment. Elsevier Science, Armsterdam, pp 57–73
Dunn SL, Vun PL, Chan AHC et al (2006) Numerical modeling of wave-induced liquefaction around pipelines. J Waterway Port Coast Ocean Eng 132(4):276–288
Dyer KR (1986) Coastal and estuarine sediment dynamics. Wiley, Chichester
Foda MA, Huang CM (2001) Fluidization model for cross-shore sediment transport by water waves. Ocean Wave Measure Anal 1356–1365
Fox GA, Wilson GV, Simon A et al (2007) Measuring streambank erosion due to ground water seepage: correlation to bank pore water pressure, precipitation and stream stage. Earth Surf Proc Land 32(10):1558–1573
Ge J, Shen F, Guo W et al (2015) Estimation of critical shear stress for erosion in the Changjiang estuary: a synergy research of observation, GOCI sensing and modeling. J Geophys Res Oceans 120(12):8439–8465
Glenn SM, Grant WD (1987) A suspended sediment stratification correction for combined wave and current flows. J Geophys Res 92:8244–8264
Henry C, Minier JP (2014) Progress in particle resuspension from rough surfaces by turbulent flows. Prog Energy Combust Sci 45:1–53
Ishihara K, Towhata I (1983) Sand response to cyclic rotation of principal stress directions as induced by wave loads. Soils Found 23(4):11–26
Jeng DS (2013) Porous models for wave-seabed interactions. Springer

Kesteren W, Kessel T (2002) Gas bubble nucleation and growth in cohesive sediments. Proc Marine Sci 5:329–341
Kirca VO, Sumer BM, Fredsøe J (2013) Residual liquefaction of seabed under standing waves. J Waterway Port Coastal Ocean Eng 139(6):489–501
Lambrechts J, Humphrey C, McKinna L et al (2010) Importance of wave-induced bed liquefaction in the fine sediment budget of Cleveland Bay, Great Barrier Reef. Estuar Coast Shelf Sci 89(2):154–162
Lavelle JW, Mofjeld HO, Baker ET (1984) An in situ erosion rate for a fine-grained marine sediment. J Geophys Res Oceans 89(C4):6543–6552
Lee GH, Dade WB, Friedrichs CT et al (2004) Examination of reference concentration under waves and currents on the inner shelf. J Geophys Res Oceans 109(C2)
Li G, Zhuang K, Wei H (2000) Sedimentation in the Yellow River delta. Part III. Seabed erosion and diapirism in the abandoned subaqueous delta lobe. Marine Geol 168(1):129–144
Li Y, Mehta AJ (1997) Mud fluidization by water waves. Cohes Sediments. Wiley, New York, pp 341–351
Maa JPY, Sanford L, Halka JP (1998) Sediment resuspension characteristics in Baltimore harbor. Maryland Marine Geol 146(1):137–145
Mehta AJ, Hayter EJ, Parker WR et al (1989) Cohesive sediment transport. I: process description. J Hydraul Eng 115(8):1076–1093
McDougal WG, Tsai YT, Liu PL et al (1989) Wave-induced pore water pressure accumulation in marine soils. J Offshore Mech Arct Eng 111(1):1–11
Merckelbach LM, Sills GC, Kranenburg C (2001) Laboratory experiments on consolidation and strength of bottom mud. In: McAnally WH, Mehta AJ (eds) Coastal and estuarine fine sediment processes. Proceedings in Marine Science 3 Amsterdam: Elsevier, pp 201–213
Mimura N (1993) Rates of erosion and deposition of cohesive sediments under wave action. In: Mehta AJ (Ed) Nearshore and estuarine cohesive sediment transport, Washington: American Geophysical Union, Coastal and Estuarine Studies, pp 247–264
Mohan R, Palermo M, Costello M et al (2011) Development of an in-Lieu Fee for sediment remediation in the Elizabeth river. In: Proceedings of the western Dredging association technical conference and Texas A&M University Dredging Seminar, Nashville, Tennessee, June 5–8
Myrhaug D, Holmedal LE, Ong MC (2014) A rational approach to seepage flow effects on bottom friction beneath random waves. Appl Ocean Res 47:322–328
Nichols RJ, Sparks RSJ, Wilson CJN (1994) Experimental studies of the fluidization of layered sediments and the formation of fluid escape structures. Sedimentology 41(2):233–253
Nielsen P (1992) Coastal bottom boundary layers and sediment transport. World Scientif Advanc Series Ocean Eng 4:324
Nielsen P (1997) Coastal groundwater dynamics Proc Coast Dynam '97, Plymouth. ASCE, pp 546–555
Nielsen P (1986) Suspended sediment concentrations under waves. Coastal Eng 10:23–31
Nouwakpo SK, Huang CH (2010) Pore water effects on soil erodibility and its implication in ephemeral gully erosion modeling. In: Joint federal interagency sedimentation and hydrologic modeling
Obhrai C, Nielsen P, Vincent CE (2002) Influence of infiltration on suspended sediment under waves. Coast Eng 45(2):111–123
Prior DB, Yang ZS, Bornhold BD et al (1986) Active slope failure, sediment collapse, and silt flows on the modern subaqueous Huanghe (Yellow River) delta. Geo-Mar Lett 6(2):85–95
Qiao SQ, Shi XF, Zhu AM et al (2010) Distribution and transport of suspended sediments off the Yellow River (Huanghe) mouth and the nearby Bohai Sea. Estuar Coast Shelf Sci 86:337–344
Rodriguez HN, Mehta AJ (2000) Longshore transport of fine-grained sediment. Cont Shelf Res 20(12):1419–1432
Sanford LP, Maa JPY (2001) A unified erosion formulation for fine sediments. Mar Geol 179(1):9–23

Sassa S, Sekiguchi H (1999) Wave-induced liquefaction of beds of sand in a centrifuge. Geotechnique 49(5):621–638
Seed HB, Rahman MS (1978) Wave-induced pore pressure in relation to ocean floor stability of cohesionless soils. Mar Georesour Geotechnol 3(2):123–150
Sekiguchi H, Kita K, Okamoto O (1995) Response of poro-elastoplastic beds to standing waves. Soils Found 35(3):31–42
Sills GC (1997) Consolidation of cohesive sediments in settling columns. In: Burt N, Parker R, Watts J (eds) Cohesive sediments: 4th Nearshore and Estuarine Cohesive Sediment Transport Conference INTERCOH'94. Wiley, Chichester, pp 107–120
Simon A, Collison AJ (2001) Pore-water pressure effects on the detachment of cohesive streambeds: seepage forces and matric suction. Earth Surf Proc Land 26(13):1421–1442
Smith AJ, Herne DE, Turner JV (2009) Wave effects on submarine groundwater seepage measurement. Adv Water Resour 32(6):820–833
Soulsby RL (1997) Dynamics of marine sands. Thomas Telford, London, p 429
Sterpi D (2003) Effects of the erosion and transport of fine particles due to seepage flow. Int J Geomech 3(1):111–122
Sumer BM (2014) Liquefaction around marine structures: (With CD-ROM). World scientific
Taki K (2000) Critical shear stress for cohesive sediment transport. Proc Marine Sci 3:53–61
Terzaghi K (1924) The theory of hydrodynamic stresses and its geotechnical applications. Proc Int Cong Appl Mech, Delft, pp 288–294
Tolhurst TJ, Black KS, Shayler SA et al (1999) Measuring the in situ erosion shear stress of intertidal sediments with the Cohesive Strength Meter (CSM). Estuar Coast Shelf Sci 49(2):281–294
Tsai CP, Lee TL (1995) Standing wave induced pore pressures in a porous seabed. Ocean Eng 22(6):505–517
Tzang SY, Ou SH, Hsu TW (2009) Laboratory flume studies on monochromatic wave-fine sandy bed interactions part 2. Sediment Suspens Coast Eng 56(3):230–243
Tzang SY, Ou SH (2006) Laboratory flume studies on monochromatic wave-fine sandy bed interactions: Part 1. Soil Fluidizat Coast Eng 53(11):965–982
Tzang SY (1992) Water wave-induced soil fluidization in a cohesionless fine-grained seabed, Ph.D. Dissertation. University of California, Berkeley, USA
Tzang SY (1998) Unfluidized soil responses of a silty seabed to monochromatic waves. Coast Eng 35(4):283–301
Van Ledden M (2002) A process-based sand-mud model. In: Winterterp JC, Kranenburg C (Eds) Fine sediment dynamics in the marine environment, vol 5. Elsevier Science, pp 577–594
Van Raaphorst W, Malschaert H, Van Haren H (1998) Tidal resuspension and deposition of particulate matter in the Oyster grounds, North Sea. J Mar Res 56(1):257–291
Verbeek H, Kuijper C, Cornelisse JM et al (1993) Deposition of graded natural muds in The Netherlands. Nearshore Estuarine Cohesive Sediment Trans: 185–204
Wang H, Liu HJ, Zhang MS (2014) Pore pressure response of seabed in standing waves and its mechanism. Coast Eng 91:213–219
Wang HJ, Yang ZS, Li GX et al (2006) Wave climate modeling on the abandoned Huanghe (Yellow River) delta lobe and related deltaic erosion. J Coast Res 22(4):906–918
Wang HJ, Yang ZS, Li YH et al (2007) Dispersal pattern of suspended sediment in the shear frontal zone off the Huanghe (Yellow River) mouth. Cont Shelf Res 27:854–871
Winterwerp JC, Uittenbogaard RE, De Kok JM (2001) Rapid siltation from saturated mud suspensions. In: McAnally WH, Mehta AJ (eds) Coastal and Estuarine Fine Sediment Processes. Elsevier Science, Armsterdam, pp 125–146
Wright LD, Wiseman WJ, Bornhold BD et al (1988) Marine dispersal and deposition of Yellow River silts by gravity-driven underflows. Nature 332(6165):629–632
Yamamoto T, Koning HL, Sellmeijer H et al (1978) On the response of a poro-elastic bed to water waves. J Fluid Mech 87(1):193–206
Yang SL (2003) Introduction to coastal environment and geomorphological process
Yang WC (2003) Handbook of fluidization and fluid-particle systems. CRC Press

Yang ZS, Ji YJ, Bi NS et al (2011) Sediment transport off the Huanghe (Yellow River) delta and in the adjacent Bohai Sea in winter and seasonal comparision. Estuar Coast Shelf Sci 93:173–181
Zen K, Yamazaki H (1990) Oscillatory pore pressure and liquefaction in seabed induced by ocean waves. Soils Found 30(4):147–161
Zhang M, Yu G, La Rovere A et al (2017) Erodibility of fluidized cohesive sediments in unidirectional open flows. Ocean Eng 130:523–530
Zhang Q (2016) Analysis of wave-induced silty seabed dynamic response. Ocean University of China

Chapter 2
Geo-Marine Environment and Sediment Properties of the Modern Yellow River Delta

2.1 Overview

The Yellow River delivers averaged 870 million tons of sediments into the Bohai Sea annually, 70% of which deposited near the estuary and formed a new continent. Approximately, 20% of the fine-grained sediments are transported into the further sea. Only a very small amount of them flows into the North Yellow Sea through the Bohai Strait. The transportation of Yellow River sediments is controlled by natural ocean dynamics. At the interaction zone between river and ocean dynamics, sediment dynamics at the Yellow River estuary are not only influenced by river dynamics, e.g., fresh water, sediment discharge, river diversions, but also greatly influenced by ocean dynamics, e.g., ocean waves, ocean currents, tides, and storm surge, etc. The Yellow River Delta forms, develops, and evolves under such a unique environment.

Deltas and estuaries are quite different in the world, because of the diversity in geographic latitude and natural environment. Among the deltas, the landform genesis and evolution process of the Yellow River Delta are very similar to that of the Mississippi River Delta. They both are the accumulated complex of the sub-deltas formed at different periods and both push forward to the sea with a tongue-shaped pattern. However, there are also significant differences between the hackly fan-shaped Yellow River Delta and the bird foot-shaped Mississippi River Delta. Moreover, it takes seven years on average for the Yellow River to form a tongue-shaped sub-delta; but it takes 100 years for the Mississippi River to form a similar sub-delta. This indicates that the Yellow River Delta has a unique geological environment that differs from any other deltas in the world.

Numerous studies have been done on the marine and geological environment characteristics, e.g., water and sediment discharges of the Yellow River, coastline evolutions, topography and geomorphology, types and distribution of bottom sediments, and the ocean dynamic environment. All these contents play a significantly important role in the investigations on the dynamic response of sediments and the geological hazards in the coastal zones of the Yellow River Delta. Therefore, these basic information about the Yellow River Delta are introduced in this chapter.

Y. Jia et al., *Wave-Forced Sediment Erosion and Resuspension in the Yellow River Delta*, Springer Oceanography, https://doi.org/10.1007/978-981-13-7032-8_2

2.2 Formation and Evolution of the Modern Yellow River Delta

2.2.1 Geographical Range

In June 1855, the dyke breach of the Yellow River at Tongwaxiang, Lankao County, Henan Province, shapes out of the modern Yellow River Delta in the north of Shandong Province, China. The onshore part of the Modern Yellow River generally refers to the vertex Ninghai, Kenli County, south to Zhimaigou, west to Tuhai River (Fig. 2.1). Till now, 154 years has passed for the Yellow River to flow through the Modern Yellow River Delta. Except for the years from 1938–1946 when the Yellow River took a different route and poured into the Yellow Sea, it has been 140 years that the Yellow River takes the current route through Modern Yellow River Delta. By 1984, the delta area is about 5, 212 km^2. According to remote sensing images in 2001, the area of the Modern Yellow River Delta is 5, 682 km^2.

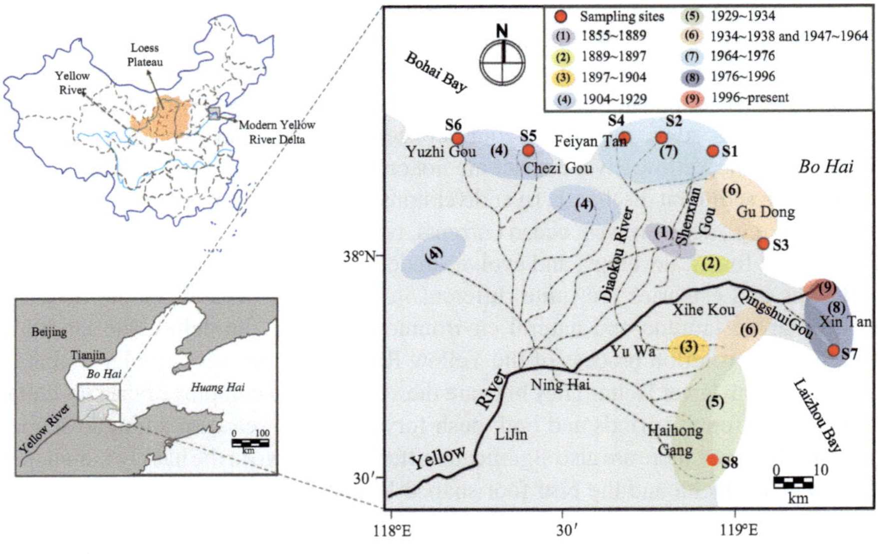

Fig. 2.1 Geographical range of the modern Yellow River Delta and historical migration of deltaic channels (modified after Xue 1994 and Fan et al. 2006). Note: (1) 1855–1889; (2) 1889–1897; (3) 1897–1904; (4) 1904–1929; (5) 1929–1934; (6) 1934–1938 and 1947–1964; (7) 1964–1976; (8) 1976–present

2.2.2 Historical Sediment Discharge

Sediment discharge rate can be regarded as the difference between the erosion amount on the whole basin and the deposition amount on the entire catchment basis. It depends on the sediment quantity of every erosion source, as well as the deposition amount of every sediment sink. The mainstream of the Yellow River is 5460 km in length and about $75,0000\,\text{km}^2$ in total catchment area. It flows through nine provinces and autonomous regions of China and empties into the Bohai Sea. Owing to the large basin area and the diverse climates, together with the intervention of human activities, the annual sediment discharge rate of the Yellow River varies greatly.

Lijin Hydrological Station locates at the top of the Yellow River Delta, about 100 km upstream of the estuary. The distance ensures that the station is not impacted by tides. This is a routine hydrological observation station set up by the Yellow River Conservancy Commission. According to the statistical data from this station, during 1951–1980, the averaged annual sediment concentration and sediment discharge rate of the Yellow River were 26.1 kg/m^3 and 1.069 billion tons, respectively. The averaged annual sediment discharge was more than that of the largest river in the world, the Amazon River, and was approximately 2.3 and 3 times of the Yangtze River and Mississippi River, respectively. It is reported that the long-term averaged annual maximum and minimum sediment concentration of the Yellow River at the downstream river are 222 and 11.3 kg/m^3, respectively; the maximum value is about 20 times of the minimum one, which is extremely variable spatially and temporally.

The sediment load (total load and averaged annual load) of different periods was summed up in Table 2.1. The sediment flux into the sea through the Yellow River drops sharply in 1996 and decreases apparently after that. In 1997, there are 226 days of zero flow at the Lijin Hydrological Station, which is the longest period of zero flow according to historical records. Correspondingly, water discharge in 1997 is the lowest in historical records. The Yellow River annual sediment flux into the sea varies greatly, and the sediment discharge rate demonstrates a continuous decline during the latest decade.

Owing to the influence of the southeast monsoon climate, the distribution of precipitation in most areas of the Yellow River Basin is also highly variable. As a result, the sediment discharge in the Yellow River and the discharge rate at the river mouth show significant variance within a year. During the flood season from July to October, the sediment discharge approximately accounts for 86% of the total annual sediment discharge. However, during the dry season from December to February, the sediment discharge only accounts for 8.2% of the total annual sediment discharge.

Sediment discharge of the Yellow River estuary shows great multiple year variations in addition to seasonal variations. The averaged sediment concentration in the flood season could be several times to ten times higher than that in non-flood season. The variation of the sediment discharge of the Yellow River estuary could be related to human activities in the Yellow River Basin, the global climate change, as well as the dynamic change of sedimentary areas.

Table 2.1 Incoming sediment of different periods at Lijin Hydrological Station (Jiang et al. 2004; Chen et al. 2009)

Period/year	July 1953–December 1963	January 1964–May 1976	June 1976–June 1996	1996–2001	2001–2007
Total incoming sediment/1 × 10^9t	129.74	134.60	128.18	10.53	11.69
Average annual incoming sediment/1 × 10^9t	12.48	10.94	6.41	1.76	1.95

2.2.3 *Channel Changes*

Owing to natural and human factors, the Yellow River has frequently changed its course and breached since it seized the Daqing River and began to empty into the Bohai Sea in 1855. According to historical records, the Yellow River diverts and breaches for more than 50 times since 1855, including larger diversions up to 10 times, as shown in Table 2.2. The distributary channel system of diversion occurred at the vertex of the delta is usually called flow path of this period. During the active period of a flow path, distributary channel breaches and shifts every year (Li and Mehta 1997), whereas only thereinafter the vertex or near the estuary of the delta. The accumulative sediments formation between two diversions near the vertex is called delta lobes, including onshore delta and subaqueous delta. The modern Yellow River Delta consists of multiple lobe bodies. There are eight lobe bodies in total in the modern Yellow River Delta since 1855, as shown in Fig. 2.1. The seven lobes formed before 1976, when Yellow River emptied into the Bohai Sea, run a time span of 112 years in total, and the average active period of one lobe is 16 years, which is considerately shorter compared with the 115–175 years of active period for lobes at the Mississippi Delta (Wells and Coleman 1987).

2.2.4 *Coastline Change*

The evolution of the Yellow River coast is shaped by the river flow, sediment conditions, and marine dynamics. The flowing channel brings sediments and coastline proceeds, while erosion takes place once a channel is abandoned. Owing to natural and human factors, the Yellow River has frequently changed its course and breached since it began to empty into the Bohai Sea in 1855. Once a channel is abandoned due to change course of the river, a large amount of sediment supply is cut off, and the

Table 2.2 History records of lower reaches of Yellow River from 1855 to 2008 (Huang et al. 2005)

Numbers	Diversion time	Diversion locations	Estuarine location	Time interval to next diversion	Duration of water line each stage	Cumulative duration of water line/a	Remarks
1	July 1855	Tongwaxiang	Mulizui in Xiaoshen temple, downstream of Tiemenguan of Lijin	33 years and 9 months	18 years and 11 months	19	Breach at Tongwaxiang
2	March 1889	Hanjiayuan	Maosituo (east of Jianlin presently)	8 years and 2 months	5 years and 10 months	25	Crevasse diversions
3	May 1897	Lingzizhuang	Southeast of Siwangkou	7 years and 1 month	5 years and 9 months	30.5	
4	July 1904	Yanwo	Laoguazui	22 years	17 years and 6 months	48	
5	June 1926	Balizhuang	Shazitou and the old course of Tiemenguan	3 years and 2 months	2 years and 11 months	51	
6	August 1929	Jijiazhuang	Nanwanghe, Songchunronggou, and Qingtuozi	5 years	4 years	55	
7	August 1934	No.1 dam in Helong	Laoshenxiangou, Tianshuigou, and Songchunronggou	18 years and 11 months	9 years and 2 months	64	
8-1	July 1953	Xiaokouzi	Shenxiangou	10 years and 6 months	10 years and 6 months	74.5	Artificial cutoff and branches merged, and forcibly diversion
8-2	1960 (branch), 1962 (river piracy)	Right bank 1 km upward No.4 Pile	Laoshenxiangou				
9	January 1964	Luojiawuzi	Between Diaokou and Walagou	12 years and 4 months	12 years and 4 months	87	Artificial diversion via dike burst
10-1	July 1976	Xihekou	Qingshuigou	20 years	20 years	107	Artificial diversion via cut flow
10-2	July 1996	Section of Qing 8	Under the azimuth angle 81.5° of Qing 8				

abandoned delta lobes are therefore suffering from serious erosion under the marine dynamic forces. The coastline retreats landwards (Ding and Dong 1988; Geng et al. 1988; Hang and Wang 1991; Ren and Zhou 1994; Chen et al. 2004; Li et al. 2004), which brings about serious threat and great economic loss on the construction of onshore and offshore oilfields and comprehensive management and exploitation of the Yellow River Delta.

Ding and Dong (1988) studied coastal line retreat by investigating the sediments characteristics around the estuarine area of the abandoned flow path of the Shenxian-gou. They found that at the first 3 years after river channel abandonment, the rate of coastline retreat landwards was about 0.93 km per year (Si and Zhang 1985). Based on the interpretation and comparison of 23 Landsat images at high tidal periods for the high tide coastal lines at 10 estuaries along the Yellow River Delta, Huang and Fan (2004) analyzed the coastline changes in the delta area since the last river course shift in 1976. The results indicated that, (1) the Yellow River Delta was currently in an erosive retreat stage overall and the length of eroded coastal line segments was twice as long as that of progressive ones, (2) human activities played an increasingly important role in the evolution of the delta, and the artificial coastline grew fast, which shortens the total coastline in length and set the coastline straight.

Yin et al. (2004) studied erosion and deposition rates and developmental evolution of the modern Yellow River Delta coast in different periods based on comparisons of nautical charts, topographic maps, satellite images and aerial photos of different times as well as field surveys and tidal flat profiles. The results of the coastline changes of the modern Yellow River Delta and Laizhou Bay were shown in Fig. 2.2. As shown in Fig. 2.2, based on coastline changes from 1855 to 2001, the estuarine delta can be divided into three periods of 1855–1934, 1934–1976, and 1976–2001. At the first period, the coastline was around Hekou and Yuwa in 1855, reached the Tiao River estuary and Taoer River estuary in 1929, and deposited southeast for 4–5 km in 1934. At the south of Songchunronggou, the silting-up extent decreased gradually from north to south. However, there was a minor erosive retreat around the Guangli estuary. At the second period, from September 1934 to July 1953, the Yellow River silted up around the Dawen River Town to the Xiaodao River areas. From July 1953 to January 1964, the Yellow River silted up to Shenxiangou Channel and this sedimentation zone was actually from the Yellow River Sea Port to Gudong Oil Field. From January 1964 to May 1976, the Yellow River changed its course to Diaokou Channel and this sedimentation zone was around the Feiyan Beach and the Yellow River Sea Port. During the 42 years of second period, the Yellow River silted up the fastest between the Tiao River estuary and the Yellow River Sea Port, and the coastline pushed 20 km toward the north. However, the coastline only pushed forward about 5 km between the Dawen River Town and the Yongfeng River estuary. The coastline changed little at other places. At the third period, the Yellow River silted up the fastest between Gudong Oil Field and Dawen River Town, and the coastline pushed about 40 km, whereas, the coast between the Tiao River estuary and the Yellow River Sea Port turned to strong erosion zone and receded for 5–7 km. Other coastlines changed only a little.

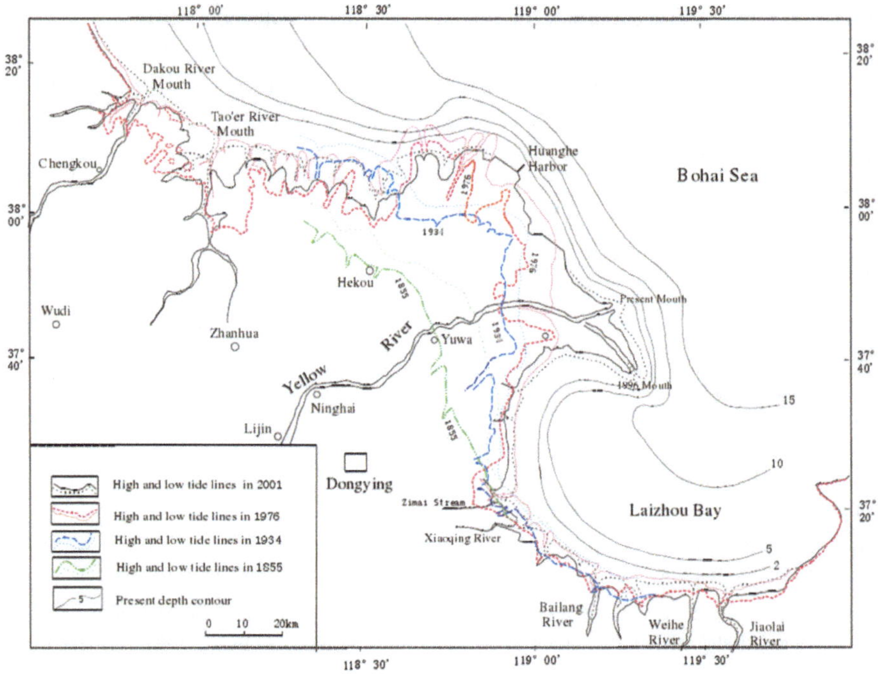

Fig. 2.2 Shoreline changes of the modern Yellow River Delta and Laizhou Bay (1855– 2001) (Yin et al. 2004)

Based on the interpretation of years of remote sensing data, Zhao (2004) identified the low tidal coastlines. His results include the entire coastlines of the Yellow River Delta in 1976, 1981, 1986 and 2001, partial coastal lines in 1996 and 1998 for Qingshuigou area. The research indicated that the main trend of the Yellow River Delta coastline since 1934 was siltation and progression, and the northern coastline has been always retreat since the Yellow River course shift in 1976. However, the coastline near the Qingshuigou River shows a progressive trend from 1976, and it goes forward the farthest in 1996.

Table 2.3 shows the interpretation based on the remote sensing data. As shown in Table 2.3, the north coastline at Diaokou has been in an erosion state and the erosion rate decreases gradually. The coastline of the Qingshuigou River is in a siltation state before 1996, with the siltation rate slowing down gradually, but it turns to be in an erosion state after 1996. The coastline of the Qingbacha River shows siltation from 1996 to 1998 and soon after it turns to be erosive, which might be related to the sharply decreased river flow and sediment discharge rate during these years. In brief, the general trend of the Yellow River Delta coastline since 1976 is as follows: the coastline near Qingshuigou estuary is progressive evidently while the north coastline retreated significantly due to erosion.

Table 2.3 Coastline changes in different regions in different periods (Zhao 2004)

Years	1976–1981	1981–1986	1986–2001	1986–1996	1996–1998	1998–2001
Distances of retreat and progress of the coastline in Diaokou	– 3	– 2.6	– 3.8	–	–	–
Rates of retreat and progress of the coastline in Diaokou	– 0.6	– 0.52	– 0.25	–	–	–
Distances of retreat and progress of the coastline in Qingshuigou	13	12	–	7	– 1	– 2.3
Rates of retreat and progress of the coastline in Qingshuigou	2.6	2.4	–	0.7	– 0.5	– 0.75
Distances of retreat and progress of the coastline in Qingbacha	–	–	–	–	6.8	– 2.1
Rates of retreat and progress of the coastline in Qingbacha	–	–	–	–	3.4	– 0.7

Note positive numbers denote siltation, and negative numbers denote erosion backward. The units of the distances and rates of retreat and progress of the coastline are km and km/a, respectively

2.3 Topography and Geomorphology of the Modern Yellow River Delta

2.3.1 Topography

The modern Yellow River Delta includes the onshore part and subaqueous part formed by the incoming sediments carried by river water, which are called, respectively, delta plain (the part above the low tidal line) and subaqueous delta (the sediments body under the low tidal line). The delta plain stretches to the Bohai Sea in a fan-shape, with the terrain of high-lying at its west-south and central portion, low-lying at its east-north part and two sides, with a gentle slope along the east-north direction. The subaqueous delta is the expansion of the onshore delta and its outer fringe expands to the isobaths line of 15–20 m from the high tidal line. The subaqueous delta, with an area of about 3000 km^2, surrounds the delta plain like a semi-ring belt. Topographically, the subaqueous delta of modern Yellow River Delta can be divided into four units: intertidal zone, steep slope zone, micro-convex submarine bed, and flat submarine bed. The intertidal zone locates at the region from the coastline to 2 m isobath curve, and it is usually passed through by major subaqueous channels and spits. The steep slope zone locates at the region from 2 m isobaths curve to 12 m isobath curve and is usually characterized by numerous irregular micromorphology. The slope angle of steep slope zone increases from 0.2 to 0.3° at the bottom to

0.4–0.5° at the leading edge of the delta. The micro-convex submarine bed located at the region of 12–18 m isobaths and sloped to the submarine with regular slope angle less than 0.1. The plat submarine bed locates at the region deeper than 20 m isobath and its slope doesn't change obviously (Prior et al. 1986).

Based on the morphology and sedimentation environment, Yellow River subaqueous delta can also be divided into pro-delta, delta front, and delta lateral area (also be called mud bay). The delta front can be further divided into estuary sandbar and distal sandbar, according to the sedimentation environment (Cheng and Xue 1997). The pro-delta locates from 17 m isobath to 18 m isobath and its submarine topography is flat. The pro-delta slopes gently, with the slope angle being only 0.01–0.02° and the width reaching 13 km. Out of the pro-delta is shelf regions of the Bohai Sea and the slope of this region is more gentle. The delta front locates at the region from the middle low tidal line out of the estuary to 12–13 m isobaths and can be further divided into delta front plat and delta front slope. The delta lateral area locates at the region from the lateral low water line out of the estuary to 12–13 m isobaths, and it is the accumulation zone of the clayey sediments. The topography of modern Yellow River Delta is shown in Fig. 2.3.

2.3.2 Geomorphology

Referencing "1:1000000 China topographic mapping standards" and geomorphology chapter in "National coastal zones and marine resources survey standards", Zang (1996) classified the Yellow River Delta into different dynamical geomorphologic types by following the multi-class classification rule of morphogenesis and based on the dynamic changes of the scope, intensity, and shoreline caused by ocean dynamic forces. The detailed result of classification is listed in Table 2.4.

According to the sediments and morphology survey conducted during 1983–1988 about distributing feature of subaqueous topography, scouring and siltation status of subaqueous nearshore and tidal flat, and formation time of the sub-deltas, the subaqueous delta of the Yellow River is generally divided into three dynamic morphology subzones, sedimentation zone, erosion zone, and dynamic-balance zone, which are shown in Fig. 2.4. The corresponding coast types of the three subzones are, respectively, siltation coast, erosion coast, and steady coast.

In Fig. 2.4, West Zone refers to the sea area in the west of Tiaohe mouth, and the Yellow River has flown out of West Zone for more than 60 years. As the serious erosion stage passed by, the subaqueous topography and dynamic condition reached a steady state, with coastline becoming stable and the terrain becoming gentle. The sediments type shows a zonal distribution pattern, parallel to the coastline. It has been in a dynamic-balance status with weak siltation and weak erosion. The morphology pattern of this zone includes tidal flat, subaqueous nearshore, small subaqueous delta, and delta plain. Middle Zone refers to the sea area between Tiaohe estuary and Shenxiangou. Since the Yellow River changed its course in 1976, this zone has experienced strong remake by the sea water dynamic forces. The coastline is

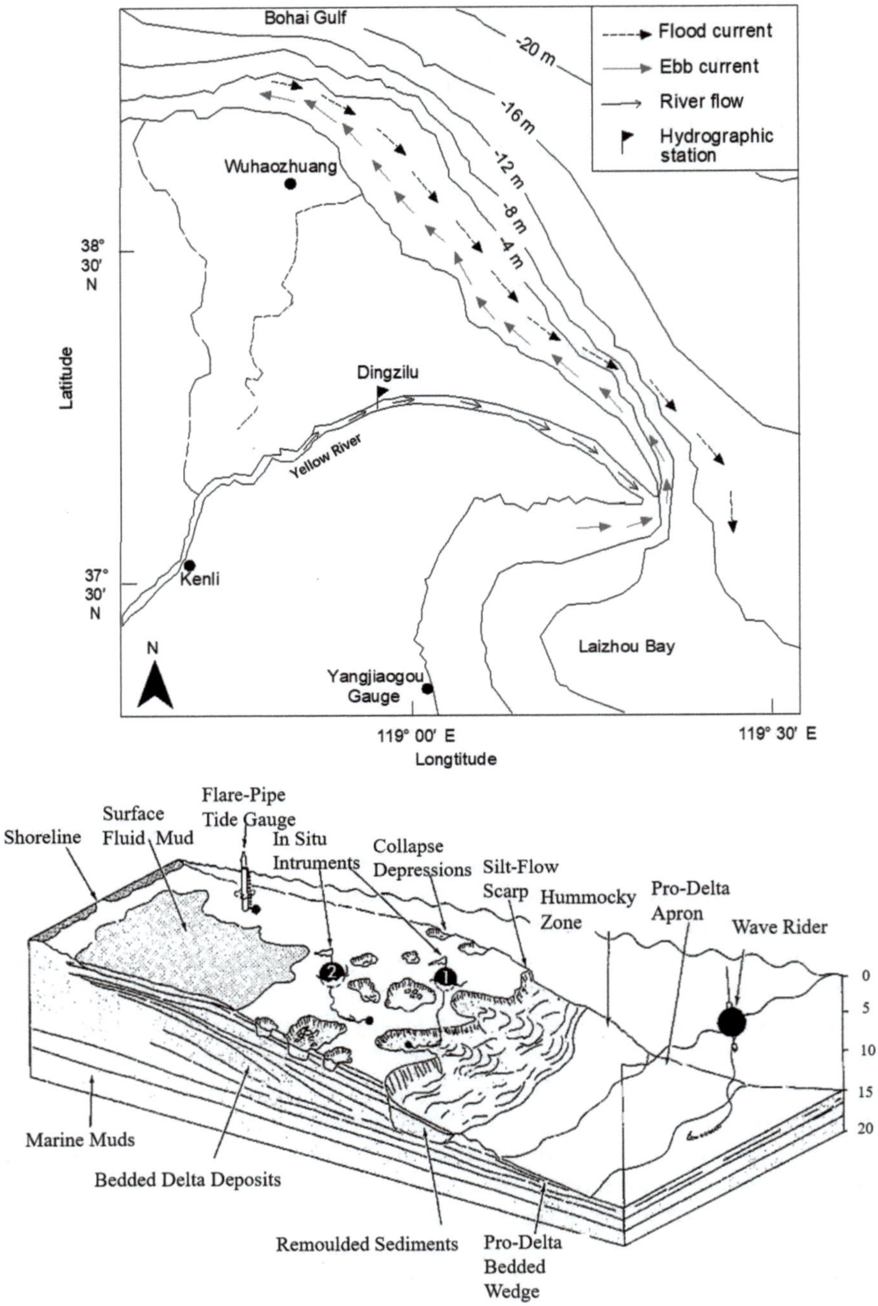

Fig. 2.3 Sketch map of topography and hydrodynamic environment in the modern Huanghe delta (Wang et al. 2007; Prior et al. 1989)

Table 2.4 Classification of marine dynamical geomorphy (Si and Zhang 1985)

Marine dynamical zone (from onshore to sea)	Dynamical geomorphy types		Distribution range	Tidemark/water depth range (m)
Intertidal zone	Tidal flat	High tidal flat	From Taoer estuary to Pile 12 located in the east of Haigang, and from Xiaodao estuary to Zhimaigou mouth	From high tide level of mean springs to the minimum low tide level
		Middle tidal flat		
		Low tidal flat		
	River mouth bar and sandspit		Current flowing river mouth	
	Modern oyster reef		Zhimaigou mouth, Tiaohe mouth, Yangkejungou mouth, and Taoerhe mouth	Close to the line of 0 m, which is the theory depth datum
Transition zone	Offshore slope	Accumulation bank slope	From Yongfeng estuary to Zhimaigou mouth	0–5
		Stability bank slope	From Tiaohe mouth to Taoer estuary	
		Erosion bank slope	From pile 11 located in the east of Haigang to Tiaohe mouth	0–12
	Active zone of sediment			5–10
	Front slope of delta		Current flowing river mouth	0–10
	Subaqueous delta of small rivers		Zhimaigou mouth, Tiaohe mouth, Yangkejungou mouth, and Taoer estuary	
Subaqueous delta plain	Subaqueous delta plain	Accumulation plain	Each segment of coastline	The range of water depth in each part of coastline is different
		Erosion plain	From Shenxiangou mouth to Diaokou estuary	
	Mud Bay		Both sides of the current flowing river mouth	

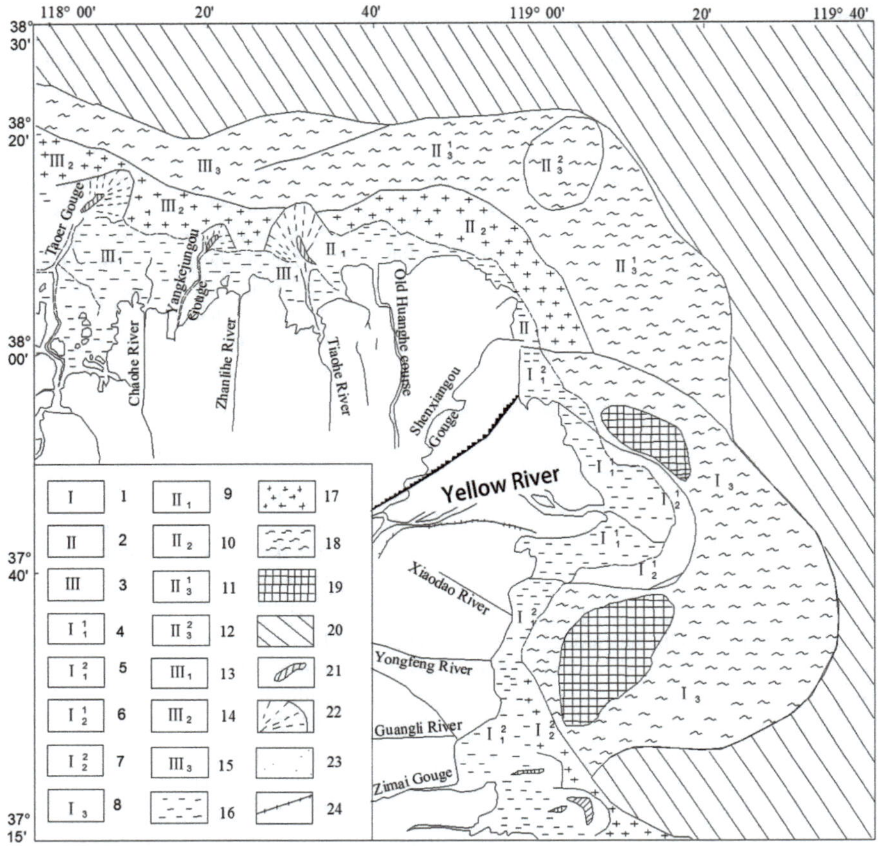

Fig. 2.4 Geomorphological map of the modern Yellow River subaquatic delta (Zang 1996). *Note* 1. Siltation area, 2. erosion area, 3. dynamic-balance area, 4. estuary sand bar, 5. silt tidal flat, 6. underwater delta front edge, 7. underwater siltation bank slope, 8. front delta underwater plain, 9. eclipse ebb tide beach, 10. underwater erosion bank slope, 11. underwater delta siltation plain, 12. underwater delta erosion plain, 13. Weak erosion and weak siltation tidal flat, 14. relatively stable underwater bank slope, 15. ancient Yellow River subaqueous delta plain, 16. intertidal zone, 17. transition zone, 18. underwater delta plain, 19. rotten mud bay, 20. shallow sea plain, 21. modern estuary oyster reef, 22. small river underwater delta, 23. beach surface river channel and tidal water ditch, 24. artificial dam

eroded backward and is straightened. Under the influence of erosion, the subaqueous terrain becomes flat and the sediment surface turned to gentle. The sediments type shows a zonal distribution pattern along the coastline. The morphology pattern of Middle Zone is mainly tidal flat, subaqueous erosion slope, active sedimentation zone, and subaqueous delta erosion plain. South Zone refers to the sea area between Shenxiangou and Zhimaigou mouth. The Yellow River flows into the sea from this area since 1976, and it takes a lot of sediments which mostly accumulates at the estuary and results in fan-shaped expansion to the sea. The subaqueous morphology shows steep slope and overlap of sandbar from different estuary. The morphology pattern of South Zone includes tidal flat, estuary sandbar, mud bay, subaqueous nearshore, and delta outer plain (Zang 1996; Feng et al. 2004).

2.4 Marine Dynamics in the Modern Yellow River Delta

2.4.1 Meteorology

The Yellow River Delta is located in the mid-latitude, in the warm temperate zone, on the back land of the sea, and is affected by the Eurasian continent and the Pacific Ocean. It belongs to the warm temperate semi-humid continental monsoon climate zone. The basic climatic characteristics are as follows: winter, cold, and hot, with four distinct seasons. Spring is dry and windy, early spring is cold and warm, often there is cold spring, and the spring is warming up quickly, often spring drought; summer, hot and rainy, warm, and humid, sometimes affected by typhoon; autumn, temperature drops, rain drops; in winter, the weather is dry and cold, the wind is blowing, the rain and snow are scarce, and the main wind direction is north wind and northwest wind. The temperature difference of the four seasons in the Yellow River Delta is obvious, the annual average temperature is 11.7–12.6 °C, the extreme maximum temperature is 41.9 °C, the extreme minimum temperature is –23.3 °C; the annual average sunshine hours are 2590–2830 h; the frost-free period is 211 days; the average annual precipitation is 530–630 mm. 70% is distributed in summer; the average evapotranspiration is 750–2400 mm.

2.4.2 Waves

The Yellow River Delta locates at the interior of the semi-enclosed Bohai Sea, east to the Laizhou Bay, and north to the Bohai Bay. At the mouth of the Bohai Sea, an island chain, Changshan Isles, blocked the outer ocean waves from entering into the inner Bohai Sea. Therefore, the waves of the Bohai Sea are mainly generated by the wind and are characterized as fast generation and disappearance. Big waves with wave period more than 10 s are seldom seen. The maximum waves, the monthly average wave height, and the largest wave period near Dongying Harbor within one year (March to December) are listed in Table 2.5.

From the wave parameters shown in Table 2.5, it can be seen that the waves in the Yellow River Delta sea area feature evident seasonal variations, which has a close relation to the wind field on the Bohai Sea. The waves are the highest in winter, medium in fall and spring, and the waves are the smallest in summer months.

In addition, based on historical observation, the directions of the strong waves and sub-strong waves are, respectively, NNE-ENE and NNW, and the direction of

Table 2.5 Statistics of wave height in the sea area near Dongying Harbor in Yellow River Delta (Yan et al. 2006)

Parameter	3	4	5	6	7	8	9	10	11	12
Max. wave height/m	3.9	3.1	3.2	3.4	2.0	2.2	2.6	4.2	5.8	3.7
Monthly averaged wave height	0.6	0.4	0.4	0.4	0.5	0.4	0.4	0.6	0.7	0.7
Max. wave period	8.3	8.9	6.7	7.5	4.8	5.9	5.6	7.6	9.0	6.8

Table 2.6 Predicted wave parameters

Water depth/m	3	4	5	6	7	8	9	10	12	14
Effective wave height once in 50 years	1.8	2.3	3.0	3.6	4.1	4.9	4.5	4.6	4.8	4.9
Effective wave height once in 5 years	1.8	2.3	3.0	3.6	4.1	3.7	3.8	3.9	–	4.2

ordinary waves is S, sometimes SSE. The wave direction in this region coincides with the frequency of wind direction in different seasons. For example, strong gust more than 8 grades mainly blows NNE in winter and SSE in summer. Therefore, the change of wave direction suggests that the wave of the Yellow River Delta sea area has obvious seasonal variations.

Another characteristic of the wave in the Yellow River Delta sea area is apparent annual variation, which is again related to the wind field on the Bohai Sea. In spring, the waves appear alternately between northeast and southeast direction. In summer, the dominant wave direction is southeast. In the fall, the northern wave is dominant. The northeast wave is the most abundant waves in the entire year and the frequency is 10.3%, with the highest wave being 5.3 m. Consequently, the northeast coast of the Yellow River Delta is the most severely eroded, and north coast eroded moderately, and erosion of the southeast coast is the lightest. On the other hand, the highest wave observed during small-wave year (1986–1989) is only 3.5 m, and the highest wave ever observed is 5.8 m during large-wave year (1984–1985). The wind field that caused big wave during this sea area is mainly northeast strong wind, and also eastern strong wind at one side of the Laizhou Bay. The weather processes that cause this kind of wind field are cold wave, typhoon, and cyclone. Based on the observations conducted by Ocean University of China, and Institute of Oceanology, Chinese Academy of Sciences, the 50-year and 5-year wave factors of the strong waves in this area are calculated and shown in Table 2.6.

2.4.3 Tide

Tide is the rise and fall of the sea levels caused by the tide-generating forces of other planets. After tidal waves entering into the Bohai Sea, reflection wave forms under the influence of shallow sea depth and topography form at the bottom of the Bohai

Sea, as well as blocking effects of the sea coast. The incident wave and reflection wave of tidal wave form standing wave node outside Shenxiangou, which is an amphidromic point. Under the influence of M2 constituent, the current location of this amphidromic point (located at N38°04′, E119°04′) can change with the variance of the Yellow River estuary, and the variance of topography and sea level in the sea area of the Yellow River Delta. The amphidromic point has the tendency of moving to the southeast.

The tide characteristic shows the obvious difference between the two sides of the amphidromic point (Zang 1996) and the tidal hours at the two sides of the amphidromic point differ greatly. Observation shows that, at the south of the Bohai Bay, west of amphidromic point, the high tide appears 5 h later than the moon's transit, and the average high tide interval is 5 h and 7 min at the east of Diaokou River. However, along the coast of Laizhou Bay, south of amphidromic point, the high tide appears 10 h later than the moon's transit, for example, the average high tide interval is 10 h and 7 min at Wuhaozhuang. Although the distance between the two sides of the amphidromic point is only 20 km, the tidal time difference is 5 h. The significant difference of tidal time causes the flooding and ebbing tides appearing from west to east sequentially. When the west of the amphidromic point is at high tide, the south of the amphidromic point is at low tide, and vice versa. The tidal range near amphidromic point is only 0.2 m, and the tidal ranges increase gradually from amphidromic point to west and south. Furthermore, the tidal range in the west is larger than that in the east. The tidal range with low in the middle and high at both ends features a saddle-shaped tidal distribution curve. According to the classification method based on the ratio of partial tide amplitude, the tide type at the sea area west of Wuhaozhuang and Dongying Harbor is informal semidiurnal tide, and the tide type at the sea area near Wuhaozhuang and Dongying Harbor is diurnal tide, however, the tide type at the sea area south of Wuhaozhuang and Dongying Harbor gradually changes into informal semidiurnal tide again. Because the tide level change caused by astronomical factors is relatively regular, the average tidal range of this sea area is small, with it being, respectively, 0.7, 1.5, and 1.7 m near the Yellow River Harbor, Guangli River mouth, and Taoer River mouth. The tide range along the coast of the Yellow River Delta is shown in Table 2.7.

The nonperiodic water level changes caused by other factors are stronger; sometimes, the nonperiodic factors play the dominant role in the change of tidal level. For example, monsoon caused water increase and reduction in the Bohai Sea, seawater density change, precipitation, and runoff are the important reasons for sea level changes. On December 14, 1986, a strong northeast wind at the harbor area causes water increase reaching more than 1.5 m, which lasts for 6 h. The highest water increase reaches 1.76 m at high tide time, which is the highest water level since the harbor is built. On January 13, 1987, the strong northwest wind causes water decrease reaching 1.77 m, which is the lowest water level since the harbor is built.

Table 2.7 Tidal change in the coastal area of Yellow River Delta (Shi and Zhao 1985)

Regions	Maximum tidal range/cm	Minimum tidal range/cm
Tanggu	294	162
Dagu River mouth	288	160
Wanwangou Castle	184	96
West of Yellow River mouth	110	70
East of Yellow River mouth	40	30
Shenxiangou mouth	26	22
South of Donglanni	22	18
Pile 5	24	8
Tianshuigou	106	46
Qingshuigou	164	72
Weihekou	128	64
Longkou	104	56

2.4.4 Currents

The observed sea current can be referred as the seawater flow under the combined action of many factors. Sea current, in general, includes two parts: one part is the relatively stable periodic flow caused by astronomical tide, which is usually called tidal current. The second part is the nonperiodic seawater flow caused by wind, uneven density of seawater, runoff to the sea, and the topography in shallow beds, which is usually called residual current. At the Yellow River Delta sea area, tidal current possesses the ability of unbalanced transmission of water and sediments out of bay mouth, and residual current is significantly important in transportation and diffusion of the sediments carried by the Yellow River into the sea.

Li (1984) analyzed 201 sea current observations conducted during 1954–1979 at 74 stations, and the 66 wind observation station data concluded that, in the sea area near the Yellow River Delta, the flow speed of the tidal current is faster than that of residual current. But, the horizontal migration distance of water particles in tidal current is limited, and it is 11.5 sea miles near the Yellow River estuary. The flood tide duration is short in the estuary. At the sea area south of 38° 10′ N in the Laizhou Bay and the Bohai Bay, the direction of the greatest speed of M2 partial tide points to bay mouth. In addition, there is a flow divergent zone at the bay mouth. Therefore, at the near Yellow River Delta sea area, the tidal current has an unbalanced capacity of transporting water and sediments out of the bay mouth. The speed of residual current is slow, which is usually no more than 10 cm/s, but it is a nonperiodic flow that can flow for a long distance in a certain direction. Therefore, the residual current has the capacity to transport sediments for a long distance.

(1) *Tidal Current*

Tidal current is the horizontal movement form of water particles in the tidal wave. Tidal current and the change of tidal level are two different manifestation forms in the same phenomenon. So the basic feature of tidal current is not only influenced by the movement of planets but also is restricted by some factors, such as the shape of the coastline and the topography of seafloor. The Yellow River estuary is located on the dividing line of the Bohai Bay and the Laizhou Bay. Most tide along the coast of the estuary area is informal semidiurnal tide and only the coast of the Bohai Straits is informal diurnal tide. The ratio between the short axis and the long axis of the tidal current oval is less than 0.1 and can be characterized as reciprocating currents flowing parallel to the coastal line (Yang and Wang 1993). The long axis of the M2 partial tidal oval basically parallels the coastal line and the directions of flood current and ebb current are, respectively, along southeast and northwest direction, which is shown in Fig. 2.3.

At the amphidromic region, the strong current flows northwest and the flow speed reaches 100–120 cm/s. At the sea area near the Yellow River estuary and amphidromic region, the horizontal distribution of flow speed is in close relation with the isobaths line. The flow speed of nearshore zones shallower than 5 m of the isobath line is smallest. At the sea area from 10 to 15 m of isobath lines, the flow speed is the biggest. At the deep water area deeper than 15–20 m of the isobath lines, the flow speed slows down again. Figure 2.5 shows the observed flow speed field of tidal current in the sea area of the Yellow River Delta. It can be seen that there are two high flow speed belts for the tidal current of the Yellow River subaqueous delta. One is outside the Yellow River estuary, and the other is outside the Shenxiangou and Wanwangou. The high flow speed belt outside the Yellow River estuary is perpendicular to the runoff direction of the Yellow River, and it flows south and north, respectively, at flood current and ebb current period. The tidal current here is typically regular semidiurnal tide, and the flow direction changes four times in a day. The high flow speed belt plays an important role in transportation of estuary sediments to both sides. The high flow speed belt between Shenxiangou and Wanwangou is close to amphidromic point, which falls within the diurnal tide zone. The ellipticity of tidal currents is very small and the long axis is usually paralleled to the coastal line. The tidal currents have the feature of reversing currents and the high flow speed belt causes strong scouring force to the seafloor, and it carries away fine sediments and leaves coarser sediments at the sea. Zhao (2004) compared and found that the two strong current zones are in correspondence with the serious erosion zones at the nearshore area north of the delta and the deposition zone at the estuary.

(2) *Residual Current*

There are many factors that influence the residual current. For the sea area of the Bohai Bay, especially the sea area near the Yellow River Delta, residual current is mainly influenced by tidal wave system, runoff to the sea, wind, uneven seawater density, sea level slope, seafloor topography, and the boundary condition of coastal line (Dong and Wang 1997). Among these factors, runoff to the sea is the main factor

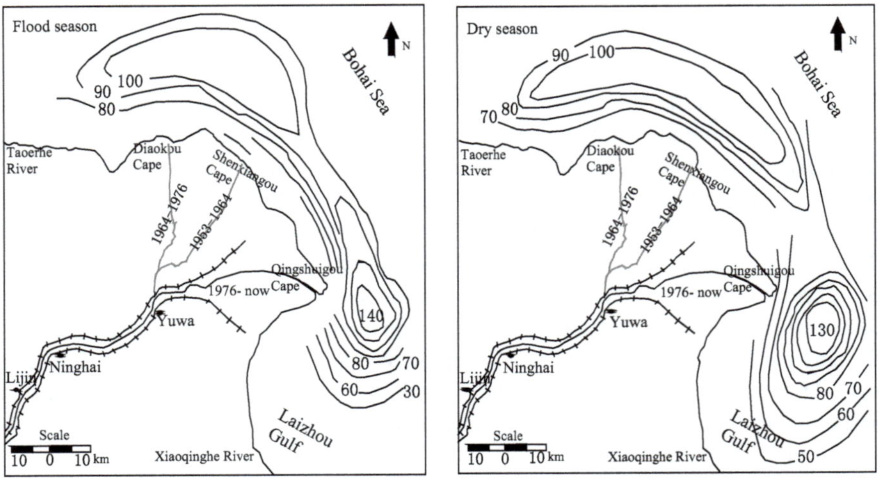

Fig. 2.5 Filed diagrams of measured longshore current (cm/s) in the modern Yellow River Delta (modified after Chu et al. 2006)

that causing density flow and inclined flow, whereas, tidal wave system usually needs the combined effects of seafloor topography and coastline boundary conditions to form the tidal residual current. Therefore, at the Bohai Bay sea area of the Yellow River estuary, the influencing factors of residual current can be categorized into three aspects: tidal wave system, wind, and Yellow River runoff into the sea. Under the condition of no influence from other factors, the above three factors can independently generate its individual residual current field, which are the tidal residual current field, wind residual current field, and residual current field caused by runoff to the sea.

The residual current field at the sea area near the Yellow River estuary is actually the overlaying of the above three residual current fields. At the different parts of the sea area and during different seasons, the overlaid residual current field perhaps differs, especially at the sea area near the Yellow River Delta. The direction of the residual current field after superposition is the diffusion direction of the sediments to the sea.

Studies by Dong and Wang (1997) indicated that the residual field caused by tidal wave system near the Bohai Bay can form anticlockwise circulation of tidal residual current field, which coincides with the direction of the strongest tidal current. The residual current field caused by wind is easily influenced by the monsoon, especially at the sea area south of the Bohai Bay and nearby the Yellow River Delta. The relatively stable residual current field caused by monsoon mainly flows west or south in spring, west or north in summer even if being scattered, northeast in fall, and southeast in winter. That is, the residual current field caused by wind flows into the Bohai Bay in summer half year and flows out of the Bohai Bay in winter half year. Thus, it can be seen that the superficial residual current of this sea area has significant seasonal changes. In case of no wind, the direction of residual current is basically from south to north, which flows from the Laizhou Bay to the Bohai Bay. The bottom residual current at deeper than 5 m isobath line is stable and mainly flows northeast. The flow speed of this residual current is generally low at about 5–11 cm/s. The residual current near the estuary is obviously influenced by runoff to the southeast direction, which is consistent with river runoff direction. The flow speed of this residual current is very high and the highest observed speed reaches 33 cm/s. There is eddy at both sides of estuary, and the flow direction is anticlockwise on the north side and clockwise on the south side.

The residual current field caused by runoff to the sea is mainly influenced by diffusion and baroclinic current. The scope of residual current field caused diffusion current is small and the diffusion residual current field mainly locates at the sea area near the estuary. Its direction is consistent with the runoff direction to the sea. The diffusion residual current field diffuses to the open sea in the shape of a tongue-shapedly diffuses to the open sea and its size depends on the quantity of water flowing into the sea. The residual current field caused by baroclinic current can influence the entire Bohai Sea area and it mainly flows east to the Bohai Strait (Dong and Wang 1997).

2.4.5 Storm Surge

Wave and storm surge play important roles in the resuspension and redeposition of the sediments in the Yellow River Delta. They are the shaping forces means dynamic force, Torphology and dynamic characteristic of the Delta area. Wave under the ordinary oceanic condition, storm surge in the extreme condition, and their interaction have become important research topics in international sediments dynamic studies.

An abnormal, sudden rise of sea level, caused primarily by strong winds offshore, or by a sharp drop in atmospheric pressure, or by the introduction a storm surge from outside the area, is called storm surge. Storm surge is a commonly seen marine natural disaster that may cause serious damage in the coastal regions, especially at the estuary delta area. The Yellow River Delta, lying between the Bohai Bay and the Laizhou Bay, is a storm surge disaster frequently occurrence area along the rim of the Western Pacific. This region is characterized as a broad coastal tidal flat area, with gentle slope and shallow water. In region like this, it is very easy to form serious storm surge, which consequently causes strong coastal erosion.

According to statistics, there are many disastrous storm surge along the Yellow River Delta in history. There were six serious storm surges during the latest hundred years, respectively, in 1845, 1890, 1938, 1964, 1969, and 1980. The storm surge disasters of the modern Yellow River Delta from 1856 to 2004 are shown in Table 2.8. Based on historical records, the storm surge disasters occur every 5.5–7.9 years, and 6.4 years on average. Compared to old Yellow River Delta, the average storm surge cycle of the modern Yellow River Delta is shortened for about 5 years. Compared to the period of 1856–1949, the average storm surge cycle is prolonged for about 2 years since 1949, and the cycle is prolonged to 7.9 years from 5.5–6.0 years. However, the

Table 2.8 Statistics of storm surge disasters of the modern Yellow River Delta (1856–2004) (Chen et al. 2007)

Dynasties	Time ranges	Statistics of years	Numbers of disasters	Frequency/(1/a)	Period/a
Five years after Xianfeng in Qing Dynasty	1856–1911	55	10	1/5.5	5.5
The Republic of China	1914–1938	24	4	1/6.0	6.0
The People's Republic of China	1949–2004	55	7	1/7.9	7.9
In total	–	134	21	1/6.4	6.4

degree of damage of storm surges increased, and the sea water invades landwards could be more than 20 km.

The wave height is no more than 1.5 m under ordinary weather condition, however, the wave height reaches more than 5.8 m under extreme weather conditions such as storm surge. The annual extreme water increase reached 1.8–3.2 m, and 2.2 m on average at the Chengbei sea area near Diaokou. Being a semi-closed shallow bay, the entire Bohai Sea water reduction (recession of sea level) is about 1– 2 m after a day of continuous northwest wind, even if the wind force is only 5–6 degrades. The Bohai Sea level is increased under the effect of the southeast wind. As a result of water increase and reduction, the average fluctuation of the Bohai Sea level can be 1–2 m. Therefore, if there is firstly southeast wind causing water increase in the Bohai Sea area, and then the wind changes to the northeast, the sea level along the Yellow River Delta will increase sharply, causing catastrophic storm surge and easily result in deformation and failure of the marine building foundation.

The storm surge occurrence of the Yellow River Delta is highly variable seasonally. The Yellow River Delta locates south of the Bohai Sea with a temperate continental climate. The weather processes, including cold wave, typhoon, and cyclone, cause this area to be one of China's coastal regions with strong winds. The climate provides the prerequisite for the formation of storm surge disaster. Three of the seven strong storm surge disasters since 1949 happened in typhoon season, three in April, and one in November—a period of frequent cold air movements (Yan et al. 2006). It is thus clear that influenced by meteorological factors, such as cold air and typhoon, the storm surge disasters in the Yellow River Delta have a seasonal pattern, which usually occur in early winter, April, and typhoon period.

2.5 Seabed Sediment Properties of the Modern Yellow River Delta

2.5.1 Sediment Types and Distribution in the Modern Yellow River Delta

Clayey silt or silty clay is the main components of the sediments in Yellow River Delta, the overall trend is that clay content is lower near coastline compared with that of offshore area. In addition, rich in calcium carbonate is the most important chemical feature of sediments in Yellow River Delta, and its content in surface sediments is generally high, with a maximum of 17.91%, the vast majority is between 10 and 14%.

According to the naming principles of primary and secondary components stipulated in the specifications for oceanographic survey, there are mainly the following types of sediments in the region according to the survey data available (Zang 1996; Feng et al. 1999):

(1) Very fine sand: It is mainly distributed in the estuary entrance bar and the ablative central area on top of sand tip; it accounts for 80–90% of components with a median particle size of about 3.6–3.4 φ. The rest is coarse silt with good sorting which is close to normal distribution and small planar distribution area.
(2) Silty sand: It is mainly distributed in the top surface of the estuary sand tip and the bottom of the escutcheon and the gullies. The median particle size is between 3.6–4 φ, the very fine sand content is 50–70% while the silt content is less which is accounting for 20–40%, the clay content is low, generally only 1–5% and no more than 10%.
(3) Sandy silt: It is mainly distributed in the area of intertidal zone and the lower part of low tide line sand wave band, which is from near the zero meter isobath to the 2 m isobath, with a median particle size of about 4.4–4.5 φ and good sorting. The coarse silt accounts for 46–70%, the content of fine-grained sand accounts for 20–40%, the clay-grained content only accounts for 3–8%.
(4) Silt: It is the main sediment type in the intertidal zone and epicontinental sea of the area, with a median particle size of about 5–6 φ and good sorting. Most of the underwater base slopes are mainly coarse silt, and the content of silt is about 70–80%, The above fractions of clay and sand are relatively low in content.
(5) Clayey silt: It is mainly distributed in the epicontinental sea area of from shore to 5 m isobath. The content of silt and clay is about 60 and 30%, respectively. The above fractions of fine sand are about 5%, and less than 10%. The median particle size is about 5–7 φ, with the sorting from moderate to poor.
(6) Silty clay: It is mainly distributed on sides of the sludge bay. The content of silt and clay are about 39–51% and 49–61%, respectively. The above fractions of fine sand are less than 5%, and the median particle size is about 6–8 φ.

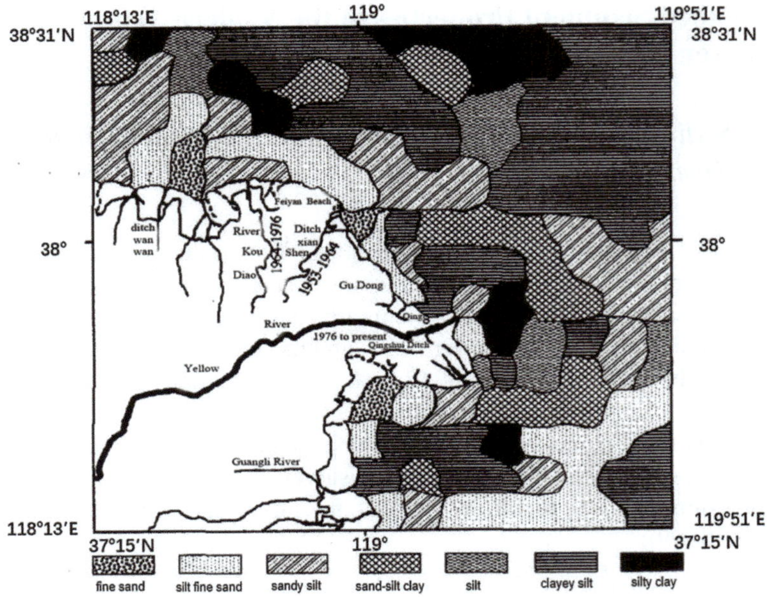

Fig. 2.6 Sediment types and distribution in the coastal areas of the Yellow River Delta (Zang 1996)

Sediment types and distribution in the coastal areas of the Yellow River Delta are shown in Fig. 2.6. It can be seen that the sediments of the modern subaqueous delta of Yellow River have obvious zonation, and the overall trend is that clay content is lower near coastline compared with that of offshore area (as shown in Fig. 2.7).

2.5.2 Geological Strata

Yang et al. (2009) and Jia et al. (2011) selected typical test area in the northwest (118.44°E, 38.11°N), the north (119.2°E, 37.8°N), the eastern (118.95°E, 38.08°N), and the southeast (118.93°E, 37.36°N) of the modern Yellow River Delta, namely Dawangbei, Qikou, Xintan, and Guangli Harbor (Fig. 2.8). By standard penetration test, static cone penetration test and sampling and testing during borehole drilling, the sedimentary layered structure of the modern Yellow River Delta was investigated, as shown in Tables 2.9 and 2.10.

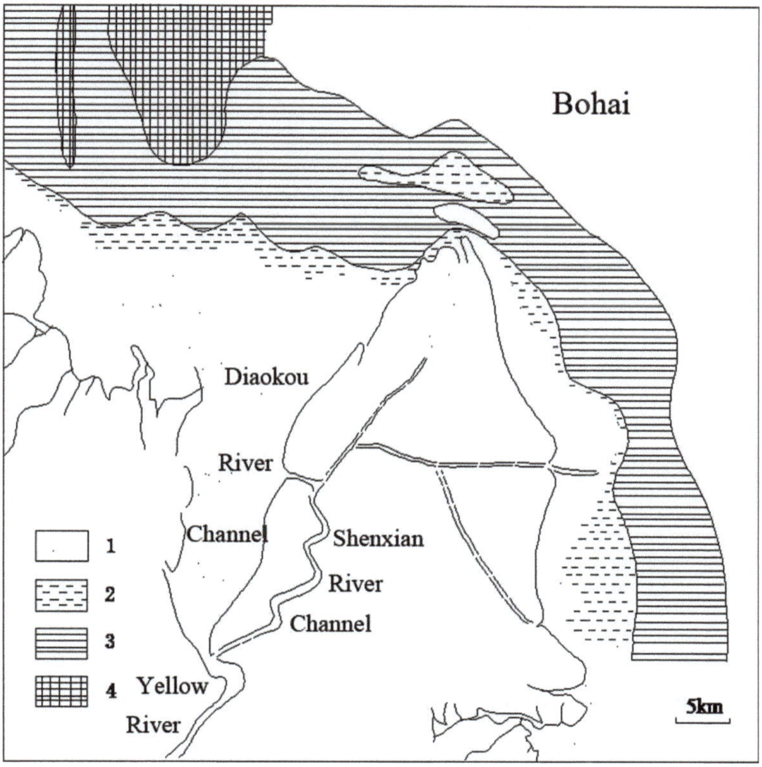

Fig. 2.7 Distribution of Clay Content of Sediments in the Diaokou Passage of the Yellow River Delta (Cheng and Xue 1997). 1: < 3%; 2: 3%–10%; 3: 10%–30%; and 4: > 30%

2.5.3 *Sediment Grain Size and Mineral Composition*

(1) ***Grain size Composition***

The granulometric composition of sediment of the modern Yellow River Delta has a characteristic of layered distribution, and the clay content varies greatly (as shown in Table 2.11).

The particle cumulative frequency curves for each site are presented in Fig. 2.9.

There is no obvious regularity in the granulometric composition characteristic of surface sediments from the coastline to the sea on vertical direction, including the sorting coefficient, skewness, kurtosis, and average particle size. But the fractal dimension of particle size shows a good positive correlation with the offshore distance. On the other hand, the dispersion of fractal dimension of sediment particle size in a weak hydrodynamic environment is larger than that under strong hydrodynamic environment (Yang 2009).

By analyzing the granulometric composition of surface sediments in the modern Yellow River Delta through dosimeters method, Meng et al. (2012) obtained the

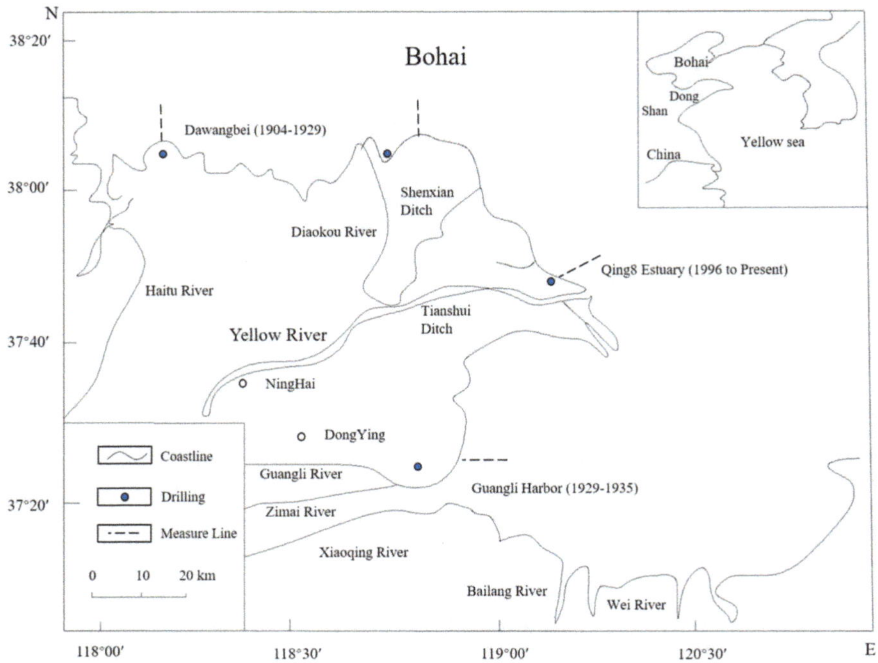

Fig. 2.8 Location of sediment layer structure survey

distribution of granulometric composition of sediments on the tidal flat in the coastal waters around the modern Yellow River Delta (as shown in Table 2.12).

A certain amount of fine sand, silt, and clay is distributed in the sediments in the stiff shell of the Yellow River Delta. The range of content of each fraction is relatively large. Generally, the sand content is relatively small, and the average is 16.72%, the content of silt is in the range of 44.43–86.99%, which is dominant. And the clay content is the least, with an average of 9.9%. According to the standard of *Specifications for Oceanographic Survey*, the stiff shell sediments of the Yellow River Delta can be classified as silt, sandy silt, and clay silt (Wang 2004).

(2) ***Mineral Composition***

Wang et al. (2004) sampled on-site on the Diaokou leaf which located on the north of Yellow River Delta, using X-ray diffraction phase analysis (Fig. 2.10), and analyzed the total mineral composition and clay mineral composition of the surface sediments (Table 2.13).

The debris minerals were predominant in the sediments of abandoned leaf of the modern Yellow River Delta, accounting for more than 75%, among them, quartz > feldspar > calcite > dolomite. Among the clay minerals, illite > chlorite > kaolinite > montmorillonite. Under strong hydrodynamic conditions, the mineral content of sediment is higher than that in the weak hydrodynamic environment and shows obvious nonuniformity. In the depth profile, the surface sediments have

Table 2.9 Typical sediment layer structure in the modern Yellow River Delta

Study site	Geological age	Layer number	Bottom elevation/m	Bottom depth/m	Thickness of Layer/m	Sediment type	Measured mean of SPT/time	Corrected mean of SPT/time	Static cone penetration	
									P_a/MPa	P_f/kPa
Dawangbei	Q_4^{al}	1	– 5.30	5.30	5.30	Silt	7.40	5.78	2.40	28.00
	Q_4^{al}	2	– 8.70	8.70	3.40	Mucky silty clay	1.61	1.70	0.40	5.00
	Q_4^{al}	3	– 15.60	15.60	6.90	Silty clay	3.25	2.61	0.70	11.00
	Q_4^{al}	4	– 18.60	18.60	3.00	Mucky silty clay	5.30	3.88	1.10	12.00
	Q_4^{al}	5	– 21.00	21.00	3.60	Silty clay	10.50	7.50	1.20	17.00
Guangli harbor	Q_4^{ml}	1	– 1.70	1.70	1.70	Plain fill	9.50	9.50	2.50	36.00
	Q_4^{al}	2	– 7.50	7.50	5.80	Silt	6.40	5.88	1.90	24.00
	Q_4^{al}	3	– 12.50	12.50	5.00	Mucky silty clay	3.30	2.72	1.00	14.00
	Q_4^{al}	4	– 16.20	16.20	3.70	Silt	8.00	5.98	1.80	21.00
	Q_4^{al}	5	– 18.90	18.90	2.70	Silty clay	6.50	4.70	0.70	11.00
	Q_4^{al}	6	– 20.50	20.50	1.60	Silt	11.67	8.20	4.80	61.00
	Q_4^{al}	7	– 21.30	21.30	0.80	Silty clay	5.00	3.50	3.10	42.00
Xintan	Q_4^{al}	1	– 5.00	5.00	5.00	Silt	5.00	4.94	2.90	24.00
	Q_4^{al}	2	– 7.30	7.30	2.30	Mucky silty clay	1.88	1.68	1.20	14.00
	Q_4^{al}	3	– 13.00	13.00	5.70	Silt	11.10	9.23	3.00	32.00
	Q_4^{al}	4	– 19.50	19.50	6.50	Mucky silty clay	3.57	2.61	0.90	12.00
	Q_4^{al}	5	– 23.00	23.00	3.50	Silt	34.00	24.00	4.20	44.00
Dongying harbor	Q_4^{al}	1	– 7.50	7.50	7.50	Silt	7.69	7.20	3.10	30.00
	Q_4^{al}	2	– 12.50	12.50	5.00	Silty clay	5.22	4.12	1.75	30.00
	Q_4^{al}	3	– 19.00	19.00	6.50	Mucky silty clay	4.15	2.99	0.90	12.00

higher detrital mineral content than that of bottom sediments, while the clay mineral content shows an opposite trend. These mineral composition characteristics of sediment indicate that the surface sediments are superior to lower sediments of the modern Yellow River Delta in the process of coarsening under hydrodynamic action.

Table 2.10 Typical sediment layer description in the modern Yellow River Delta

Study site	Layer number	Sediment type	Description of soil layer
Dawangbei	1	Silt	High moisture and slight density, and it is in quick response to shaking. This layer is interbed with thin clay layer
	2	Mucky silty clay	Composed of dust-colored mucky silty clay, ranges from fluid plastic to soft plastic. This layer is wet and a bit smooth. with moderate dry intensity, contains a little organic matter
	3	Silty clay	Composed of dust-colored mucky silty clay, ranges from soft plastic to plastic. This layer is wet and a bit smooth with moderate dry intensity; contains a little organic matter; contains tan silt interlayer, high moisture and slight density with low dry intensity, in quick response to shaking; and contains organic matter and shell fragments and clay silt interlayer
	4	Mucky silty clay	Dust-colored, range from soft plastic to plastic, wet and a bit smooth, with moderate dry intensity, with plenty of silt blocks and sludge blocks
	5	Silty clay	Cinereous, plastic, wet and a bit smooth, with moderate dry intensity, contains small amounts of shell fragments
Guangli harbor	1	Plain fill	Tan to dust color, moderately dense. It is mainly made up of silt, mixed with a little of coarse sand and coal cinder
	2	Silt	The response to shaking is rapid, doesn't have shiny reflection and the dry strength is low. The organic matter content is low but with some shell fragments and localized silty fine sand

(continued)

Table 2.10 (continued)

Study site	Layer number	Sediment type	Description of soil layer
	3	Mucky silty clay	Dust-colored, soft plastic, wet, a bit smooth, with moderate dry strength and toughness, contains a little organic matter and shell fragments but plenty of silt blocks, mingled with silt thin layer in the depth of 8.9–9.8 m
	4	Silt	Dust to tan color, wet, with moderate density. Its reaction is rapid when shaking. It doesn't have shiny response and the dry strength is low, low toughness,low content of clay, contains mica and localized interbeds of thin silty clay layer. Along the profile, the layer from 13.5 to 14.5 m is fine silty sand
	5	Silty clay	Soft plastic to plastic, wet, a bit smooth, with moderate dry strength and moderate toughness, low contents of organic matter and shell fragments
	6	Silt	Wet, moderate density, doesn't have shiny reflection, low dry strength, contains iron oxides and mica, along the profile, the layer from 19.5 to 20.0 m is fine silty sand
	7	Silty clay	Dust-colored plastic, a bit smooth, with moderate dry strength and moderate toughness, low contents of organic matter and shell fragments
Xintan	1	Silt	Tan color, slight dense, and low in dry strength, in quick response to shaking, containing iron oxide and mica, localized with silty sand and mucky silty clay thin layers
	2	Mucky silty clay	Dust-colored fluid plastic to soft plastic state, saturation, a bit smooth, containing low content of organic matter
	3	Silt	Tan color, slight dense, wet to saturation, low dry strength, in quick response to shaking, contains iron oxide and mica. Along the profile, at the depth of 7.5–7.7 m, 9.5–9.7 m and 12.0–12.2 m are silty clay thin layers, and the layer of 10.5–11.0 m is a gravel interlayer

(continued)

Table 2.10 (continued)

Study site	Layer number	Sediment type	Description of soil layer
	4	Mucky silty clay	Dust-colored, fluid plastic to soft plastic, saturation, smooth, contains low content of organic matter and local areas are interlayer with thin silt layers
	5	Silt	Tan, slight dense, wet to saturation, low dry strength, in quick response to shaking, contains iron oxide and mica. The bottom is fine silty sand
Dongying harbor	1	Silt	Tan, wet to saturation, slight to moderate dense, in quick response to shaking, contains iron oxide and mica, mingled with silty clay and silty sand thin layers
	2	Silty clay	Dust-colored, soft plastic, wet, a little smooth with low content of organic matter, mingled with multiple silt soil interlayers
	3	Mucky silty clay	Dust-colored, soft plastic to plastic, wet, contains a bit organic matter

2.5.4 Sediment Microstructure

The sediments of the modern Yellow River Delta are mainly composed of silt particles and the microstructure is mostly framework structure. The framework is formed by silt particles, which is loose, relatively homogenous, and very porous. There was a small amount of clay particles unevenly distributed in the microstructure. The clay particles distributed at the surface of single particles either form as thin films, or were scattered, and sometimes were located at the contact points of the silt particles, playing a role of interconnection. The connected force of this structure was weak.

By sampling on-site, Jia et al. (2011) conducted microstructure observation of the soil samples from different sea areas of the modern Yellow River Delta in vertical and horizontal direction. The test results show that the sediment microstructures in different study areas exhibit the same structural characteristics.

Surface silt sediments (within 1 m depth), in vertical direction, the particles were large in size and no significant directionality trend was found. The structure is also blocky aggregation and very porous. There was a little clay grains adhering on the surface of the sample and correspondingly, the psephicity was low (Fig. 2.11). In horizontal direction, the particles were also large in size and no significant directionality trend (Fig. 2.11). Within 5 m depth, in vertical direction, the particles were large in size and no significant directionality trend was found. The structure is also blocky aggregation and very porous. There was a little clay grains adhering on the surface of the sample and correspondingly, the psephicity was low (Fig. 2.12). In

Table 2.11 Sediment clay content distribution characteristics in the modern Yellow River Delta

Soil depth/m	Study site	Partition characteristics				General characteristics			
		Maximum	Minimum	Average	Standard deviation	Maximum	Minimum	Average	Standard deviation
0–6	Dawangbei	15.34	7.07	10.23	3.06	38.17	4.42	12.72	7.68
	Guangli harbor	20.07	4.42	12.75	6.55				
	Diaokou	19.33	6.15	11.38	6.14				
	Xintan	38.17	8.87	16.25	12.48				
7–20	Dawangbei	60.55	13.53	42.58	12.35	68.05	3.00	29.06	17.07
	Guangli harbor	35.67	3.00	17.57	10.21				
	Diaokou	56.52	8.33	23.42	14.04				
	Xintan	68.05	5.76	31.61	19.95				

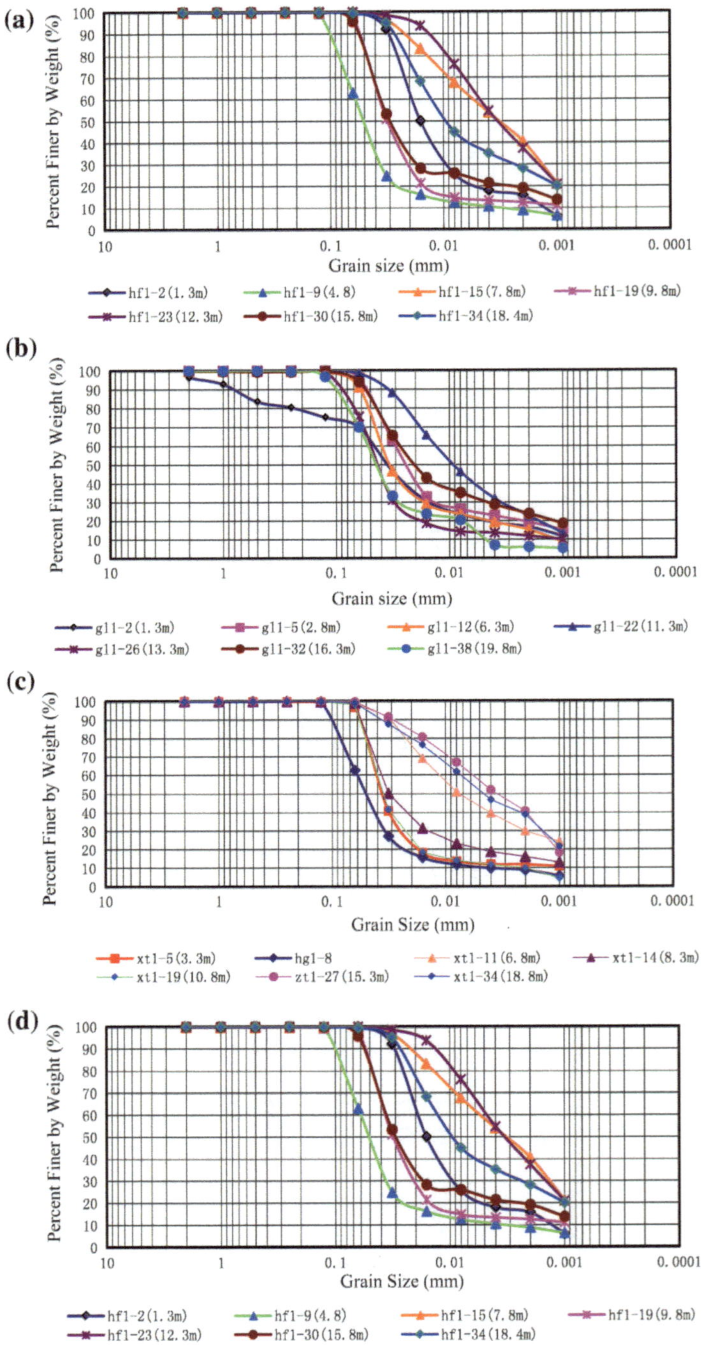

Fig. 2.9 Grain size distribution curve of sediment samples at (a), (b), (c), and (d)

Table 2.12 Sediment grain size composition in the modern Yellow River Delta

Study site	Latitude and longitude	Average particle size/mm	Sand content/%	Clay content/%
Dongying harbor	38°04.042′N, 118°56.499′E	0.037	3.5	12.6
Dongying harbor	38°04.045′N, 118°56.468′E	0.043	9.1	14.8
Dongying harbor	38°04.043′N, 118°56.494′E	0.042	4.9	13.6
Dongying harbor	38°04.083′N, 118°56.407′E	0.025	2.8	18.6
Dongying harbor	38°04.082′N, 118°56.407′E	0.020	0.9	23.7
Dongying harbor	38°04.080′N, 118°56.418′E	0.027	4.1	24.2
Dongying harbor	38°04.077′N, 118°56.402′E	0.027	3.5	18.8
Pile 106	38°08.110′N, 118°46.816′E	0.016	3.2	33.2
Gudong	37°51.174′N, 119°05.829′E	0.044	3.4	17.5
Feiyan beach	38°08.379′N, 118°42.634′E	0.042	3.8	18.7
Chezigou fishing village	38°06.129′N, 118°27.872′E	0.051	10.8	7.8
Mines 4th team 7th	38°06.131′N, 118°27.870′E	0.040	8.6	13.9
Xintan	38°08.054′N, 118°12.773′E	0.029	1.4	13.3
Guangli harbor	37°42.137′N, 119°16.457′E	0.058	35.4	9.2

horizontal direction, The particle size was small with plenty of well-proportioned fine particles that enriched at the surface of the blocky grains and there was no significant directionality but with good psephicity (Fig. 2.12). Within 15 m depth, silty clay sediments, in vertical direction for sample, the particles were small in size with no significant directionality trend. There were abundant pores and fine particles in the sample. The fine particles adhered on the surface of big grains. The psephicity was low (Fig. 2.13). In horizontal direction, the particles size was large with no significant directionality. The sample is mostly in blocky aggregation structure. There was a small amount of clay grains adhering on the surface of the low perspicuity debris particles (Fig. 2.13).

Our research results showed that under strong hydrodynamic conditions, the fine-grained matter of sediments within 3 m depth showed a trend of little-aggregation-

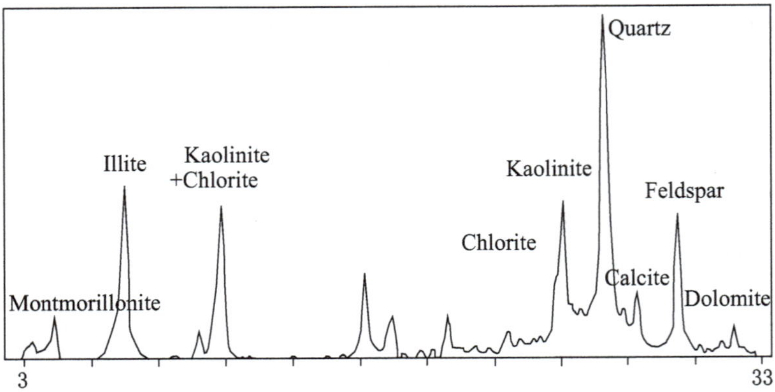

Fig. 2.10 X-ray diffraction spectrum of sediment samples at all study areas

reduction-normal, and correspondingly the filling form was transformed into the attached form; there is no corresponding change law under the weak hydrodynamic environment. On the other hand, the uniformity of sediment particles under the strong hydrodynamic environment is generally higher than that under the weak hydrodynamic environment and exhibits periodic changes along the depth. But for both the two hydrodynamic environments, the flaky particles on the surface of sediment of the modern Yellow River Delta are mostly vertically inserted into the combination of lower soil particle, and the long axis direction is relatively uniform. The vertical inclination of the particles is concentrated at 50–100°, with a trend of vertical alignment. Through the analysis of fractal characteristics of microstructures, Wang et al. (2007) found that in the zone with strong hydrodynamic action, the fractal dimension of tidal flat sediments is small, the dispersion is large, and showing an "S" type change with depth, and the directional fractal dimension of sediment particles is small on average, large in dispersion, and also shows an "S" type change with depth. In the weak hydrodynamic environment, the sediment shows a large fractal dimension with good consistency and monotonously increases with depth. The directional fractal dimension of sediment particles shows a high degree of fractal dimension with high consistency, showing a gentle anti "S" type change with depth.

2.5.5 *Physical and Mechanical Properties of Sediment*

(1) ***Physical Properties***

The water contents of the sediment of the modern Yellow River Delta is generally in the range of 20.8–45.8%. The water content of sediments in the northwest is generally in the range of 20.8–45.8%, that in the north is generally in the range of 23.8–43.2%, and in the eastern part is generally 23.2–43.1%, and generally 21.9–31.6% in the south.

Table 2.13 Mineral composition of surface sediment in Diaokou lobe of the modern Yellow River Delta

Sample depth/cm	Quarta/%	Feldspar %	Calcite/%	Dolomite/%	Illite/%	Chlorite/%	Kaolinite/%	Montmorillonite/%	Detrital mineral/%	Clay mineral/%
2	38–50	17–26	9–11	1–4	11–18	3–6	4–9	0–1	70–81	19–29
40	32–46	18–24	6–11	1–2	12–21	4–8	6–9	0	64–77	23–36

Fig. 2.11 Representative SEM micrograph of silt sediments within 1 m depth in horizontal and vertical directions

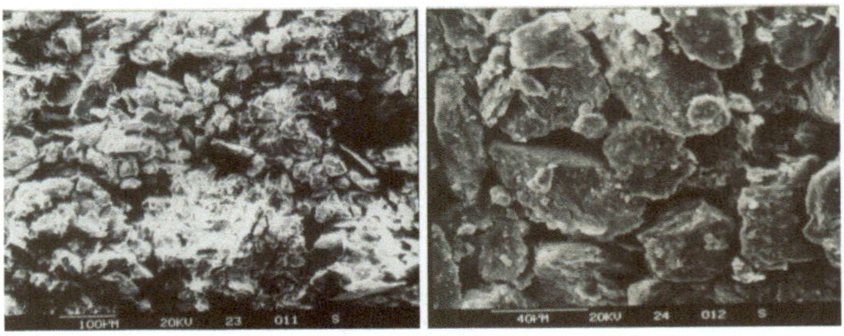

Fig. 2.12 Representative SEM micrograph of silt sediments within 5 m depth in horizontal and vertical directions

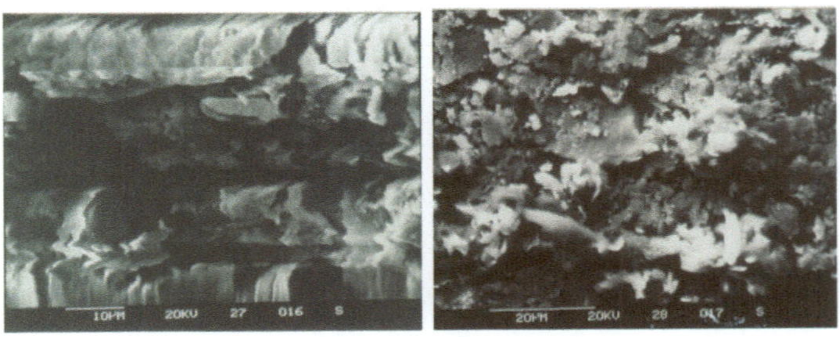

Fig. 2.13 Representative SEM micrograph of silt sediments within 15 m depth in horizontal and vertical directions

The wet density of the sediment of the modern Yellow River Delta is generally in the range of 1.71–2.09 g/cm^3. The wet density of sediments in the northwest is generally in the range of 1.71–2.06 g/cm^3, that in the north is generally in the range of 1.83–2.09 g/cm^3, and in the eastern part is generally 1.86–2.03 g/cm^3, and generally 1.89–2.05 g/cm^3 in the south.

The dry density of the sediment of the modern Yellow River Delta is generally in the range of 1.21–1.96 g/cm^3. The dry density of sediments in the northwest is generally in the range of 1.21–1.68 g/cm^3, that in the north is generally in the range of

1.32–1.66 g/cm^3, and in the eastern part is generally 1.31–1.50 g/cm^3, and generally 1.45–1.69 g/cm^3 in the south.

The pore ratio of the sediment of the modern Yellow River Delta is generally in the range of 0.60–1.25. The pore ratio of sediments in the northwest is generally in the range of 0.60–1.25, that in the north is generally in the range of 0.69–1.15, and in the eastern part is generally 0.64– 1.10, and generally 0.64–1.08 in the south.

The liquid limit of the sediment of the modern Yellow River Delta is generally in the range of 22.0–50.7, while the plastic index is generally in the range of 6.7–18.5. The liquid limit and plastic index of sediments in the northwest is generally in the range of 27.0–40.0 and 9.6–15.8, respectively, that in the north is generally in the range of 25.4–35.4 and 8.7–14.5, and in the eastern part is generally 22.0–47.0 and 7.8–14.4, and generally 23.0–35.0 and 7.2–13.8 in the south.

The regional variation of sediment water content, void ratio, and liquid plastic limit basically increased with the increase of depth of water and the distance from the estuary of the Diaokou flow. In areas within 10 m water depth, the sediment water content is basically within 30%, and the water content is higher in areas where water depth is higher than 10 m, which can up to 60%.

(2) ***Mechanical Properties***

The mechanical properties of sediments in the modern Yellow River Delta are related to granulometric composition, mineral composition, texture and structure, and the stress history. Jia et al. (2011) obtained the variation range of mechanical properties of sediments in four typical study areas of the modern Yellow River Delta through consolidation experiments and undrained static triaxial tests (Table 2.14). Meng et al. (2012) conducted field tests on the physical and mechanical properties of sediments on tidal flat along the modern Yellow River Delta, as shown in Table 2.15.

2.6 Summary

This chapter first analyzes the transporting characteristics of the Yellow River estuary and its material composition based on observations at Lijin Hydrological Station. Then it introduces the formation and evolution, the topography and morphology, and the ocean dynamic environment of the Yellow River Delta. Finally, the seabed sediment properties including sediment types and distribution, geological strata, sediment grain size and mineral composition, microstructure, as well as physical and mechanical properties of the Modern Yellow River Delta are demonstrated.

Table 2.14 Sediment mechanical properties in the modern Yellow River Delta

Study site	Compressibility/MPa^{-1}	Compression modulus/MPa	Effective angle of internal friction/°	Cohesion/kPa	Pre-consolidation pressure/kPa	Over-consolidation
Northwest	0.0983–0.771	2.79–19.30	8.3–27.4	9–38	62–155.9	0.4–10.3
North	0.078–0.720	3.06–22.30	13.1–35.8	10–35	82.1–158.3	0.5–6.0
East	0.089–0.643	2.65–18.48	12.0–36.5	7–36	40.9–207.0	0.5–1.7
South	0.077–0.295	5.99–21.48	12.5–38.9	6–34	87.1–169.2	0.5–2.6

Table 2.15 Tests of sediment mechanical properties in the modern Yellow River Delta (Meng et al. 2012)

Study site	Latitude and longitude	Penetration resistance/N	Shear strength/kPa	Compressibility/MPa^{-1}
Dongying harbor	38°04.042′N, 118°56.499′E	0.7	8.0	0.122
Dongying harbor	38°04.045′N, 118°56.468′E	1.0	5.0	0.133
Dongying harbor	38°04.043′N, 118°56.494′E	1.1	1.0	0.130
Dongying harbor	38°04.083′N, 118°56.407′E	4.1	23.0	0.117
Dongying harbor	38°04.082′N, 118°56.407′E	3.3	10.3	0.130
Dongying harbor	38°04.080′N, 118°56.418′E	2.9	10.0	0.122
Dongying harbor	38°04.077′N, 118°56.402′E	4.4	9.0	0.130
Pile 106	38°08.110′N, 118°46.816′E	0.3	0	0.179
Gudong	37°51.174′N, 119°05.829′E	0	7.7	0.151
Feiyan beach	38°08.379′N, 118°42.634′E	0.5	6.5	0.153
Chezigou fishing village	38°06.129′N, 118°27.872′E	0.8	4.7	0.120
Mines 4th team 7th	38°06.131′N, 118°27.870′E	1.6	8.8	0.169
Xintan	38°08.054′N, 118°12.773′E	1.9	7.2	0.149
Guangli harbor	37°42.137′N, 119°16.457′E	0.4	3.2	0.128

References

Chen JQ, Li XJ, Yang JR et al (2009) Analysis on the reclamation of Xinkou gate in Yellow River between 1996 to 2007 31(3): 101–102

Chen SL, Zhang GA, Gu GC (2004) Mechanism of heavy coastal erosion on Yellow River delta and its countermeasures. J Hydraulic Eng 7:1–6

Chen SL, Gu GC, Wu SY (2007) Storm surge disasters on the Yellow River delta and countermeasures. Geogr Geo-Inf Sci 23(3):100–112

Cheng GD, Xue CT (1997) Sedimentary geology of the yellow river delta. Geology Publishing House, Beijing

Chu ZX, Sun XG, Zhai SK et al (2006) Changing pattern of accretion/erosion of the modern Yellow River (Huanghe) subaerial delta, China: Based on remote sensing images. Mar Geol 227(1):13–30

Ding D, Dong W (1988) A preliminary study on the erosion of the modern Huanghe River delta. Mar Geol Quat Geol 8(3):53–60
Dong NH, Wang GY (1997) Residual current analysis of the Yellow River mouth area in Bohai gulf. J Ocean Huanghai Bohai Seas 15(1):64–69
Fan H, Huang HJ, Thomas QZ et al (2006) River mouth bar formation, riverbed aggradation and channel migration in the modern Huanghe (Yellow) River delta, China. Geomorphology 74:124–136
Feng XL, Qi HS, Wang T, et al (2004) Geomorphological evolution and geological disasters analysis in Chengdao sea area of the Yellow River delta. Rock Soil Mech 25(supp): 17–20
Feng XL, Lin L, Zhuang ZY (1999) Relationship between geotechnical engineering parameters and sedimentary environment of the modern Yellow River underwater delta since the Holocene [J]. Coast Eng 4:1–7
Geng XS, Zhang QN, Fu MZ et al (1988) Quantitative analysis and evaluation on the erosion and deposition of Haigang in Yellow River mouth. Acta Oceanol Sin 10(6):712–719
Huang HJ, Fan H (2004) Monitoring changes of nearshore zones in the Huanghe (Yellow River) delta since 1976. Ocean Limnol Sin 35(4):306–314
Huang HJ, Le KT, Li F (2005) Interaction between Yellow River delta and Bohai and Huanghai sea. Science Press, Beijing
Jiang MX, Yang FD, Huo RJ et al (2004) On the evolution of the Huanghe River delta and its relation to the riverway and sand entering the sea. Trans Ocean Limnol 3:6–15
Jia YG, Shan HX, Yang XJ et al (2011) Sediment dynamics and geologic hazards in the Estuary of Yellow River. Science Press, Beijing, China
Li AL, Li GX, Cao LH et al (2004) The coast erosion and evolution of abandoned lobe of the Yellow River delta. Acta Geogr Sin 59(5):731–737
Li Y, Mehta AJ (1997) Mud fluidization by water waves. Cohesive sediments, Wiley, New York
Li ZG (1984) Analysis of tidal currents in the sea area near the Yellow River delta. Mar Sci Bull 3(5):12–16
Liu DS (1964) Loess in middle course of Yellow River. Sciences Press, Beijing
Meng XM, Jia YG, Shan HX (2012) An experimental study on erodibility of intertidal sediments in the Yellow River delta. Int J Sedim Res 27(2):240–249
Prior DB, Suhayda JN, Lu NZ et al (1989) Storm wave reactivation of a submarine landslide. Nature
Prior DB, Yang ZS, Bornhold BD et al (1986) Active slope failure, sediment collapse, and silt flows on the modern subaqueous Huanghe (Yellow River) delta. Geo-Mar Lett (2)
Ren YC, Zhou YQ (1994) Morphological features and evolution of the abandoned Huanghe River delta. Mar Geol Quat Geol 14(2):19–28
Shi MC, Zhao JP (1985) The analysis of hydrographical characteristics in the nontidal region M2 near delta of the Huanghe River. J Ocean Univ China 15(1):129–131
Si SH, Zhang GQ (1985) Variation in depth at area of Huanghe River delta and assumption for choicing a more suitable site for deepwater port constructing. Coast Eng 4(1):18–24
Wang ZY, Wang WL, Tian SM (2007) Mineral composition and distribution of the sediment in the Yellow River basin. J Sediment Res 5:1–8
Well JT, Coleman JM (1987) Wetland loss and the subdelta life cycle. Estuar Coast Shelf Sci 25(1):111–125
Xue CT (1994) Division and recorgnition of modern Yellow River delta lobes. Geogr Res 13(2):59–66
Yan TS, Pu GJ, Zhang JH (2006) Costal erosion and protection of Shengli Oilfield in Yellow River Delta. Yellow River Conservancy Press, Zhengzhou
Yang XJ, Jia YG, Liu HJ (2009) Discussion on consolidation pressure of silty soil in the Yellow River Delta [J]. J East China Norm Univ (Natural Sciences) 03:146–152
Yang ZS, Wang T (1993) Chengdao oilfield exploration and development. Ocean University of China Press, Qingdao
Yin YH, Zhou YQ, Ding D (2004) Evolution of modern Yellow River delta coast. Mar Sci Bull 23(2):32–39

Zang QY (1996) Onshore sediment in Yellow River Delta. China Ocean Press, Beijing
Zhao DB (2004) The study of the erosion of Diaokou course coast in Yellow River delta (Master's thesis). Qingdao, Ocean Unversity of China, pp 1–75

Chapter 3
Erosion Survey of the Modern Yellow River Delta

3.1 Overview

Delta areas are not only the world's most economically developed regions, but also the most complex areas of the Earth's environmentally dynamic systems. In recent years, due to the impact of human activities in the watersheds, and the sediment brought by rivers to the sea was significantly reduced, resulting in a strong retreating delta coast (Komar 2000; Chen and Chen 2002). For example, for the Nile in Egypt in the twentieth century, the amount of sediment entering the sea was (1.2–1.4) × 108 t/a; but since the construction of the Aswan Dam in 1964, due to the intercepting and storing the sediment further upstream, 98% of the sediment was trapped in the dam, and the delta suffered a strong coastal erosion (Fanos 1995). Similar cases are the Mississippi Delta in the United States (Trenhaile 1997), Ebro River Delta in Spain (Guillen and Palanque 1997) and Luanhe Delta in China (Qian 1994), and the Yellow River Delta (Chen et al. 2004), especially in the Yellow River Delta.

Due to its protruding morphology into the Bohai Sea and the decrease in water and sediment supply of Yellow River Delta in recent years, most of the Yellow River Delta Region, with the exception of a small area around the river mouth, are facing varying degrees of coastal erosion under the dynamics of currents, waves, tides, and storm surges. Once the world's largest piece of newly formed coastal wetlands, the delta area is in gradual degradation of land resources, the wetlands suffer a gradual loss of its function to buffer storm surges, therefore, the development of modern progradational delta and the economic developmental environment is deteriorating, detrimental not only to the nearshore platforms on the Shengli Oil Field, and pipeline infrastructure, but also to the entire Yellow River Delta's national and efficient eco-economic model of wetland areas and their natural environment has also brought serious consequences (Yang and Wang 1995; Wu et al. 2000). Therefore, the research on erosion characteristics around the Yellow River Delta would be applicable to the design and construction of nearshore oil field engineering and development.

Y. Jia et al., *Wave-Forced Sediment Erosion and Resuspension in the Yellow River Delta*, Springer Oceanography, https://doi.org/10.1007/978-981-13-7032-8_3

This chapter summarizes the study on the erosion status of the modern Yellow River Delta. Field measurements conducted on the north abandoned the lobe of the Yellow River Delta, which has become a kind of typical beach strongly eroded by waves after the Yellow River estuary changing to Qingshuigou flow path. Historical topographic data are also collected to demonstrate the erosion process of the subaqueous delta.

3.2 Erosion Survey of a Typical Coast

3.2.1 Methodology

(1) ***Study Area***

(1) Survey sites

The research area of Chezigou fishing village is above the leaflets of the Yellow River Sub-Delta, located in the northern part of the modern Yellow River Delta (38°05.981'N, 118°27.815') and west of the Qingkou River, here is the Yellow River inlet from 1904 to 1929 (Fig. 3.1) with no beach protection measure as a nature open beach. The range of diffusion of sediment was limited since the Yellow River rerouting to the Qingshuigou flow path in 1976, its affecting north boundary not exceeding the Yellow River's seaport, and the sediment flux of Qingshuigou flow path at the discarded leaflets which are located in the north of the Yellow River Delta was 0.1 mm/a (Li et al. 2005). In the absence of sediment supply, coupled with the lack of effective revetment measures, the erosion of the shoreline is very serious under the influence of ocean dynamics such as tides, waves, and storm tide. The original Chezigou fishing village was unable to continue living under severe erosion and has now been relocated.

Through the research on the beach surface, many scholars have found that erosion and recession first manifested as narrowing of the intertidal zone, the erosion rate of the low-tide line was greater than that of the high-water line, the beach surface was roughened with many kinds of recession microtopography appearing, such as erosion scarp, eroded bad land, mud gravel, and erosion pits. This phenomenon also exists in the study area (Fig. 3.1a–d). The elevation difference of the erosion scarp in the study area is 20–50 cm, with mud gravels and collapsed large-scale soil blocks below the scarp, showing a strong erosion geomorphology. In the study area, there are special geomorphologies such as ripple marks and erosion scarps on the tidal flats, which show that the tidal flats of Chezigou are erosion–sedimentation tidal flats, and the waves have a significant shaping effect on beaches.

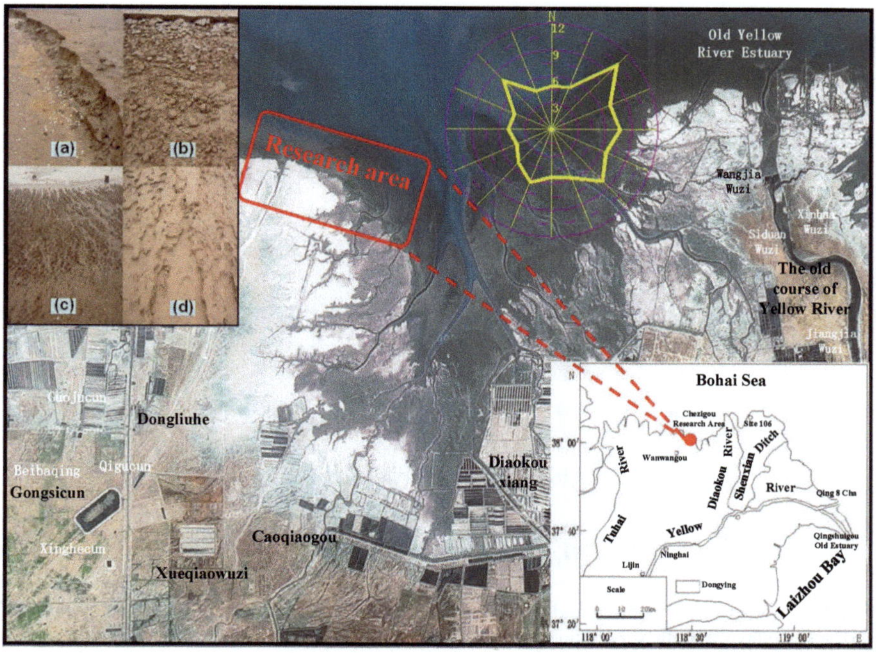

Fig. 3.1 Location map of research area (Among them: The upper left corner is the recession geomorphology in the study area: **a** erosion scarp; **b** eroded bad land; **c** mud gravel; **d** erosion pit. The upper part is a rose diagram of wave direction)

As the frontier of sea–land interaction, the tidal flat is the most active area of sea–land interaction. Its characteristics largely represent the nature of the coastal area. The silt and erosion changes in the Yellow River Delta directly affect the development of tidal flats and ditch. However, there are relatively few researches on muddy silt tidal flats with a large width, muddy beach, and tidal ditches.

(2) ***Data Acquisition and Analysis***

For a long time, there is no uniform standard of the demarcation of the shoreline. Because of the different research objects and perspectives of different scholars, they differ greatly in the demarcation of land and sea, and even contradict each other (Lu 2005). However, for sandy beaches with steep shores, people usually use the steep shore to define the shoreline. Beaches with steep shores generally do not develop berm, and the beach directly connects with scarp, where vegetation development often occurs above and basically no vegetation under. A unified understanding of the shoreline is basically reached, that is the intersection of the beach and the scarp is the shoreline (Han 2008). The development of the steep shore in the selected study area in this chapter is very obviously and meets the conditions for the demarcation of the shoreline. Therefore, the steep shore is selected as the shoreline. The shoreline survey uses high-precision GPS to select multiple points above the steep shore for

Fig. 3.2 Layout of survey line and conditions of wind towers at different times (The upper left corner is the layout of the survey line in the study area, there are 4 survey lines: A–D, using wooden piles for positioning; the other pictures in the picture are the changes of the beach around the wind tower and its base in different periods, and the benchmarks for line elevation measurement are selected to measure the wind. The measurement is based on the base of the wind tower)

measurements. From December 2008 to December 2009, a total of 10 measurements were performed, and the actual distance between the wind tower and the shoreline was used for the correction of result from GPS.

In addition, through field trips of tidal flats near Chezigou fishing village, four profiles (Fig. 3.2) were placed on the site. In order to make the location of the profiles relatively constant during the survey, piles were buried at the observation points of the profile. The wooden is the first choice for the materials of piles because of its light so that it can be less affected by self-settling. Each pile has a length of 1 m and a diameter of about 4 cm. Engrave a groove at 10 cm from the top, put the pile until the groove. The distance between the piles is 20 m. Using this method, four survey lines were placed on the tidal flat in the study area. The location of the survey line is shown in Fig. 3.2. From November 2008 to December 2009, 11 elevation measurements were taken along the profiles. The elevation measuring instrument is a level gauge with an accuracy of 0.1 mm. The measurement is based on the base of the wind tower, which is the only structure on the tidal flat, as the reference point, and a closed leveling line measurement is performed on the elevation of the section.

3.2.2 Results

(1) *Evolution of the Shoreline*

From December 2008 to December 2009, a total of 10 measurements of shoreline changes were performed. Using GIS software, all measured latitude and longitude coordinate data were converted to geodetic coordinate and then data points were plotted. Connect each measurement point in order with a smooth curve to obtain the shoreline for each measurement, and then the measured results of the distance between the shoreline and the wind tower were used to regulate the shoreline. Take the result of the first shoreline survey in December 2008 as the baseline, and then calculate the difference between the shoreline and the baseline to obtain the change in the shoreline from the baseline (Fig. 3.3). As can be seen from the figure, during the period from December 2008 to December 2009, the shoreline continued to erode, and the maximum erosion distance was nearly 130 m. The erosion was very strong. At the same time, it can also be seen that there is a great difference in the rate of

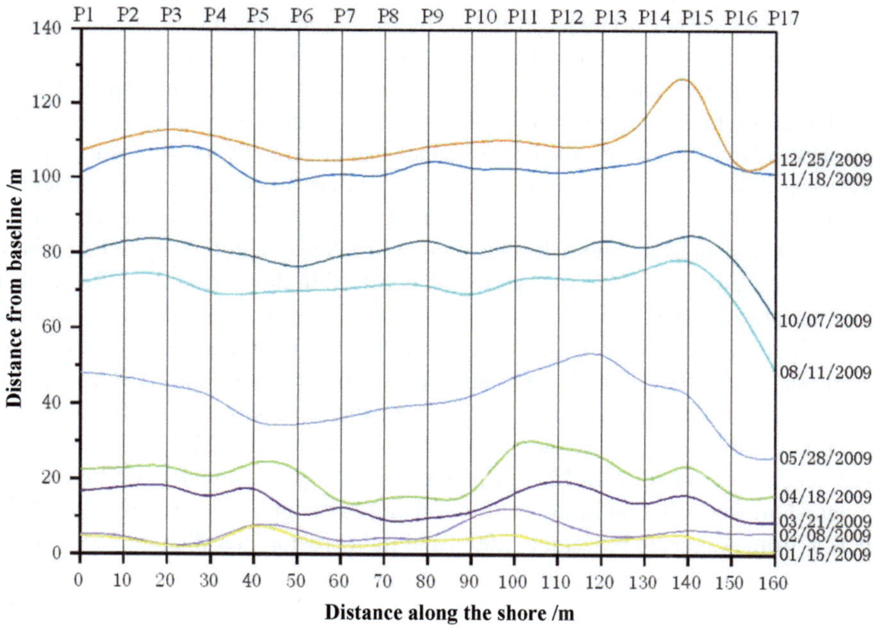

Fig. 3.3 The relative change of the position of the shoreline over time (In the figure, the x coordinate axis is distance along the coast, and the starting point for the shoreline measurement is zero point, the unit is m; the y coordinate axis is the distance between the measured shoreline and the baseline in different periods, that is, the variation of the shoreline relative to the baseline. The unit is m; the right side of the drawing is the measurement date; the upper side of the drawing is the profile perpendicular to the shoreline, P1–P17 are used to calculate the daily average erosion rate of the shoreline)

erosion of shoreline during different observation periods; even in the same period, the rate of erosion shown by different locations is not the same.

(2) ***Evolution of the Beach Elevation***

In order to facilitate the analysis, some conversions of 11 elevation measurements taken along the profiles from October 2008 to December 2009 were made, the data of the measured beach (N, E, Z) was converted to km network (X, Y, Z), to plot the profile of the survey line (Fig. 3.4). At different periods, the elevation changes significantly on the tidal flat, and the erosion and deposition phenomenon is obvious.

As can be seen from figure, the wind tower is located on the tidal flat in the study area. The interface between its foundation and surrounding soil can intuitively reflect the erosion or deposition of the beach. From October to December 2008, the base of the wind tower was exposed on the beach surface which was originally under the beach surface, exposed about 22 cm, the beach surface was eroded; during the period from December 2008 to February 2009, the interface between the foundation and the surrounding soil changed little. The erosion phenomenon was not obvious. The wind tower was broken off and fell on the beach surface during the period, indicating that there was windy weather during this period. From February to March of 2009, the interface was significantly reduced and the base exposed about 35 cm, strong siltation occurred on the beach surface; during March–April 2009, the interface was moved up greatly, and the foundation was covered again under the soil; during April–August 2009, the interface changed slightly. However, the change was not obvious; during October to December 2009, the beach surface was covered by ice, and when the ice layer was thickest at the wind tower, it was more than 80 cm above the foundation. The thickness of the ice layer exceeded that of the foundation of the wind tower.

As can be seen from figure, the elevation of line A was significantly reduced from October to December 2008, strong erosion occurred in this period; from December 2008 to January 2009, only slight erosion occurred within a short distance near the coast; there was no change in elevation from January to February 2009, basically; the elevation of tidal flat reached a minimum from February to March 2009, and strong erosion occurred; from March to April 2009, the shoreline was strongly silted and the maximum sedimentation thickness up to 35 cm; from April to May 2009, the beach was generally eroded; from May to August 2009, it was characterized by erosion in nearshore and siltation in the offshore; there was slight erosion in the nearshore section from August to November 2009, and light erosion and siltation occurred on the beach.

The change of line B is basically the same as line A, but in January to February 2009, siltation occurred and the maximum sedimentation thickness was as large as 72 cm from March to April 2009.

The overall condition of line C is consistent with line B, and the maximum sedimentation thickness is 51 cm from March to April 2009.

As for line D, erosion occurred in the nearshore within 180 m from October and December of 2008, and significant siltation occurred in the subsequent section. The change from December 2008 to November 2009 was the same as that of line B.

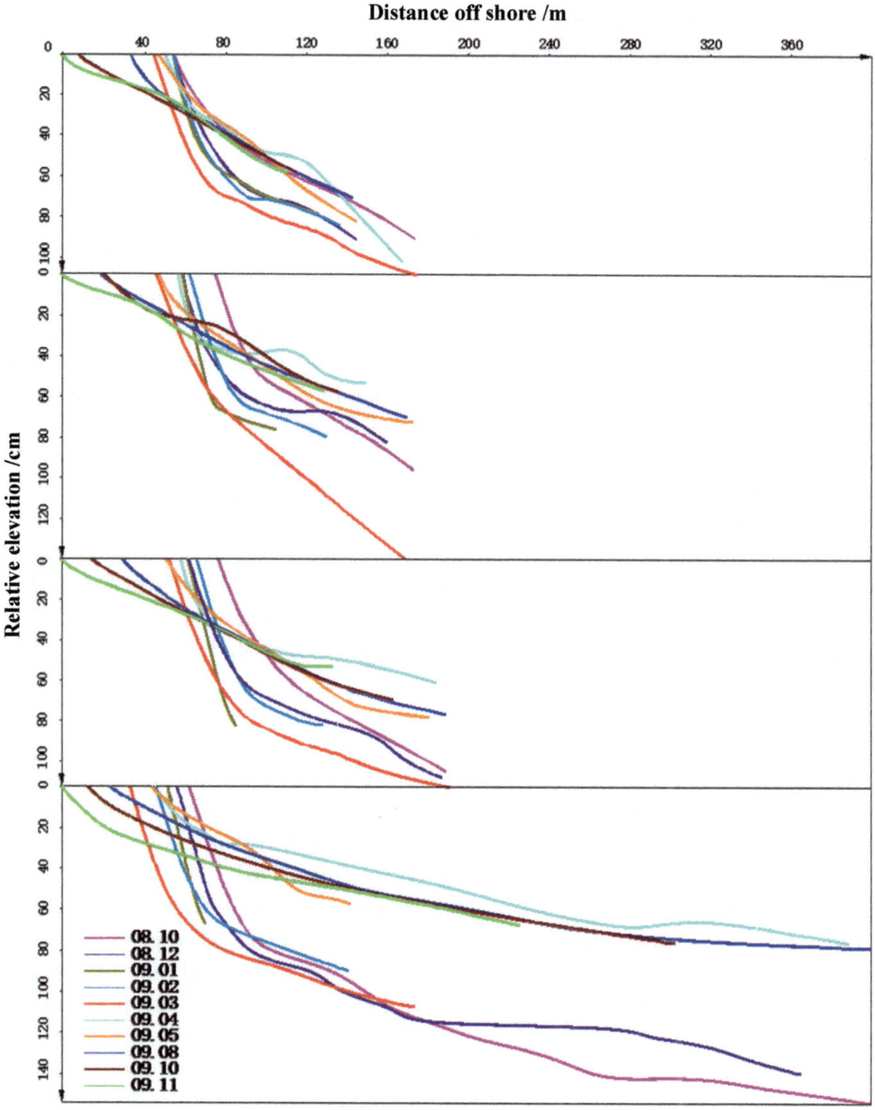

Fig. 3.4 Measurement results of elevation of beach

Among them, from March to April 2009, the maximum sedimentation thickness was 76 cm.

During the measurement of elevation of the beach, because of the serious erosion and recession of shoreline, the starting points of the 4 survey lines were continuously extended to the land along the survey line. The last measurement start point in November 2009 was taken as zero point, and the data in other periods are converted accordingly; in addition, the measurement length of the survey line is limited due to different conditions of ebb tides, sea ice in winter and other factors, and the survey line cannot be completely measured in some periods.

3.2.3 Analysis of Coastal Erosion

(1) ***Factors Influencing the Coastal Evolution***

Based on the shoreline change data in 39 coastal countries, the famous book of Bird (1981) "Change of the Shoreline of a Hundred Years" found that coastal erosion is a common phenomenon across the world in the past 100 years. Erosion and sedimentation are two basic processes for shaping coastal profiles, and these two processes are caused by many factors (including internal forces and external forces). These two processes can be quantified by the amount of sediment balance. For a given coastal unit, if the loss (output) is greater than the income (input), there will be a net loss, which shows a decrease in sediment amount, therefore, erosion of the coast. Conversely, if input exceeds output, the coast will progradation. In this way, coastal morphology has always been evolving over time. For the Yellow River Delta, there are two main factors affecting the evolution of the coast: (1) normal sea conditions, e.g., wind waves, tidal currents, river sediment discharge, sea level change, and sea ice; (2) extreme factors, mainly storm surge.

(a) **Waves**

The waves in the Bohai Sea are mainly wind waves which are obviously influenced by the seasonal winds in the Bohai Sea. Therefore, the distribution of wave elements has obvious seasonal characteristics, with the frequent wave direction of NE and the frequency of 10.3%. Strong wave mainly comes from NNE–ENE, especially from NE to strongest, followed by partial NW. As can be seen from the statistical data (Table 3.1), wave elements such as maximum wave height, average wave height, and maximum wave cycle are generally highest in spring and autumn, followed by winter, and weakest in summer, that is, wave energy is relatively high in spring and autumn.

Wave is one of the main dynamic factors of offshore sediment movement and coastal geomorphology. The same is true for the Yellow River Delta, where the sediment movement in the coastal areas is mainly caused by waves and water flow. After the wave propagates to the nearshore, it will be deformed due to changes in the topography, such as shallow water bottom friction deformation, wave refraction, diffraction, wave crashing, etc. These changes of waves in the nearshore are accompanied by the release of large amounts of wave energy, which caused erosion of the shore beach (Fig. 3.5), disturbing and lifting of sediments on the seabed, and causing changes in the direction and intensity of sediment movement. Longitudinally, the oblique incidence of waves forms a coastal current parallel to the shore, causing sediment transport along the coast. Horizontally, due to the asymmetry of the nearshore wave motion and the underwater bank slope having a certain slope, lateral

Table 3.1 Monthly statistics of waves and tidal ranges (Yan 2006)

Parameter	Month												Year
	1	2	3	4	5	6	7	8	9	10	11	12	
Max. wave height (m)	–	–	3.9	3.1	3.2	3.4	2.0	2.2	2.6	4.2	5.8	3.7	3.41
Ave. wave height (m)	–	–	0.6	0.4	0.4	0.4	0.5	0.4	0.4	0.6	0.7	0.7	0.51
Max. wave period (s)	–	–	8.3	8.9	6.7	7.5	4.8	5.9	5.6	7.6	9.0	6.8	7.11
Max. tidal range (m)	1.79	1.54	1.57	2.12	1.43	1.6	1.52	1.4	1.42	1.98	2.32	1.56	1.69
Ave. tidal range (m)	0.93	0.92	0.84	0.89	0.86	0.93	0.94	0.8	0.85	0.86	0.94	0.96	0.89

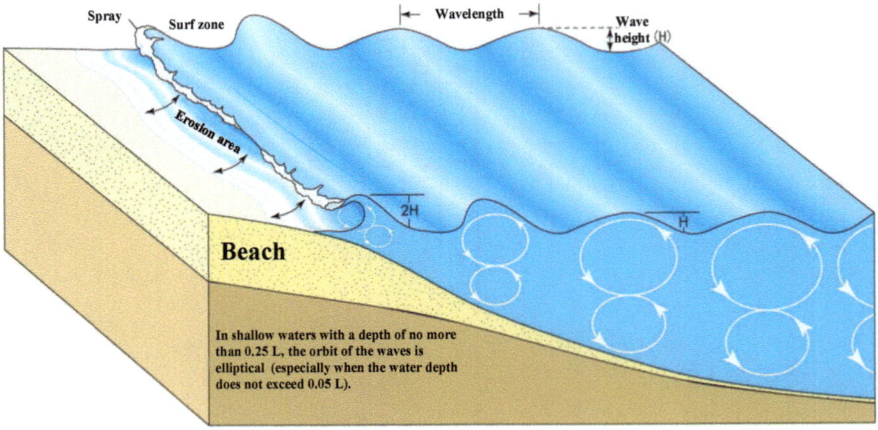

Fig. 3.5 Schematic diagram of the effect of waves on the beach

movement of the sediment near the shore is caused. The direction of motion has a greater relationship with wave steepness. Generally speaking, a small wave steepness is favorable for the migration of sediments to the shore. Larger wave steepness is more conducive to the offshore movement of sediments.

(b) **Tidal Current**

Tidal current is the main and eternal driving force for transporting sediment from the Yellow River Delta. The waters in the northern part of the Yellow River Delta are irregular semidiurnal tides. The tidal current pattern in the sea area in which the study area is located is controlled by M2 amphidromic point positions of tidal waves (119° 04′ E, 38° 04′N) in the northeast. From the amphidromic point to the west, the tidal range gradually increases, and the maximum tidal velocity gradually decreases. The average high tidal range of the study area is 1.84 m, the average small tidal range is 0.96 m, the semidiurnal tide age is 2d 18 h, the average high tide gap is 4 h, 27 m, and the average low tide gap is 9 h, 45 m. The base level of the tide is Dagu, with an average high tide of 249 cm, an average low tide of 122 cm, and an average tidal range of 127 cm. In this area, there are two rising tides and two falling tides each day. The direction of the rising tide is SN direction, the direction of the falling tide is NW direction, and the mainstream is parallel to the shoreline and flows reciprocally. The tidal range is one of the important signs of tidal strength in a sea area. According to statistics, the maximum tidal range reached the maximum of 2.32 m in November, followed by April, and the summer tidal range was the smallest, only about 1.45 m (Table 3.1).

The reciprocating motion of the nearshore fluctuation tide has an important influence on the transport and distribution of fine-grained sediments. When tidal currents interact with local wind-driven currents, rivers, and ocean circulations, they tend to cause net transport of sediments in a certain direction. As the water depth changes along the coast, the seabed friction increases and the tidal flow velocity decreases toward the shore. The transport of fine-grained sediments is controlled by the "lagging effect of erosion" and "lagging effect of sedimentation", showing the tendency of transport and deposition to the shore. For the Yellow River Delta, the tidal current is also one of the important driving forces for the evolution of the beach.

(c) **River Sediment Discharge**

The transport of sediments by rivers and the sediment transport caused by marine dynamic erosion of the coast are the two aspects of the development of the material base of the delta and the sediment discharge from the Yellow River is the most fundamental factor affecting the evolution of the delta. In 1976, the Yellow River diverted from the Qingshuigou Channel into the sea, and the northern Yellow River Delta beach lost its direct supply from the Yellow River; while the sediment diffusion scope from Qingshuigou was limited, affecting the northern boundary no more than the Yellow River seaport, and the depositional flux of Qingshuigou channel sediment in Feiyantan is about 0.1 mm/year. The northern beach of the Yellow River Delta lost its supply from the Yellow River after 1976, and lacked material base supplies for the development of the delta. Therefore, for the northern beach, the amount of sediment from the Yellow River has little effect on the recent evolution of the beach.

(d) **Sea Level Rise**

According to the "2009 China Sea Level Bulletin" issued by the State Oceanic Administration, the average rate of sea level rise in the Bohai Sea is 2.3 mm/y. In 2009, the sea level in the Bohai Sea was 53 mm higher than the average year, which is basically the same as in 2008. Among them, the relative rising height of sea level from June to September is more obvious than that of the normal year, and the rising height is about 30 cm. From January to April and from November to December, the relative height of sea level was lower than normal. Although sea levels have risen in May and October, the increase has been modest. The monthly changes in sea level are shown in Fig. 3.6.

Sea level change is one of the important factors of coastal evolution. When the sea level rises, the coastal dynamic activity belt moves to the land, resulting in coastal erosion, sand-bar migration, and lowland flooding. Each type of terrain unit responds specifically to sea level rise values and rates because of its different resistances. People have conducted extensive discussions on the coastal beach effect of rising sea level from different regions and natures of the coast. Although there is a consensus on the retreat of the shoreline, there is a great deal of disagreement on the section shaping.

(e) **Sea Ice**

Regarding the impact of sea ice on the evolution of the beach in the Yellow River Delta, there are currently few relevant studies and no literature has been found in this area. However, the Bohai Bay and Laizhou Bay on the verge of the Yellow River Delta belong to the high latitude sea area, and sea ice will appear every winter. According to the 2009 China Ocean Disaster Bulletin issued by the State Oceanic Administration, it is pointed out that the ice conditions in the Bohai Sea and the

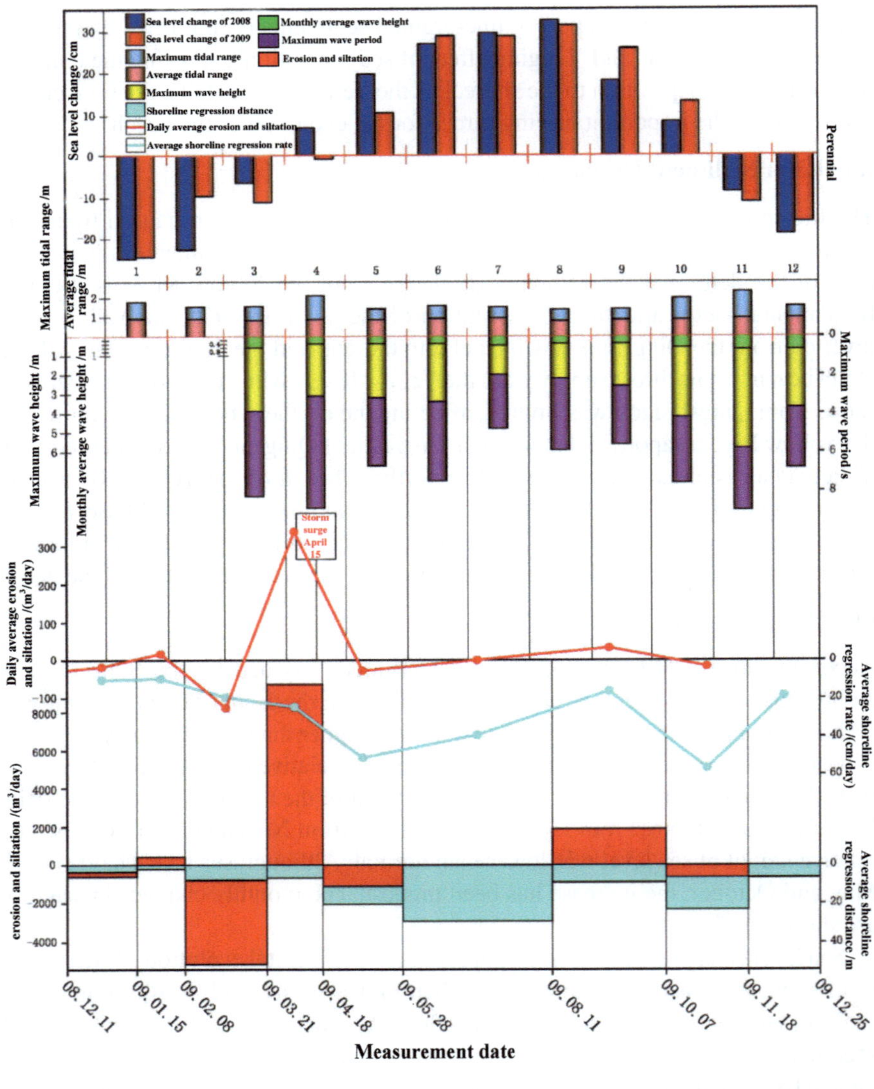

Fig. 3.6 Correspondence between the average rate of erosion on the shoreline and elements of each month (The uppermost part of the figure is the average of the relative increase of sea level in the Bohai Sea in each of the 2008 and 2009 months; next is the statistical value of the maximum tidal range and average tidal range for each month; the central part is the statistical value of monthly wave elements, including the maximum wave height, average wave height, and maximum wave cycle; the curves are the average daily erosion volume of the beach and the average daily shoreline erosion; the bottom part of the figure are beach erosion and shoreline erosion)

Fig. 3.7 Sea ice conditions in the study area in January 2009

northern part of the Yellow Sea in the 2008/2009 winter are generally light, and the ice conditions throughout the winter are approximately the same as in the previous year. The period from the end of January to the beginning of February 2009 was the most serious period of ice this year. The largest ice floe range was 18 nautical miles which occurred in late January in the Bohai Bay. The sediment content in the ice layer whose thickness was about 40–50 cm was extremely high and showed a distinct layered appearance (Fig. 3.7).

(f) **Storm Surge**

The Bohai Sea is a semi-enclosed sea area, and the delta area is exactly the south coast of Bohai Bay and the west coast of Laizhou Bay. The area has a large area of mudflats, a gentle slope, and a small water depth in the sea. Therefore, under the influence of strong northeast winds, the storms increase water obviously, especially after the strong winds in the southeast after turning to the northeast winds, it is easy to form storm surges. During the study period, an extraordinary storm surge occurred on April 15, 2009. Affected by the cold air in the northwestern Bohai Sea and the inland low pressure in the southern Bohai Sea and the low pressure of the Yellow Sea through the Shandong Peninsula, there were 11 levels of gales and 12 levels of gusts on the coast of Cangzhou; the storm surge caused by the strong winds at Huangye tide gauge station caused the greatest storm surge of 176 cm, and it rose to 5.14 m at 6h23, surpassing the warning level of 34 cm; the maximum wind speed of 21.9 m/s was observed in the 0357–1403 sea area near Weifang Port; the maximum wind speed in the 0633 sea area is 35.8 m/s, and the Weifang Port tide station is 3.51 m high and 21 cm below the warning water level (3.72 m). The maximum storm surge was measured at 176 cm in the tide station along the Bohai Sea. At 5:00–8:00 on April 15th, it was the period of astronomical high tide along the coasts of Bohai Bay and Laizhou Bay. Under the influence of strong winds, high tides, and giant waves, this severe storm of temperate regions was formed (Table 3.2).

Table 3.2 The maximum stormwater increase and the highest tide level at the coastal stations on April 15, 2009

Site	Measured maximum stormwater increase and highest tide level and time				Warning water level (cm)	Measured super-warning water level (cm)
	Maximum water increase (cm)	Time (days, hours, minutes)	Highest tide level (cm)	Time (days, hours, minutes)		
Bayuquan	106	15–16:00	291	14–07:00	470	–
Huludao	90	15–16:00	340	15–20:02	405	–
Jingtanggang	101	15–07:00	235	15–05:46	290	–
Caofeidian	127	15–07:00	345	15–06:51	380	–
Huangwei	176	15–08:00	514	15–06:23	480	34
Tanggu	173	15–09:00	484	15–06:49	490	–
Longkou	141	15–11:00	207	15–10:28	270	–
Penglai	93	15–13:00	299	15–12:40	310	–
Yantai	72	15–13:00	379	15–13:08	390	–

(2) ***Mechanism of the Shoreline Evolution***

As can be seen from Fig. 3.3, the shoreline shows a continuous erosion recession, but the rate of erosion recession is different at different locations. In order to calculate the average rate of erosion recession of the shoreline accurately, 17 sections (P1–P17) are arranged in the direction perpendicular to the shoreline, and the section spacing is 10 m. The erosion recession distance of the 17 shore sections of each interval is obtained, and then the average value was taken and divided by the corresponding observation interval, which is the daily average erosion recession rate of the shoreline, shown in Table 3.3. During the period from December 2008 to April 2009, the rate of erosion recession is relatively slow, and the rate of erosion recession is faster during the period from April to November 2009. The slowest rate of erosion recession occurred between January 15 and February 8, 2009, which was only 9.9 cm/d. The fastest occurred during the period from October 7 to November 18, 2009, with a rate of 56.6.cm/d, followed by the period from April 18 to May 28, 2009, at a rate of 51.2 cm/d.

It's an open natural coast without any protective measures near the abandoned village of Chezigou fishing village in the northern part of the Yellow River Delta. We can see from the observation result that the shoreline has been in an erosive state as time changes, which is mainly due to the influence of the topography of the Bohai Sea area. There is the longest wind zone in the NNE–NE–ENE direction, as a result, waves can grow fully in the wind zone before spreading to the shallow waters near the shore. However, there are differences in the rate of shoreline erosion in different periods. The average rate of shoreline erosion is only 9.9 cm/d, and the highest is 56.6 cm/d. In general, the shoreline has a large erosion rate in spring and

Table 3.3 Average line erosion rate

Observation date	09.01.15	09.02.08	09.03.21	09.04.18	09.05.28	09.08.11	09.10.07	09.11.18	09.12.25	Total
Time interval (d)	35	24	41	28	40	75	57	41	36	377
Average distance of erosion recession (m)	3.69	2.37	7.99	6.85	20.48	29.32	9.28	23.20	6.65	109.83
Average rate of erosion recession (cm/d)	10.5	9.9	19.5	24.5	51.2	39.1	16.3	56.6	18.5	29.1

autumn, and the erosion rate in summer and winter is small, which is consistent with the season characteristics wave of the area. The waves in this area are mainly wind waves, and the seasonal influence of the wind in the Bohai Sea is obviously, so the distribution of wave elements has significant seasonal characteristics. In the spring, due to the transition monsoon and the disordered wave direction, the wave direction is mainly eastward, and the total frequency is 61.7%, of which the northeastward wind wave accounts for 16.6%; in autumn, the northward wave appears as high as 67.7%, of which the northeastward wave accounts for 33.1 and 13.0% for northwest. The effective wave height and period of the northeast wave are larger than other directions (Table 3.3), that is to say, the wave energy is larger. And the direction of the terrain in study area protruding toward the sea is also northeast, which contributes to a strong interaction with the waves from the northeast, resulting in a very high rate of shoreline erosion, above 50 cm/d, forming a typical erosive beach. However, in the summer, controlled by the southeast wind, the waves are mainly ESE and SE, and the frequency of occurrence is 16.6% and 15.6%, respectively, with a relatively small wave height and small rate of shoreline erosion; In the winter, the waves controlled by the northeast wind are also mainly in the northeast direction. The collapse of the wind tower also indicates that there have been strong winds, which means in theory, the shoreline should have strong erosion recession, but in fact, the rate of winter shoreline erosion is relatively slower, only about 10 cm/d. The above phenomenon is due to the fact that the sea surface freezes during the 3 months, and the existence of sea ice significantly impedes hydrodynamic conditions such as tides and waves, which greatly weakens the hydrodynamic conditions on the shoreline.

Many studies have shown that most of the coast with a sea level relatively rising is eroded. And the sea level relative rising usually leads to an increase in the nearshore water depth, which increases the wave action on the shore and causes coastal erosion (Bird 1996). According to the results of the *2009 China Sea Level Gazette* published by the State Oceanic Administration (Fig. 3.6), in 2008 and 2009, the sea level of Bohai Sea rose significantly only in summer compared to the perennial, and the increase in other seasons is not obvious or at a downtrend. However, hydrodynamic conditions such as waves and tides have little effect on the abandoned leaf petals in the northern part of the Yellow River Delta in summer, which weakens the influence of relative sea level rise on the coastal erosion in the northern part of the Yellow River Delta.

(3) ***Mechanism of the Beach Evolution***

The elevation data of the survey line were further sorted out. ArcInfo 3D Analyst in GIS platform was used to interpolate the data, encrypt the data points, and the resolution of the data points was selected as 0.5 m. Then, the logarithmic data points were meshed to form a curved surface. In order to increase the comparability of erosion deposition in different measuring periods, the rectangular frame range 200 × 160 m was selected for the boundary of the surface. According to this method, the beach surface of each period is made separately, and the volume between the two surfaces is the erosion and deposition amount of the beach surface of the corresponding period, as shown in Table 3.4, where the positive value represents deposition and the negative value represents erosion.

As can be seen from Table 3.4, the beach surface presents an alternate state of erosion and deposition. The most serious erosion occurred from February 8 to March 21, 2009, and the amount of erosion was as high as 5150 m^3, followed by the period from October 1 to December 11, 2008. The most serious deposition occurred from March 21 to April 18, 2009, and the deposition was as high as 9460 m^3. During the period from May 28 to August 11, 2009, the erosion and deposition amount was the smallest, only −28 m^3, and the average daily erosion and deposition amount were almost 0, indicating that the income and expenditure of the beach surface basically reached a balance. It can be seen from the above results that the beach surface is always in a state of constant erosion and deposition. The change of the interface between the wind tower base and the beach surface, which is located on the tidal flat, can roughly indicate and verify the erosion and deposition of the beach surface. As can be seen from the changes in the interface, erosion mainly occurred in two periods, namely, October 2008–December 2008 and February 2009–March 2009. The deposition occurred mainly from March to April 2009, and it can be seen that the erosion and deposition of the beach surface are consistent with the change of the upper survey line.

During the period from October 1 to December 11, 2008, the elevation decreased significantly, and the beach surface experienced a great degree of erosion, with the amount of erosion being 1859.3 m^3. According to the existing research results, the critical starting shear stress of tidal flat in this research area is 0.143–0.175 Pa, with an average of 0.159 Pa (Meng 2009). Wang et al. (2010) studied the evolution and dynamic mechanism of the abandoned leaf petals in the modern Yellow River Delta. It was found that the wave-induced bottom shear stress increased sharply due to the shallow effect and wave breaking after the wave entered shallow water, especially in the abandoned delta front. The wave-induced undercut stress is as high as 0.25 Pa, which is greater than the critical starting shear stress of the tidal flat in the research area, indicating that the wave plays a very important role in the beach erosion process. Wave statistics in the Bohai area show (Fig. 3.6) that the maximum wave height, the monthly average wave height and the maximum wave period all occur in November, which also indicates that waves can provide enough force on the tidal flat during this period, resulting in strong erosion of the beach surface.

From December 13, 2008 to January 15, 2009, there was a certain degree of erosion on the beach surface, but the amount of erosion was small, about 618.7 m^3. Since this period has entered the ice age, the beach surface is covered with sea ice. Although the waves still have an effect on the beach surface, the presence of sea ice weakens the wave force, so the beach surface does not have a large degree of erosion. From January 15 to February 8, 2009, the beach surface was silted up in the nearshore section, and the silting amount was about 420.8 m^3. This period was exactly the most serious ice condition of this year, with the thickness of ice cover being the largest and the phenomenon of sea ice stacking, which greatly weakened the erosion effect of waves on the beach surface. Reimnitz et al. (1990) pointed out that the stacking of sea ice can cause sediment transport to the land, which will lead

Table 3.4 Estimation of erosion and deposition amount of beach surface

The observation date	08.12.11	09.01.15	09.02.08	09.03.21	09.04.18	09.05.28	09.08.11	09.10.07	09.11.18
The time interval (d)	71	35	24	41	28	40	75	57	41
The amount of erosion and deposition (m^3)	−1859.3	−618.7	420.8	−5150.7	9460.6	−1122.5	−28.6	1884.9	−665.0
The average daily amount of erosion and deposition (m^3/d)	−26.2	−17.7	17.5	−125.6	337.9	−28.1	−0.4	33.1	−16.2
The amount of erosion and deposition on the unit surface ($\times 10^{-3}$ $m^3/m^2 \cdot d$)	−0.82	−0.55	0.55	−3.93	10.56	−0.88	−0.01	1.03	−0.51

to deposition of the nearshore beach surface. Therefore, the deposition of the beach surface in the research area may also occur during this period.

During the period from February 8 to March 21, 2009, the beach surface elevation was reduced to the lowest value of the whole measurement period. The beach surface was strongly eroded, the erosion amount was as high as 5150.7 m^3, and the daily average erosion amount was 125.6 m^3/d. This is mainly because during this period, the temperature in the Bohai Sea increased, the ice age ended, the beach surface was no longer covered by sea ice, and the protective layer was lost. This caused the waves to directly act on the beach surface. In addition, the wave action began to strengthen in the Bohai Sea during this period. The wave elements such as the maximum wave height, the monthly average wave height, and the maximum wave period have higher values during this period. In this case, the beach surface is strongly eroded.

From March 21 to April 18, 2009, the entire beach surface not only did not erode, but it was severely silted, with a thickness of up to 76 cm. The elevation was much higher than that measured in the first measurement in October 2008. The deposition amount is as high as 9460.6 m^3, and the daily average deposition amount is 337.9 m^3/d. However, the wave conditions in March and April are not much different. The sea level in April is basically unchanged from the normal year. The only difference is that a major storm surge occurred on April 15th. Therefore, it is believed that this unusually intense deposition was caused by the storm surge. It is generally believed that storms will cause extensive beach erosion and serious coastal property losses (Kelley et al. 1989). Although storms sometimes affect coastal zones in this way, the storm process is complex and plays a significant role in the remodeling and reconstruction of tidal flats. For example, on the coast of Atlantic, the northeast storm generates downflow, which leads to the transport of sediments offshore, while the upwelling caused by the southwest storm leads to sediment deposition (Wright et al. 1994), which is also proved by the effect of this huge storm surge on the northern shore of the Yellow River delta on April 15.

From April 18 to May 28, 2009, the overall beach surface erosion was relatively serious, with the amount of erosion being about 1122.5 m^3. There may be two reasons for this: (1) the action of waves on the beach surface is still the key factor of beach surface erosion; (2) the newly deposited sediments on the beach surface caused by storm surge are not fully consolidated and have low strength, so they are easy to be eroded under the action of waves.

From May 28 to August 11, 2009, the beach surface showed erosion in the nearshore segment and deposition in the offshore segment. On the whole, the beach surface did not change much, and the average daily erosion was only 0.4 m^3/d. Compared with the hydrodynamic conditions in this period, it can be seen that the wave conditions in this period are the weakest throughout the year, and the tidal range is the smallest in this period. Therefore, only slight erosion occurs in the nearshore segment. At this time, hydrodynamic conditions cannot provide enough power to carry away the eroded sediments, resulting in sediment accumulation in the offshore segment. From August 11 to October 7, 2009, the beach surface only suffered slight erosion in the nearshore section, again due to its weak hydrodynamic conditions. From October 7 to November 18, 2009, the beach surface was eroded, mainly due to the strengthening of wave action in this period.

3.3 Erosion Survey of the Subaqueous Delta

3.3.1 Methodology

Since the 1970s, the Hydrology and Water Resources Survey Bureau of the Yellow River estuary have been observing the fixed sections in the Yellow River Delta. It was usually conducted after the flood seasons. During the measurement period, there were short-term and temporary tide stations to observe the tide level for correcting the measured water depth. Based on the topographic profile data of 22 lines on the subaqueous delta (Fig. 3.7) in nearly 30 years from 1976 to 2004, the erosion process was analyzed. First, Kriging interpolation is performed on these data, so that the scattered data points are equally spaced, and then the data of different years are selected for the difference calculation, and the variation is obtained, and the variation is the erosion during the period. Based on this, the erosion process of the Yellow River subaqueous delta and the effect of storm surge on erosion control are analyzed.

The Yellow River subaqueous delta can be divided into two parts in general, as shown in Fig. 3.8. The first part is the area north of the abandoned delta lobes in the northern part of the Yellow River Delta, where the line CS1–CS8 is located. The second part is the area on both sides and east of the Qingshuigou flow path, where the line CS14–CS27 is located. For the convenience of description, the first part is defined as the I zone, and the second part is defined as the II zone. Since the Yellow River was divagated from Qingshuigou flow path to Qing-8 branch in 1996, in this chapter, when studying the erosion process of the underwater delta, the erosion of the subaqueous delta in the two periods from 1976 to 1996 and 1996 to 2004 was discussed with 1996 as the dividing line. Among them, for the amount of erosion, a negative value represents erosion and a positive value represents deposition.

3.3.2 Historical Erosion and Deposition Evolution of the Yellow River Delta

(1) *Survey Results from 1976 to 1996*

According to the contour map of the Yellow River subaqueous delta from 1976 to 1996 (Fig. 3.8), it can be seen that the 0 line of the selected area in the Zone I is located near the Y coordinate of 4240, and the north of the 0 line is mainly characterized by deposition, and the south is characterized by erosion. The area of the eroded area

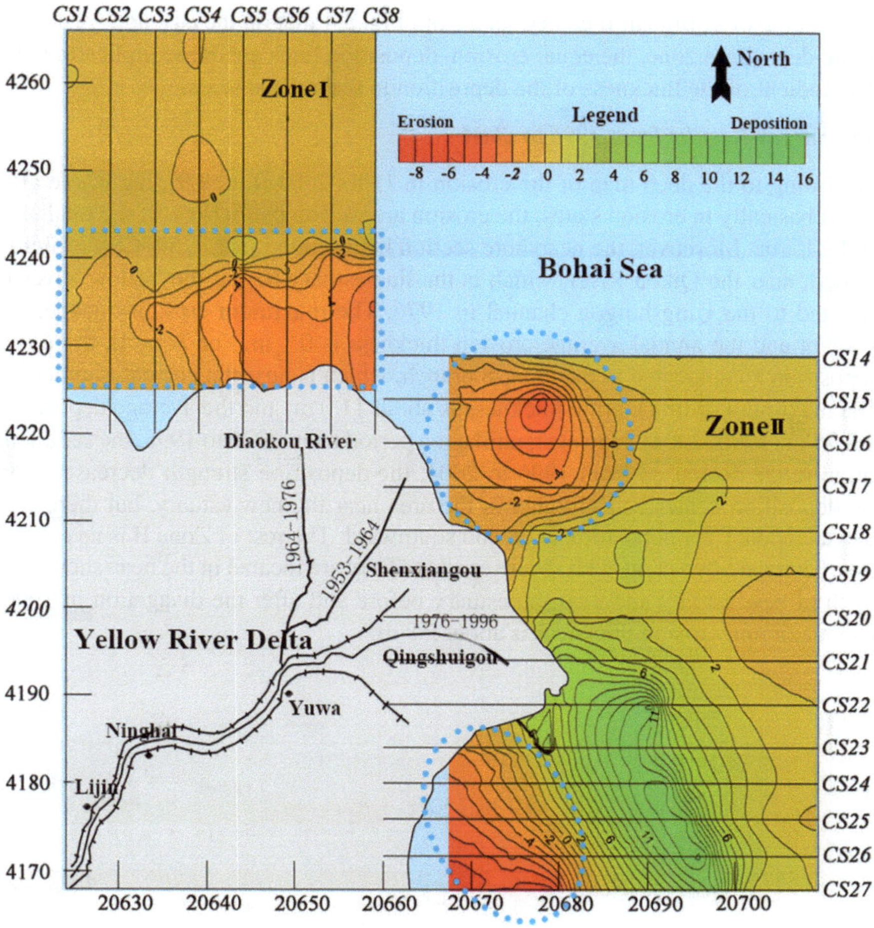

Fig. 3.8 Variation of the subaqueous delta of the Qingshuigou flow path from 1976 to 1996

accounts for about 38% of the area of the I area, while the deposition area accounts for about 62%. The whole area of the II area is mainly silt, and there are two strong erosion areas, which are located in the nearshore section of the test line CS14–CS18 and the nearshore section of the test line CS23–CS27. The area of the erosion area only accounts for about 8.3% of the total area of the study area. The deposition area is mainly concentrated near the estuary, but the deposition center is on the south side of the estuary. The maximum deposition thickness is 14.9 m, the average deposition is 0.93 m, the average deposition thickness of the I area is about 3.58 m, and the average annual deposition is 0.22 m. The total sedimentation in Zone II is about 12.4 billion tons. During the period, the Yellow River Lijin Station has a total of 10.3 billion tons of sand, and its sedimentation capacity is 120% of the amount of sand coming. This also shows that sea sediment carried into the sea by the river is not the only source of sediment. It can be seen from Fig. 3.4 that on the east and west sides of the deposition zone, the equal erosion–deposition lines are dense, indicating that the gradient of the thickness of the deposition in this area is steep.

(2) ***Survey Results from 1996 to 2004***

According to the deep map of the erosion in 1996–2004 (Fig. 3.9), the whole area of I is basically in erosion states, the erosion area accounts for 96% of the total area of the I area. Moreover, the nearshore section between the lines CS4–CS8 is highly eroded, near the Qikou River, which is the line estuary before the Yellow River is diverted to the Qingshuigou channel in 1976. The maximum erosion thickness is −3.2 m and the annual average erosion thickness is 0.1 m. For Zone II, there is a deposition zone located at the Qing-8 branch, which is near the current Shuihekou. The maximum sedimentation thickness is about 11.0 m, and the average deposition is 2.75 m (1996–2004). However, unlike the period from 1976 to 1996, the sedimentation range decreased from 1996 to 2004, the deposition strength decreased, and the deposition center concentrated in the area near the new estuary, but there was also a tendency for deposition to extend southward. The rest of Zone II is an erosive area. There are two areas with severe erosion. They are located in the near-shore area north of new estuary and near the estuary before and after the divagation in 1996. The maximum erosion thickness is about 2.8 m.

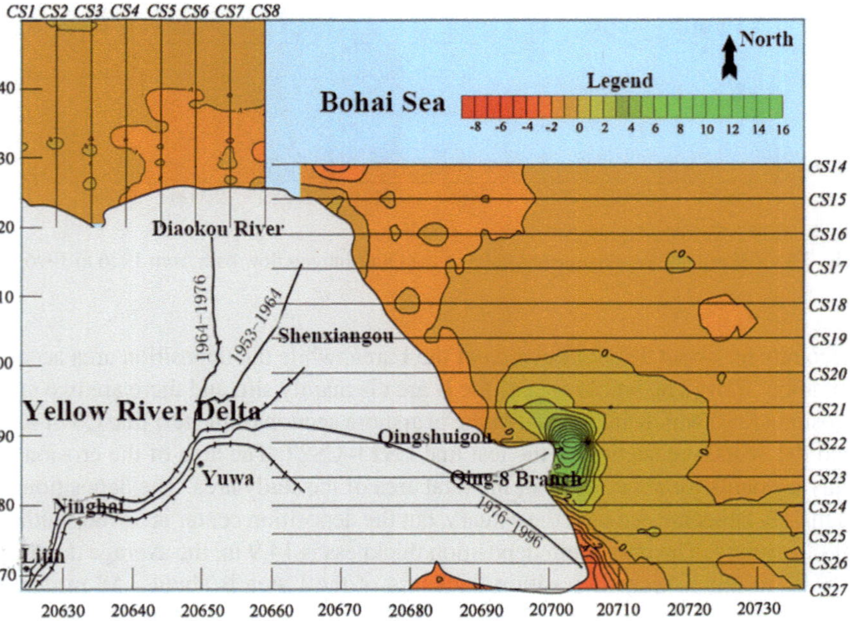

Fig. 3.9 Deposition–erosion variation of the subaqueous delta of Qingshuigou flow path from 1996 to 2004

(3) ***Erosion and Deposition Analysis During 1976–1996 and 1996–2004***

The erosion area, deposition area, total deposition–erosion amount, annual average deposition–erosion amount and annual average deposition–erosion amount per unit area of Yellow River subaqueous delta and sediment amount of the Yellow River into the sea were calculated for the two periods, as shown in Table 3.5. It can be clearly seen from Table 3.5 that during the period from 1976 to 1996, the Yellow River subaqueous delta was dominated by deposition, and the area of the deposition area was much larger than that of the erosion area. Although the total amount of deposition–erosion in Zone I is $-0.36 \times 108\ m^3$, which is generally erosion, the total amount of deposition–erosion in Zone II is as high as $68.9 \times 108\ m^3$, and the deposition is very severe. The deposition amount is higher than the total deposition amount from the Yellow River to the sea. The deposition amount is more than the amount of deposition from the Yellow River to the sea, which indirectly proves that there are other sources of sediment in Zone II. The sediments eroded on the abandoned delta lobes in the northern part of the Yellow River Delta are transported southward under the influence of tidal currents and deposited in the southern part of the Yellow River underwater delta. From 1996 to 2004, the entire Yellow River subaqueous delta was changed from the previous deposition to the erosion oriented, and the erosion area was much larger than the deposition area. For Zone I, compared with 1976–1996, the eroded area expanded to the entire Zone I. The area of the eroded area accounted for 96% of the Zone I, and the total deposition–erosion amount was $-7.1 \times 108\ m^3$, which was 59 times the annual average deposition–erosion amount per unit area from 1976 to 1996. For Zone II, during the period 1976–1996 the area of the deposition area was reduced from 76 to 30%, and the total deposition–erosion amount was also changed from the previous 68.9×108 to $-3.77 \times 108\ m^3$. Considering the element of sediment flow from the Yellow River to the sea, during the period from 1996 to 2004, the sediment volume of the Yellow River into the sea is sharply reduced. The annual average sediment volume into the sea is only $0.9 \times 108\ m^3$, which is less than one-third of the sediment volume from 1976 to 1996. The serious shortage of sediment from the Yellow River into the sea may be one of the main factors leading to the erosion of the Yellow River underwater delta.

3.3.3 Impact of Storm Surge on Subaqueous Delta Erosion

According to the "*China Marine Disaster Bulletin*" and related literature published by the State Oceanic Administration in each year, since the 1950s, there have been more than 10 serious storm surges in the Yellow River Delta, most of which were caused by typhoons, and a few were caused by the northern temperate system. This section uses the three major storm surges that occurred in the Yellow River Delta since the 1990s to study the effects of storm surges on the subaqueous delta deposition–erosion of the Yellow River.

Table 3.5 Statistics on the deposition–erosion variation amount of the Yellow River subaqueous delta and the sediment amount of the Yellow River into the sea

		1976–1996	1996–2004
Zone I	Erosion zone (%)	38	96
	Deposition zone (%)	62	4
	Total erosion–deposition amount ($\times 10^8$ m^3)	−0.36	−7.1
	The annual average erosion–deposition amount ($\times 10^7$ m^3/y)	−1.8	−8.9
	The annual average erosion–deposition amount per unit area ($\times 10^{-2}$ $m^3/m^2 \cdot y$)	−0.15	−8.8
Zone II	Erosion zone (%)	24	70
	Deposition zone (%)	76	30
	Total erosion–deposition amount ($\times 10^8$ m^3)	68.9	−3.9
	The annual average erosion–deposition amount ($\times 10^7$ m^3/y)	34.4	−4.96
	The annual average erosion–deposition amount per unit area ($\times 10^{-2}$ $m^3/m^2 \cdot y$)	13.3	−1.3
The amount of sediment from the Yellow River into the sea	Total amount ($\times 10^8$ m^3)	57.2	7.2
	The annual average amount ($\times 10^7$ m^3/y)	2.86	0.9

(1) ***Typical Storm Surge***

The storm surged on September 1, 1992. The storm surge was caused by a strong tropical cyclone and was more serious in the Yellow River Delta and adjacent areas. The continuous water increase time in the Yellow River Delta and adjacent areas is more than 50 h. The shortest duration of continuous water increases of 50 cm or more is more than 10 h, the average water increase is above 60 cm, and the highest tide level is 5.93 m. The NEE in the Shandong Peninsula and the central and western Bohai Sea has a NEE to the windy level of 8–9. The maximum wave height H1/10 of the storm surge is 5.9 m, the wave direction is SE, and the period is 8.7 s.

The storm surged on August 19, 1997. The storm surge was caused by the typhoon 9711. When the typhoon 9711 affected Shandong, it coincided with the astronomical tide, and the storm surge was superimposed on the astronomical climax of the day, causing many areas to exceed the local warning level. Affected by the storm surge, the tide level of Tianjin Tanggu is as high as 5.59 m, the maximum water increase is 1.93 m; the Bohai Sea has a 4–5 m huge wave area; the central wind exceeds 12, the wind direction is NE direction, and the average precipitation in the Yellow River Delta area is above 150 mm.

The storm surged on October 11, 2003. The storm surge was caused by the temperate storm surge, which was caused by the combination of high-altitude cold air and ground southwest inverted trough. When the storm surge occurred, it was the tide of the astronomical astronomy in Bohai. The northeast wind of the ground reaches 24.6 $m{\cdot}s^{-1}$, and the maximum instantaneous wind force measured by the offshore platform is 40 $m{\cdot}s^{-1}$, and the 10–11-grade east wind is maintained at sea for more than 20 h. Affected by the continued strong coastal wind and astronomical tide, coupled with the impact of heavy rain, the water level in the western coast of the Bohai Bay has increased sharply, and the highest tide level has reached 5.54 m.

The disasters caused by these three major storm surges are shown in Table 3.6.

(2) ***Survey Results in the Years with Storm Surges***

The arrangement of the water depth measurement line of the Yellow River subaqueous delta is shown in Figs. 3.10, 3.11 and 3.12. The difference between the two water depth measurements before and after the storm surge on each line is the difference, and the difference represents the evolution of deposition and erosion in the subaqueous delta in the year of the storm.

A typhoon storm surge occurred on September 1, 1992. The evolution of erosion and deposition in the Yellow River subaqueous delta from October 1991 to October 1992 is shown in Fig. 3.10. It can be seen that, except for the test line CS26, the entire Yellow River subaqueous delta is dominated by erosion, and the maximum erosion depth occurring on other lines is 1.9 m. For Zone I, the area of the erosion area is still larger than the area of the sedimentation area, but the difference is that the center of the strong erosion area is not located in the nearshore section of the Diaokou River, but is offset to the northeast. A large deposition zone appeared between the test lines CS1 to CS3. This "the eastern eroded and the western deposited" situation is inconsistent with the distribution of strong tidal currents under normal sea conditions in the northern delta. For Zone II, it is also dominated by erosion. Contrary to the situation described in Sect. 3.2.2, the location of the estuary during the period 1992–1993 was not only without deposition, but also with severe erosion, with a maximum erosion depth of −9.3 m.

A major typhoon storm surge occurred on August 19, 1997. The evolution of the deposition and erosion of the Yellow River subaqueous delta from October 1996 to October 1997 is shown in Fig. 3.11. The measurement data of Zone I in 1997 is lacking, and the contour map of the Zone I is not drawn. For Zone II, it can be seen that the area of the eroded area is larger than that of the deposition area, and it is basically bounded by the Qing-8 branch (the current estuary after the divagation in 1996), and "the northern eroded and the southern deposited". The maximum sediment thickness is 1.8 m; at this time, the Yellow River has been divagated to Qing-8 branch, and there is a thick silt near the estuary, with a thickness of 3.2 m. The change was

Table 3.6 Disasters caused by storm surge

Time	Disaster situation	Degree of disaster
1992.9.1	On September 1, 1992, the east coast of Dongying was hit by a violent storm surge. The highest tide level was 3.5 m, the seawater intrusion was 10–20 km inland, and the direct economic loss of local and oil fields was more than 500 million yuan. 1992 China Marine Disaster Bulletin: Dongying suffered the largest storm surge since 1938. The seawater rushing into the inland, the maximum distance is 25 km, and the inundation area is 960 km^2 from the high tide line. The tidal disaster destroyed 50 km of tidal dikes, 350 hydraulic structures, 5,388 collapsed houses, more than 1,000 damaged ships, more than 15,000 hm^2 of salt fields, and more than 5,000 hm^2 of artificial grassland. Twelve people in the city were drowned by the sea, with a direct economic loss of 359 million yuan. In this extraordinary tidal disaster, Shengli Oilfield suffered huge losses, flooding 105 oil wells, causing serious losses in drilling, oil production, power supply, communication, transportation, production, and living facilities. 21 people in the oilfield died of this tidal disaster. The direct economic loss is 150 million yuan	Extraordinary typhoon storm surge
1997.8.19	On August 19, 1997, the typhoon storm surge of No. 9711 hit Dongying, and the coastal area was 1417 km^2. In the 61 villages of Hekou District and Lijin, 12,000 households entered the water, 6,000 people were trapped by water, 6 people died, 60 km of tidal dikes were damaged, 32,450 houses were damaged, and 9 436 houses were collapsed. The communication was scraped, 3,575 power supply poles, 145 km of damaged roads, and 1,259 bridges and culverts. The affected area of crops was 109,000 hm^2, and the shrimp and crab ponds were destroyed by 0.56 million hm^2, 158 fishing boats were damaged, and 1230 pieces of nets were destroyed. The damage to salt fields was 10.9 million hm^2, and the direct economic losses amounted to 700 million yuan, of which oilfield industry lost 520 million yuan	Extraordinary typhoon storm surge
2003.10.11	2003 China Marine Disaster Bulletin: On the night of October 11th–12th, the phenomenon of three tides and zero pressure appeared in the Yellow River mouth. The maximum water increase of the Yangjiaogou tide station is 300 cm, and its highest tide level is 624 cm (the third highest tide in history), exceeding the local warning water level of 74 cm. The direct economic loss of Shandong Province was 613 million yuan. The five districts and counties in Dongying were affected by the disaster. The affected population was 0.56 million, the aquaculture damage area was 35,000 hm^2, 180 damaged houses, 40 km of the seawall, 38 km of the roadbed, 1 bridge, 36 ships, and the direct economic loss was 140 million yuan. The local water conservancy department monitored that from the 11th to the 12th, the high tide water level continued to retreat, 4.2 m higher than the normal tide level, which is basically close to the 1997/8#200 extraordinarily large stormwater level. The storm surge caused the newly built 3800 m access road in Diaokou County, Lijin, to be destroyed by tidal water. The storm-proof dam was 4300 m, 4000 hm^2 forest land was flooded, 1300 hm^2 shrimp and crab were washed away by water, the Sihe Bridge was broken, and 20,000 tons of raw salt was dissolved, and directly lost economic losses of more than 40 million yuan	Extra temperate storm surge

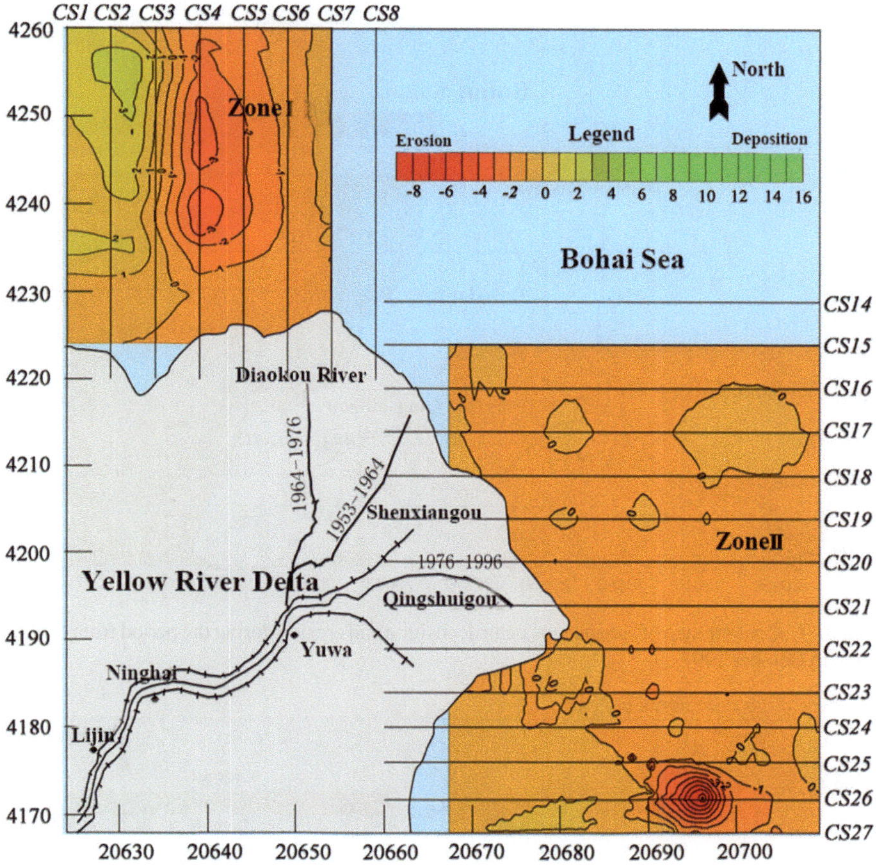

Fig. 3.10 Contour map of subaqueous delta deposition and erosion during the period from October 1991 to October 1992

basically consistent with the overall change from 1996 to 2004, but in the early stage of the divagation, there was no obvious erosion near the Qingshuigou flow path in the original estuary.

A typhoon storm surge occurred on October 11, 2003. The evolution of the deposition and erosion of the Yellow River subaqueous delta from August 2003 to August 2004 is shown in Fig. 3.12. The Zone I is dominated by erosion and the Zone II is dominated by deposition. For Zone I, the area of the erosion area accounts for 85% of the total area of Zone I. The erosion phenomenon is obvious, which is consistent with the overall change from 1996 to 2004. For Zone II, the area of the deposition area accounts for 82%. Excluding the line CS22 located at the estuary location, the maximum erosion thickness of other lines is 1.5 m, and the maximum sediment thickness is 1.4 m, while the line CS22 at the estuary shows a large silt layer near the estuary 10 km, with a maximum sedimentation thickness of 5.4 m. The locations

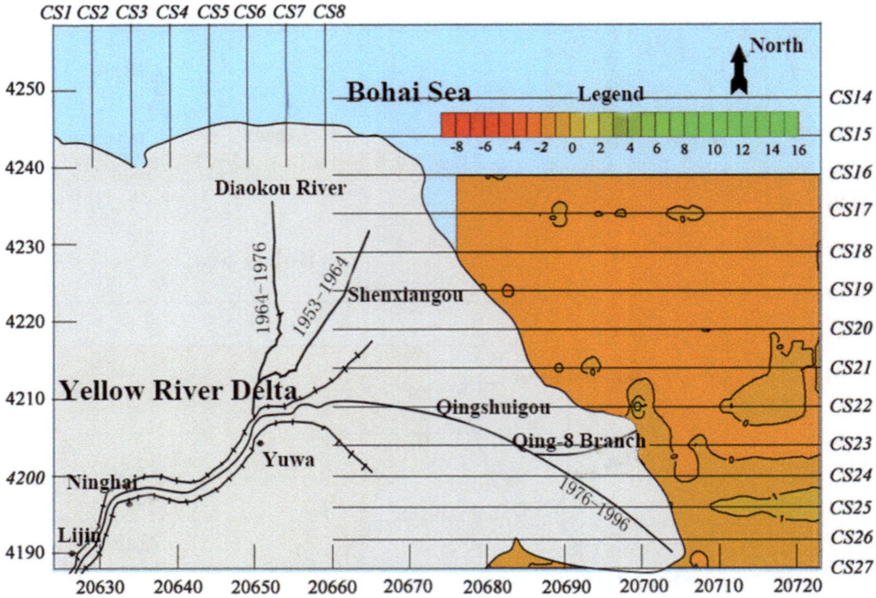

Fig. 3.11 Contour map of subaqueous delta deposition and erosion during the period from October 1996 to October 1997

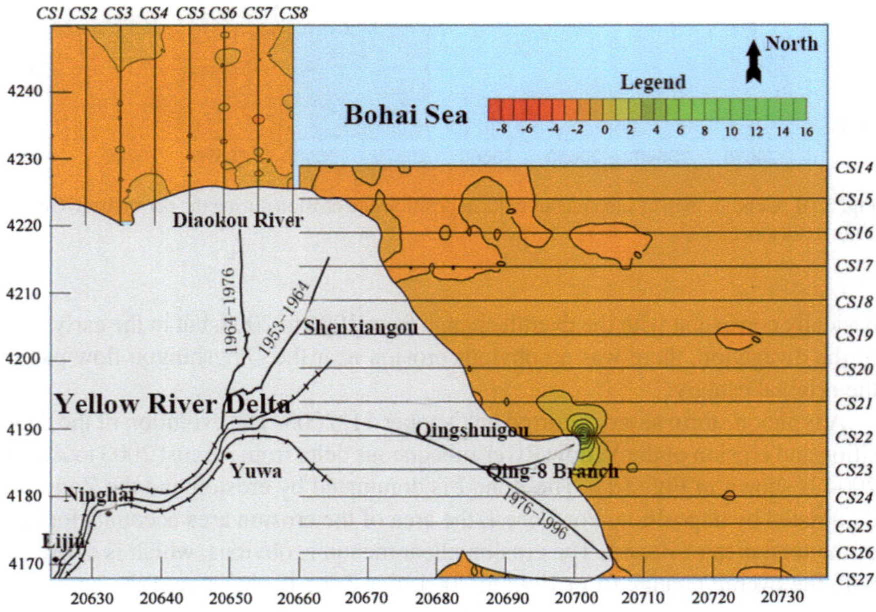

Fig. 3.12 Contour map of subaqueous delta deposition and erosion during the period from August 2003 to August 2004

of the two strongly eroded areas present in the area are basically the same as those described in 3.2.2, but the area of the deposition area is larger than the overall change from 1996 to 2004.

(3) ***Influence of Storm Surge on Erosion and Deposition of the Subaqueous Yellow River Delta***

The area of the eroded area, the area of the deposition area, the annual average erosion–deposition per unit area of the Yellow River subaqueous delta in the storm surge year, and the sediment amount of the Yellow River into the sea are counted as shown in Table 3.7.

Between 1992 and 1993, the total amount of erosion and deposition of the entire subaqueous delta (sum of Zone I and Zone II) was $-11.56 \times 108\ m^3$, during which the amount of sediment from the Yellow River to the sea was comparable to the multi-year average during the period 1976–1996. Different from the period from 1976 to 1996, during the period from 1992 to 1993, the subaqueous delta was dominated by erosion, and the most severe part of the eroded area occurred at the estuary, which is an anomaly. This anomaly may be related to the effect of storm surge.

During the period from 1996 to 1997, Zone II was still dominated by erosion, and the total amount of erosion and deposition was $-2 \times 108\ m^3$. During this period, the sediment volume of the Yellow River into the sea was $1.2 \times 108\ m^3$, while the average annual sediment volume into the sea was $0.9 \times 108\ m^3$ from 1996 to 2004. The annual average sediment amount into the sea in 1996 and 1997 is still slightly higher than the multi-year average, and its erosion intensity is higher than the multi-year average of 1996–1997 by 6 times. The occurrence of this phenomenon may also be related to the role of the storm surge.

During the period from 2003 to 2004, the total erosion of the Yellow River II area reached $10.4 \times 108\ m^3$, while the sediment volume of the Yellow River into the sea was only $1.75 \times 108\ m^3$. If the total amount of sediment in the I area is added, it is still less than one-half of the sedimentation in the II area. It is obviously impossible for such large numbers of unclear sources of sediment to be transported only by the tides under normal sea conditions. Only extreme sea conditions such as storm surges can carry out large-scale long-distance sediment transport.

In summary, the storm surge played a very important role in the evolution of the erosion and deposition of the Yellow River underwater delta. Large-scale and long-distance sediment transport will cause huge changes in the subaqueous delta terrain of the Yellow River, and even change the direction of the current transporting sediment under normal sea conditions, resulting in abnormal erosion. However, due to the lack of data, there is no measurement data of the subaqueous delta before and after the storm surge in the strict sense. Therefore, the accurate quantification of the evolution of the erosion and deposition of the Yellow River subaqueous delta by storm surge has yet to be further studied.

Table 3.7 Statistics on the variation of the Yellow River delta during the storm surge and the sediment amount of the Yellow River

Year	Zone I (North)				Zone II (Estuary)				Sediment load of Yellow River	
	Erosion zone (%)	Siltation zone (%)	Total amount of erosion (×108 m^3)	Annual average ecliptic per unit area (× 10–2$m^3/m^2 \cdot y$)	Erosion zone (%)	Siltation zone (%)	Total amount of erosion (×108 m^3)	Annual average ecliptic per unit area (× 10–2$m^3/m^2 \cdot y$)	Total (× 108 m^3)	Annual average (× 108 m^3)
1992–1993	60	40	−3.66	−49.4	82	18	−7.9	− 30	–	2.5
1996–1997	–	–	–	–	64	36	−2	−8	–	1.2
2003–2004	85	15	−2.8	− 28	18	82	10.4	35.6	–	1.75

3.4 Summary

In this chapter, a field survey has well detected the serious erosion status of the Yellow River Delta. It shows that the shoreline of the abandoned leaf petals in the northern part of the Yellow River Delta is still in a strong erosion recession state, and the average daily erosion rate is up to 56.6 cm/day. The shoreline erosion has obvious seasonal characteristics, which is consistent with the seasonal characteristics of the waves. The wave direction in spring and autumn is consistent with the shoreline direction, and the wave energy is larger at this time, causing strong erosion of the shoreline recession. In summer, waves are mainly to the southeast and the wave energy is small, and the rate of coast erosion recession is relatively slow. In winter, sea ice has a certain protective effect on the beach, and the rate of coastal erosion is only about 10 cm/d, besides, the accumulation of sea ice leads to siltation of the beach surface. In addition, the effect of the relative rise of sea level to the erosion of abandoned leaf blade in the northern part of the Yellow River Delta is not obvious very much. During the measurement period, the beach surface showed an alternate change between erosion and sedimentation, which was closely related to hydrodynamic conditions and did not conform to the traditional law of "winter rushing summer depositing". The changes in the erosion of the shoreline and the beach surface are not synchronized, because the shoreline is in a state of continuous erosion recession, and the beach surface is sometimes eroded and sometimes deposited. The storm surge has a controlling effect on the monolithic evolution process of erosion and deposition in the subaqueous Yellow River delta.

References

Bird ECF (1981) World-wide trends in sandy shoreline changes during the past century. Géog Phys Quatern 35:241–244

Bird ECF (1996) Coastal erosion and rising sea-level. In: Milliman JD, Haq BU (eds) Sea-level rise and coastal subsidence. Coastal systems and continental margins, vol 2. Springer, Dordrecht

Chen JY, Chen SL (2002) Estuarine and coastal challenges in China. Chin J Oceanol Limnol 20(2):174–181

Chen SL, Zhang GA, Gu GC (2004) Mechanism of heavy coastal erosion on Yellow River delta and its countermeasures. J Hydraul Eng 13(7):1–6

Fanos AM (1995) The impacts of human activities on the erosion and accretion of the Nile Delta coasts. J Coastal Res 11(3):821–833

Guillen J, Palanque A (1997) A historical perspective of the morphological evolution in the lower Ebro River. Environ Geol 30(3–4):174–180

Han X (2008) The sdudy on shoreline evolvement in Hebei province during the past century. Dissertation, Hebei Normal University, Shijiazhuang, Hebei, China

Kelley JT, Belknap DF, Shipp RC (1989) Sedimentary framework of the southern Maine inner continental shelf: influence of glaciation and sea-level change. Mar Geol 90:139–147

Komar PD (2000) Coastal erosion-Underlying factors and human impacts. Shore Beach 68(1):3–16

Li G, Wang H, Dong C (2005) Numerical simulation of the process of transport and deposition of sediment from the Yellow River. Acta Geogr Sin 60(5):707–716

Lu H (2005) Study on the delimitation between the modern Yellow River Delta and the sea. Dissertation, Ocean University of China, Qingdao, Shandong, China
Meng X (2009) Study on the erosion process of coastal sediment of the modern Yellow River Delta. Dissertation, Ocean University of China, Qingdao, Shandong, China
Qian CL (1994) Effects of the water conservancy projects in the Luanhe River basin on Luanhe River Delta. Hebei Province Acta Geogr Sin 49(2):158–166
Reimnitz E, Barnes PW, Harper JR (1990) A review of beach nourishment from ice transport of shoreface materials, Beaufor Sea, Alaska. J Coastal Res 6:439–470
Trenhaile AS (1997) Coastal dynamics and landforms. Clarendon Press, Oxford
Wang H, Yuan X, Wang Y et al (2010) Evolution of abandoned Shenxiangou-Diaokou delta lobe: procesesand mechanism. J Sediment Res 4:51–60
Wright LD, Xu JP, Madsen OS (1994) Across-shelf benthic transports on the inner shelf of the Middle Atlantic Bight during the "Halloween storm" of 1991. Mar Geol 118:61–77
Wu SY, Shi XF, Liu YG (2000) Influence of the Yellow River zero flow on delta petroleum and gas exploitation environment. Coast Eng 19(4):33–40
Yan S (2006) Coastal erosion and protection of Shengli oilfield in the Yellow River Delta. The Yellow River Water Conservancy Press, Zhengzhou
Yang MQ, Wang GL (1995) The Incipient Motion Formulas for Cohesive Fine Sediments. J Basic Sci Eng 3(1):99–109

Chapter 4
Erodibility of Seabed Sediments in the Modern Yellow River Delta

4.1 Overview

The modern Yellow River Delta is subject to serious processes of coastal erosion like most estuarine deltas in the world (Li et al. 2000). Processes of sediment erosion are affected not only by marine hydrodynamic conditions but also by sediment self-feature attributes including sediment particle compositions, physical and mechanical properties, and mineral compositions (Endler 2009). The sediment critical shear stress can be viewed as an important parameter that describes levels of sediment erodibility. Sediment with levels of high critical shear stress is difficult to erode, and sediment with low levels of critical shear stress can be eroded easily. Theoretically, when the shear force induced by a combination of waves and currents exceeds the sediment shear stress, sediment erosion occurs (Teisson et al. 1993).

Two methods can be used to determine the degree of sediment critical shear stress: direct and indirect measurements. Direct measurement techniques can involve using circulating water flumes (Bale et al. 2006), washdown scouring flumes (Zheng et al. 2013), an acoustic Doppler velocimeter (Andersen et al. 2007), or a cohesive strength meter (Watts et al. 2003). Indirect measurement techniques can involve using empirical formulae based on the relationship between sediment critical shear stress and sediment composition, bulk density, water content, etc. (Winterwerp et al. 2012). Direct measurement techniques are widely applied, as there are many limitations to empirical formulae constructed based on the sediment properties of a local study area. In the modern Yellow River Delta, sediment erosion occurs heterogeneously. Different sedimentary areas and different locations within the same sedimentary area undergo different forms and levels of sediment erosion and include different erosional landforms (Zheng et al. 2011). At present, the literatures about studying the erodibility of intertidal sediments in the Yellow River delta through direct measurements are sparse. Shi et al. (2006) conducted laboratory experiments to research into the erodibility of the core samples fetched from the high tidal flat in the north of the Yellow River delta. Nevertheless, the sediment erodibility certainly will change because of disturbance when the samples are sampled and transported. Therefore,

Y. Jia et al., *Wave-Forced Sediment Erosion and Resuspension in the Yellow River Delta*, Springer Oceanography, https://doi.org/10.1007/978-981-13-7032-8_4

conducting field tests to investigate the sediment erodibility in the Yellow River delta is necessary.

In this part, two kinds of direct techniques were used to measure the sediment erodibility (i.e., critical shear stress) found in lines running perpendicular to the coast in high tidal flats, middle tidal flats, and low tidal flats in different sedimentary areas formed in different depositional histories of the modern Yellow River Delta. The spatial distribution of sediment critical shear stress over different spatial scales and resulting contributions to the erosional microrelief were analyzed, and influencing factors and formation mechanisms were also considered. In addition, the heterogeneity and mean levels of sediment critical shear stress for the modern Yellow River Delta were compared with those of other estuarine deltas around the world.

4.2 Flume Measurements of Sediment Erodibility

4.2.1 Methodology

(1) ***Layout of Measurement Points***

The study was carried out at 20 tidal flat experiment sites along the seashore of the Yellow River delta (see more details about the study site in Fig. 2.1). The location of each experiment site was determined using Magellan GPS 315 (Table 4.1). Seasonal wind prevails on the Yellow River delta, generally blowing from north in winter and from south in summer. The wave direction is closely consistent with wind direction, taking on seasonal changes. The tide patterns are irregular semidiurnal tide along most sections of the Yellow River delta, and irregular diurnal tide at the local section off the eastern Shenxiangou promontory, where there is an amphidromic point of M2 tide (Zang 1996). The gradient of intertidal zone in the Yellow River delta is gentle. When the tide ebbs, the wide tidal flat is exposed, and it facilitates the work.

(2) ***Measurement of Sediment Physical-Mechanical Properties***

Surficial samples were taken to determine the bulk density, water content, grain size distribution, plasticity, mechanical strength (penetration resistance and shear strength), and compressibility for each location. The bulk density was measured using a steel ring sampler (Φ 6 cm, height 2 cm) which was pushed into the seabed. The sediment core was weighed to determine the bulk density. It was further dried in an oven for 8 h at a temperature of 105 °C and re-weighed to determine the water content. Sieving and hydrometer was used to get the grain size parameters, such as median grain size, sand content, and clay content. Liquid and plastic limits were measured using liquid-plastic limit combined tester to determine the plasticity of the surficial sediment. Penetration resistance and shear strength was measured using an electronic digital micro-penetrometer (measuring range 20 N) and a portable torsional vane shear device (measuring range 50 kPa), respectively. The compressibility

Table 4.1 The general situation of experiment sites

Date	Site	Location	The characteristics of sand ripples and biogenic features on the tidal flat
2008.7.17	S1-1	38°04.042′N, 118°56.499′E	No sand ripples; zoobenthos (mainly mussel and crab) burrow 49/dm^2, burrow diameter 1–10 mm, mostly about 5 mm
2008.7.19	S1-2	38°04.045′N, 118°56.468′E	Sand ripples, 1 cm height, 6 cm width; zoobenthos (mainly mussel and crab) burrow 58/dm^2, burrow diameter 1–10 mm, mostly about 5 mm
2008.7.20	S1-3	38°04.043′N, 118°56.494′E	Sand ripples, 1 cm height, 3–7 cm width; zoobenthos (mainly mussel and crab) burrow 45/dm^2, burrow diameter 1–10 mm, mostly about 5 mm
2008.7.23	S1-4	38°04.083′N, 118°56.407′E	Flat sediment surface
	S1-5	38°04.082′N, 118°56.407′E	Crab burrow 2/dm^2, burrow diameter about 5 cm, rugged sediment surface
2008.7.24	S1-6	38°04.080′N, 118°56.418′E	Crab burrow 2/dm^2, burrow diameter about 5 cm, rugged sediment surface
2008.7.27	S1-7	38°04.077′N, 118°56.402′E	Sand ripples, 1 cm height, 3–7 cm width
2008.7.21	S2-1	38°08.110′N, 118°46.816′E	Dense zoobenthos (mainly crab and snail) burrowing, big burrow diameter 1.5–4 cm, mostly about 4 cm, small burrow diameter 1–2 mm
2008.7.29	S3-1	37°51.174′N, 119°05.829′E	Sand ripples, 1 cm height, 4.5 cm width; zoobenthos (mainly crab and earthworm) burrow 11/dm^2, burrow diameter 1–5 mm, mostly 2-3 mm
	S3-2	37°51.174′N, 119°05.832′E	
	S3-3	37°51.175′N, 119°05.821′E	
2008.7.30	S4-1	38°08.379′N, 118°42.634′E	Sand ripples, 1 cm height, 3.5-4 cm width; zoobenthos is mainly *Bullacta exarata*
2008.10.1	S5-1	38°06.129′N, 118°27.872′E	Sand ripples, 0.8–0.9 cm height, 5.5 cm width; zoobenthos is mainly *Bullacta exarata*
	S5-2	38°06.131′N, 118°27.870′E	
2008.10.3	S6-1	38°08.051′N, 118°12.771′E	Sediment contains many shells
	S6-2	38°08.054′N, 118°12.773′E	
	S6-3	38°08.056′N, 118°12.776′E	
2008.10.4	S7-1	37°42.137′N, 119°16.458′E	Sand ripples
	S7-2	37°42.131′N, 119°16.457′E	
2008.10.5	S8-1	37°22.276′N, 118°56.869′E	Sand ripples; sediment contains shells

(usually measured in compression coefficient) can be obtained through laboratory consolidation test.

(3) ***Measurement of Sediment Erodibility***

Measurements of sediment erodibility were made using a recirculating flume (Fig. 4.1). It was designed and constructed according to the mobile recirculating flume (MORF) (Black and Cramp 1995). The flume consists of a 0.10 m wide race-track shaped channel in which water is continuously recirculated by means of a rotating paddle wheel. The flume is 1.20 m long and 0.50 m wide. The channel of the

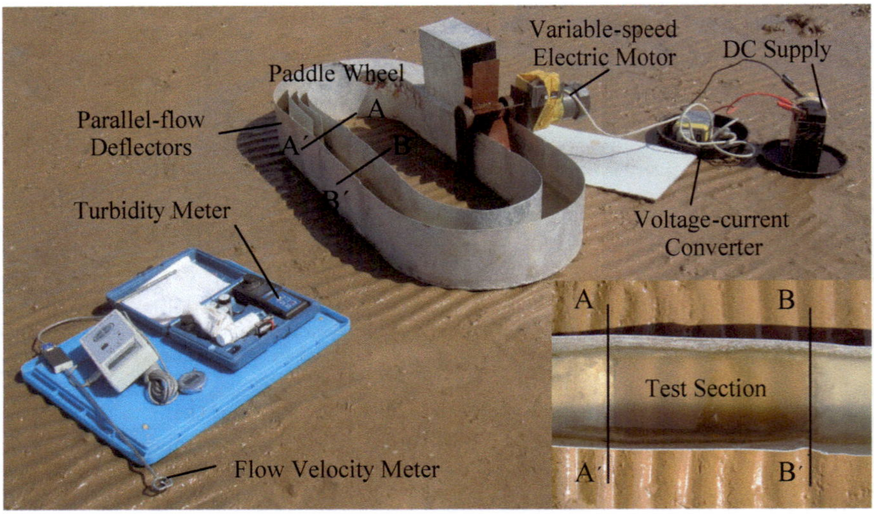

Fig. 4.1 The flume designed and constructed in this study and other instruments for erodibility measurements

flume is constructed of 1 mm thick aluminum. The walls of the channel are 0.15 m high. In the curved part of the channel, upstream of the test-section, is two parallel flow deflectors to straighten the flow, i.e., minimize cross-stream flow nonuniformities. They consist of two aluminum plates spaced at 0.03 m intervals between the channel walls. The channel has a solid base for most of its length, except at the downstream section where a rectangular hole (measuring 0.08 m $\times$ 0.25 m) has been cut in the floor. This is the test-section (showed in the bottom-right part of Fig. 4.1), and when the flume is placed directly on the mud surface it delimits an experimental bed area of 0.02 m^2. Directly underneath the open test-section is a 0.04 m deep rectangular lip, which forces its way down into the sediment upon flume placement, preventing seepage.

The flume measurements were carried out just after ebb water had vanished from the sediment surface. When the flume is properly resting on the seabed, the surface of the seabed is in line with the bottom of the flume. Local seawater was dribbled slowly into the flume from above. The working water depth was 0.08 m, and the volume of seawater used in each experiment was 0.02 m^3. The flow in the flume is generated by a 6 blade, undershot iron paddle wheel attached to a variable-speed 220 V motor. Each paddle measures 0.125 m deep by 0.08 m wide. Field measurements of flow velocity, recorded 15 mm above the bed at the entrance to the test-section, were made using a LGY-II type flow velocity meter. The concentration of suspended sediment in the flume was determined using a TURB 335IR type nephelometer through taking water samples from the flume, 20 mm above the bed 0.08 m downstream of the test-section. After the flume, variable-speed electric motor and paddle wheel was settled, the electrocircuit was connected and background turbidity was measured. Turn on the motor and fix the running speed, and start to measure the turbidity in ten seconds.

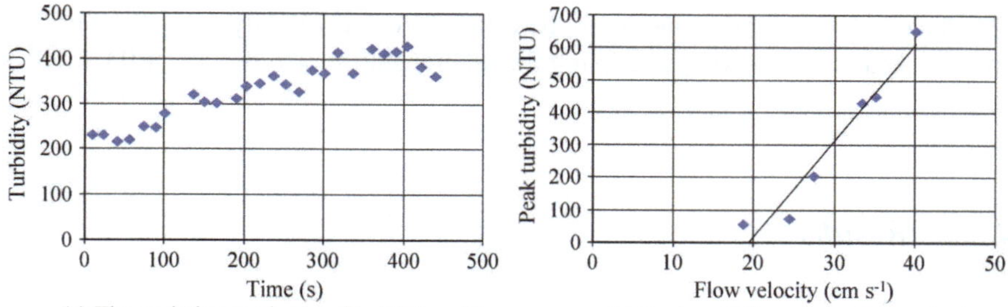

(a) The variation tendency of turbidity with time under certain flow velocity

(b) Peak turbidity versus flow velocity

Fig. 4.2 Example data from one of the erosion experiments

Continuously measure and record the turbidity and time when the water samples were taken. At the same time, measure and record the flow velocity. When the readings of nephelometer reached peak value, change the running speed of the motor. Repeat the previous steps until the readings of nephelometer reached peak value again. The flume measurements were started from the lowest running speed of the motor. If the turbidity under the flow velocity fluctuates around the background turbidity, change the running speed until the turbidity exceeds the background turbidity. Change the running speed that can stir the sediment particles up three times at least, and then cease the experiment.

It is found that the turbidity increases rapidly at the beginning and then stops increasing, even starts to decrease as time goes on when the erodibility measurement was conducted under certain flow velocity (Fig. 4.2a). For a certain site, the peak turbidity increases with the flow velocity (Fig. 4.2b). To determine the erosion threshold a linear function is fitted to the data (Fig. 4.2b). The background turbidity was substituted into the function, and the flow velocity u_c was solved. Once flow velocity exceeds u_c, the turbidity will be greater than background turbidity. It means there will be more sediment particles suspended from the surface of seabed. The flow velocity u_c is defined as the critical erosion velocity.

The critical erosion shear stress is an important parameter in sediment transport mechanics. It is obtained by the following calculation procedure. The flume belongs to open channel flume. For open channel

$$R_{ec} = \frac{v_c R}{\upsilon} = 580 \tag{4.1}$$

where R_{ec} is critical Reynolds number, v_c is critical velocity, υ is kinematic coefficient of viscosity, $\upsilon = 10^{-6}\ \mathrm{m^2\ s^{-1}}$, R is hydraulic radius, $R = A/\chi$, χ is wetted perimeter, $\chi = 0.26$ m, A is cross section of flow, $A = 0.008\ \mathrm{m^2}$. The known parameters are substituted into formula (4.1), and v_c is solved, $v_c = 1.89\ \mathrm{cm\ s^{-1}}$. The flow velocities measured in the flume range from 17.00 to 47.88 cm $\mathrm{s^{-1}}$, which exceed the critical velocity v_c. The flow is thus considered turbulent. Shear velocities (u_*) are calculated using the Karman–Prandtl logarithmic relation with $\kappa = 0.4$ (Cardoso et al. 1990).

The shear velocity corresponding to the critical erosion velocity is u_{*c}. The critical erosion shear stress is computed using the relationship $\tau_c = \rho u_{*c}^2$, $\rho = 1.03 \times 10^3$ kg m^{-3}.

4.2.2 Results

The physical-mechanical sediment properties and erosion thresholds of every experiment site are shown in Table 4.2. The experiment sites S3-1, S3-2, and S3-3 are next to each other. The heterogeneity of sediment was not taken into account, and the physical-mechanical properties were measured at only one site. It is the same with the sites S5-1, S5-2, sites S6-1, S6-2, S6-3, and sites S7-1, S7-2. Grain size parameters (median grain size, sand content, and clay content in Table 4.2) and plasticity indexes (I_p in Table 4.2. The plasticity index is the difference between liquid and plastic limits) indicate along-seashore differences. The sediment at S1-1, S1-4, S3-1, S3-2, and S3-3 is silty clay. It is sandy silt at S5-1, S5-2, and S8-1. The sediment at other experiment sites is all clayey silt. The sediment type is classified according to Chinese standard GB 50021-2001. The bulk density at the sampling sites varied from 1.82 to 2.06 g cm^{-3}, with the minimum value found at S2-1. The maximum value of bulk density was found at site S1-4. The water content ranged from 22.81 to 41.07%. It was maximum at S2-1 and minimum at S1-4. The penetration resistance and shear strength had the higher values at S1-4, S1-5, and S1-6, where the bulk density values were higher and the water content values were lower. The surficial sediment at all sites was medium compressibility. The compression coefficient varied from 0.117 to 0.179 M Pa^{-1}, with the minimum value found at S1-4. The surficial sediment at S2-1 was most easily compacted and the compression coefficient was maximum.

The critical erosion shear stress ranged between 0.088 Pa and 0.254 Pa, with the minimum value found at S1-3 and the maximum value found at S1-4 (Table 4.2), which were within the representative values given by Black et al. (2002) for estuarine tidal mudflats. As is shown in Fig. 4.3, the critical erosion shear stresses of intertidal sediments in the Yellow River delta are different in different sites. They vary widely even if in the same intertidal area. For example, the critical erosion shear stress at S1-4 is obviously higher than the other sites (S1-1, S1-2, S1-3, S1-5, S1-6 and S1-7) in the same intertidal area. The critical erosion shear stress of the sediment in the Yellow River delta measured by Shi et al. (2006) varied from 1.091 to 21.854 Pa. By comparison, the critical erosion shear stress here is much lower. The factors that make the difference may include: (1) different erosion devices; (2) the different criteria used to define the erosion threshold; (3) Shi et al. (2006) did the erosion experiments in the laboratory, and the samples used for the measurements would inevitably suffer disturbance (vibration, compaction and water loss) during transport to the laboratory; (4) different sediment properties. It seems it is difficult for the results to compare.

Table 4.2 The physical-mechanical sediment properties and erosion thresholds of experiment sites

Site	Sediment type	Median grain size (mm)	Sand content (%)	Clay content (%)	I_p	Bulk density (g cm^{-3})	Water content (%)	Penetration resistance (N)	Shear strength (kPa)	Compression coefficient (100–200 kPa) (M Pa^{-1})	Critical erosion shear stress (Pa)
S1-1	Silty clay	0.037	3.5	12.6	11.9	1.99	28.00	0.7	8.0	0.122	0.129
S1-2	Clayey silt	0.043	9.1	14.8	9.6	1.99	27.30	1.0	5.0	0.133	0.113
S1-3	Clayey silt	0.042	4.9	14.3	6.8	1.98	25.10	1.1	1.0	0.130	0.088
S1-4	Silty clay	0.025	2.8	18.6	10.9	2.06	22.81	4.1	23.0	0.117	0.254
S1-5	Clayey silt	0.020	0.9	22.1	8.6	2.00	23.39	3.3	10.3	0.130	0.098
S1-6	Clayey silt	0.027	4.1	24.2	8.5	2.04	23.28	2.9	10.0	0.122	0.115
S1-7	Clayey silt	0.027	3.5	18.8	9.7	1.97	26.67	4.4	9.0	0.130	0.090
S2-1	Clayey silt	0.016	3.2	33.2	8.8	1.82	41.07	0.3	0	0.179	0.130
S3-1	Silty clay	0.044	3.4	18.5	10.4	1.92	32.52	0	7.7	0.151	0.110
S3-2	–	–	–	–	–	–	–	–	–	–	0.121
S3-3	–	–	–	–	–	–	–	–	–	–	0.109
S4-1	Clayey silt	0.042	3.8	20.7	7.3	1.89	29.58	0.5	6.5	0.153	0.196
S5-1	Sandy silt	0.051	10.8	7.8	7.7	1.97	30.40	0.8	4.7	0.120	0.143
S5-2	–	–	–	–	–	–	–	–	–	–	0.175
S6-1	Clayey silt	0.040	8.6	13.9	5.3	1.99	26.60	1.6	8.8	0.169	0.153
S6-2	–	–	–	–	–	–	–	–	–	–	0.167
S6-3	–	–	–	–	–	–	–	–	–	–	0.192
S7-1	Clayey silt	0.029	1.4	13.3	8.2	2.01	26.90	1.9	7.2	0.149	0.177
S7-2	–	–	–	–	–	–	–	–	–	–	0.146
S8-1	Sandy silt	0.058	35.4	9.2	7.7	2.00	28.70	0.4	3.2	0.128	0.123

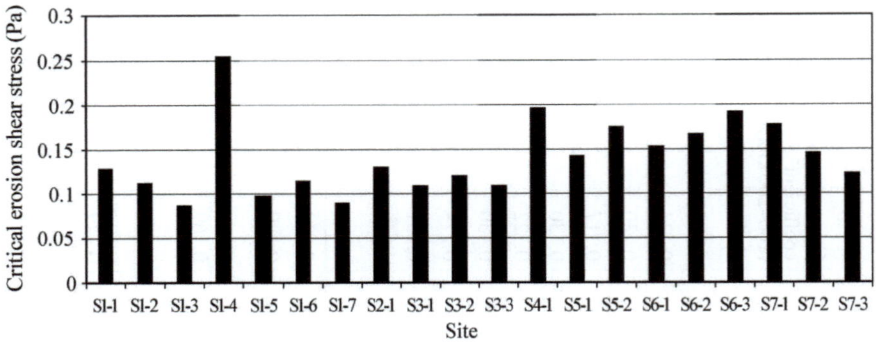

Fig. 4.3 The critical erosion shear stress at each experiment site

4.2.3 The Spatial Difference of Sediment Erodibility

Figure 4.3 shows the spatial difference of sediment erodibility. For the experiment sites that are next to each other, though the heterogeneity of sediment was not taken into account, and the physical-mechanical properties were measured at only one site, the erodibility are different from each other. The heterogeneity of sediment may not be negligible because of the uncertainty of bioturbation, etc. (Table 4.1). Therefore it is suggested that the physical-mechanical properties at every single site should be measured to study its influence on erodibility (it will be discussed below). For the same intertidal area, the surficial sediment of which the erosion resistance is poor and is easily eroded under hydrodynamic forces, but the surficial sediment of which the erosion resistance is good and is reserved. It may lead to the formation of scour pits on the tidal flat (Fig. 4.4).

4.2.4 The Effect of Sediment Physical-Mechanical Properties on Erodibility

It appears logical to think that a correlation exists between erodibility and soil properties. The importance of physical sediment properties such as grain size, bulk density, water content, and plasticity index on erodibility has long been recognized (Amos et al. 1997; Hjulstrom 1939; Postma 1967; Einsele et al. 1974; Fukuda and Lick 1980; Kamphuis and Hall 1983; Mitchener and Torfs 1996; Jepsen et al. 1997; Torfs 1997; Roberts et al. 1998; Houwing, 1999). However, Table 4.3 shows that there was no significant correlation between mean grain size, sand content, clay content, plasticity index, bulk density, and water content with critical erosion shear stress suggesting that those variables were not dominant factors influencing sediment erodibility at experiment sites. The research on relationship between the mechanical sediment properties and the erodibility is few. It is found in Table 4.3 and Fig. 4.5 that the

Fig. 4.4 The scour pits on the tidal flat near S1-5

critical erosion shear stress is positively correlated to the shear strength whereas no clear relationship is found between penetration resistance, compression coefficient and critical erosion shear stress. By contrast, Meng et al. (2010) finds that there are good correlations between critical erosion shear stress and physical-mechanical properties of the newly deposited sediment on the Yellow River delta. The probable reason is that the sediment has underwent physical and biological reworking for a long time since it deposited, and the factors influencing surficial sediment erodibility have became complicated. Nevertheless, there is a positive correlation between the critical erosion shear stress and shear strength. Watts et al. (2003) plotted sediment shear strength determined using the fall cone against critical erosion shear stress, and drew the same conclusion. The shear strength of soil is a result of friction and interlocking of particles, and possibly cementation or bonding at particle contacts. It is a macroscopic measure of the undrained sediment shear strength integrated over the upper 0.5 cm of the sediment, in contrast to the measurements provided by the flume experiment which relate to erosion within the upper few millimeters at most (Black et al. 2002). Although critical erosion shear stress cannot be determined directly from geotechnical shear strength, these measurements may provide a useful quantitative indicator of critical erosion threshold.

Table 4.3 Pearson correlation coefficient (r) of the correlation analysis of the data in Table 4.2

	Mean grain size	Sand content	Clay content	Plasticity index	Bulk density	Water content	Penetration resistance	Shear strength	Compression coefficient
Critical erosion shear stress	−0.031	−0.080	−0.096	0.005	0.147	−0.086	0.115	0.604[a]	0.051

[a]Correlation is significant at the 0.05 level (2-tailed)

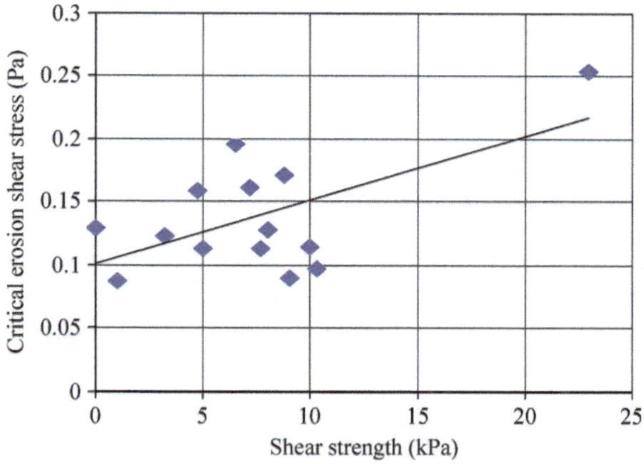

Fig. 4.5 Correlation between geotechnical shear strength and critical erosion shear stress

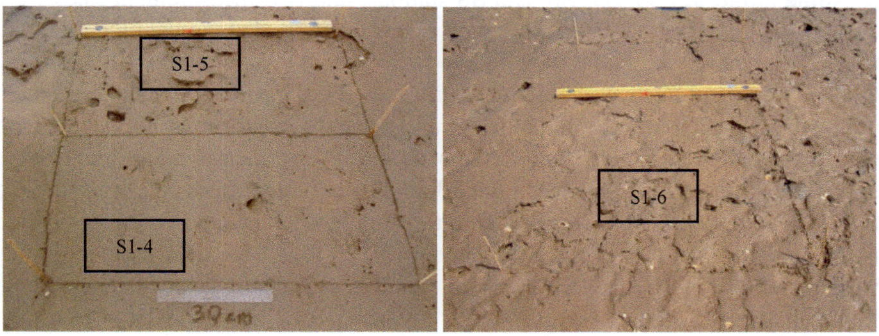

Fig. 4.6 Photos of S1-4, S1-5 and S1-6

4.2.5 The Effect of Crab-Burrows on Erodibility

Destabilization of sediments by animal tubes is indicated by observations of scouring around single tubes (Scoffin 1970; Gage 1977). Flume experiments (Eckman et al. 1981) have shown that below a certain threshold density, animal tubes may cause sediment destabilization through a sufficiently high transfer of turbulent kinetic energy to the bed. Sediment microtopography may alter entrainment thresholds by affecting near-bed flow through increasing the roughness height parameter and enhancing shear stress, thereby decreasing threshold values. The surficial sediment at S1-5 and S1-6 clearly had greater roughness than at S1-4 (Fig. 4.6), and that is one possible reason except the difference of physical-mechanical sediment properties for the lower critical erosion shear stress at S1-5 and S1-6. It is suggested that the influence of crab burrowing on erodibility should be taken into account in sediment transport models by changing the bottom roughness (increasing the bed shear stress).

4.3 CSM Measurements of Sediment Erodibility

4.3.1 Methodology

(1) ***Layout of Measurement Points***

Eight measuring stations in different sedimentary areas along the modern Yellow River Delta were selected, as indicated by S1, S2, S3, S4, S5, S6, S7, and S8 in Fig. 2.1. As it is shown in Fig. 2.1, the stations chosen to conduct measurements are nearly located at different delta lobes formed in different periods. Previous literatures revealed that sediment in different sublobes is characterized by different composition in the modern Yellow River Delta (Jia et al. 2011). Therefore, different stations with unique sediment characteristics were chosen to conduct the field investigations in this paper. Sediment critical shear stress levels were measured perpendicular to the coastline at the S2, S3, S5, S6, and S8 stations, and two measurement lines were used at station S8 with two measurement points positioned along the high tidal flat, two positioned in the middle tidal flat, and three positioned along the low tidal flat. One measurement line was applied at stations S2, S3, S5, and S6. One measurement point was applied to the high tidal flat, two were applied to the middle tidal flat, and two were applied to the low tidal flat at station S2; three measurement points were applied to all three tidal flats at station S3; and one measurement point was applied to the high tidal flat, two were applied to the middle tidal flat, and two were applied to the low tidal flat at station S5. At stations S1, S4, and S7, measurement points were positioned close to the coast, as the tidal flat sediment was not strong enough to sustain the measurement work.

(2) ***Measurement of Critical Shear Stress***

The cohesive strength meter (CSM, Partrac Ltd., Glasgow, UK) is used to measure the critical shear stress of the surface sediment on the tidal flat (Fig. 4.7). The tool generates a water jet induced by air pressure that creates a shear force that can erode and resuspend seabed surface sediment. Before measuring the sediment critical shear stress, the experimental module was inserted into seabed sediment 2 cm below the surface, and then distilled water was injected into the experimental module through needle tubing. The device was powered on, air in the tube was eliminated, and the test mode was selected before the experiment was started. During the experiment, the testing time, injection intensity, and light transmittance were recorded, and testing was stopped when the light transmittance did not vary significantly. After testing, the experimental module can be rinsed using air pressure in the device. The remaining air can then be released, and the device can then be closed.

(3) ***Data Processing***

Resuspended sediment from surface sediment lifted by shear stress induced by injected water changes the light transmittance (T) of the experiment module. The

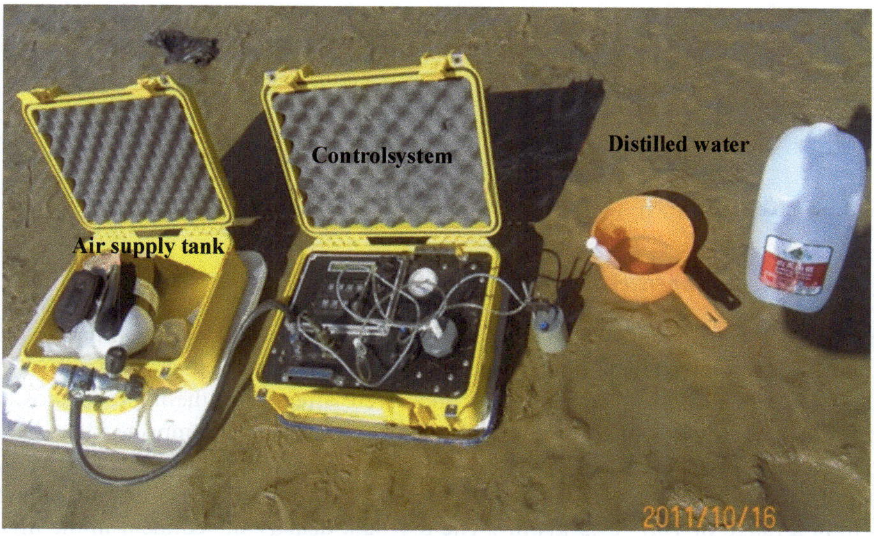

Fig. 4.7 Photo of a cohesive strength meter working in situ

suspended sediment concentration (C, g/l) can be calculated according to the relationship between T and C shown in Eq. (4.2), and the shear force can be calculated according to Eq. (4.3) (Tolhurst et al. 1999).

$$C = 4.9352e^{(-0.0953T)} \tag{4.2}$$

$$\tau = 67 \times (1 - e^{(-P/310)}) - 195 \times (1 - e^{(-P/1623)}) \tag{4.3}$$

In Eq. (4.3), P represents the injecting intensity (kPa), and τ represents the shear force induced by injected water (kPa). Variations in suspended sediment concentrations with horizontal shear force were obtained from Eqs. (4.2) and (4.3). As the injecting intensity level increases, the light transmittance decreases. The critical injecting intensity is defined as the intensity corresponding to a light transmittance that is reduced by 90% (Tolhurst et al. 1999). In this experiment, the shear force that corresponds to a suspended sediment concentration of 1 kg/m^3 can be considered equivalent to the sediment critical shear stress level.

(4) ***Measurements of Geotechnical Properties***

For each sediment critical shear stress measurement point, wet bulk density, water content, penetration strength, and shear strength levels were measured in situ; dry bulk density, pore ratio, and saturation degree levels were calculated; and sediment particle compositions were analyzed in a laboratory. Wet bulk density and water content levels were measured using an electronic balance and oven, respectively; the penetration strength was measured using an electronic digital display tiny penetrometer; the shear strength was measured using a small vane shear apparatus; the sediment particle composition was analyzed using screening and densimeter methods according to British Standard Procedure (British Standards Institution 1990). The dry bulk density, pore ratio, and saturation degree were calculated (Liu et al. 2013b).

4.3.2 *Results*

(1) ***Sediment Properties Perpendicular to the Coast***

Several bed ripples oriented roughly parallel to the coastline developed in the low tidal flat according to the field observations. Many microbenthos were active in the middle tidal flat, and various study areas were affected by the microbenthos to differing degrees. Sediment in the high tidal flat was highly compacted to a hard plastic state with low water content levels. Wet bulk density levels did not vary significantly in different locations of the tidal flat, with all measurements falling within a range of 18.7–19.5 kN/m^3. The dry bulk density of sediment in the high tidal flat was higher than that in the middle and low tidal flats, and it also featured a lower pore ratio and higher shear and penetration strength (Table 4.4). Although the saturation degree of the high tidal flat was slightly higher than that of the low tidal flat, as the high tidal flat was covered with sea water for less time and was thus exposed to air for more time than the low tidal flat, saturation degrees for both flats reached 90%. Trends found for the modern Yellow River Delta are consistent with those of other estuarine deltas in that sediment particles become finer from the sea to the land surface. Sediment particles are finer, sand content levels are lower, and clay content levels are higher in the high tidal flat than in the low tidal flat. No obvious regularities of distribution could be found in the middle tidal flat due to the presence of benthos (Table 4.4). According to principles of traditional sedimentology, sediment grows coarser from the land to the sea because hydrodynamic conditions weaken toward the shore. Thus, coarse particles settle in the low tidal flat where marine processes are significant, and fine particles settle in the high tidal flat where marine processes are weak. Studies on sediment dynamics occurring under waves show that fine sediment particles can be transported toward the sediment surface through wave forces and can then be transported away through currents, promoting the coarsening of the original sea area (Liu et al. 2013a, 2017; Jia et al. 2014). Masses of fine particles that can be transported toward the sediment surface increase with increasing wave forces, and hydrodynamic conditions are more active in the low tidal flat than in the high tidal flat. The resulting change in coarseness is thus more significant in the low tidal flat.

As shown in Fig. 4.8, which provides distributions of sediment critical shear stress for six measuring lines and five of the study sites, the sediment critical shear stress in the high tidal flat is significantly higher than that in the low tidal flat. Middle tidal flat measurement points significantly affected by benthos, such as S2 and S5, present lower levels of sediment critical shear stress than points that benthos affects less, such as S8 and S6. Both sets of measurement points in middle tidal flat present lower critical shear stress values than those derived from high tidal flat measurements. Our

Table 4.4 Sediment properties perpendicular to the coast

Study site	Location of tidal flat	Number of measuring points	Wet bulk density (kN/m^3)	Dry bulk density/(kN/m^3)	Water content/%	Pore ratio	Saturation degree/%	Shear strength/kPa	Penet-ration/N	Sediment particle composition			
										sand/%	silt/%	clay/%	mean grain size/μm
S8	Low	S8-1-1	18.7	14.1	32.5	0.92	95.7	3.6	1.0	9.3	80.8	9.9	27
	Middle	S8-1-4	19.1	14.5	31.5	0.87	98.1	6.9	1.2	17.1	74.8	8.1	36
	High	S8-1-6	19.0	14.6	29.8	0.86	91.6	4.7	1.7	16.0	76.4	7.6	34
	Low	S8-2-1	19.3	14.5	33.0	0.90	99.4	4.0	1.4	34.1	61.5	4.4	46
	Middle	S8-2-4	18.9	14.3	32.1	0.90	96.7	6.0	2.9	39.1	56.5	4.4	45
	High	S8-2-6	19.2	14.8	29.4	0.82	97.2	4.3	2.5	17.0	76.6	6.4	37
S2	Low	S2-1-1	19.3	14.8	30.5	0.83	99.9	4.0	0.8	44.3	48.2	7.5	48
	Middle	S2-1-4	19.1	14.4	31.1	0.88	96.2	3.8	1.0	33.8	56.9	9.3	39
	High	S2-1-6	19.5	15.1	28.8	0.79	96.7	4.7	1.9	11.3	85.1	3.6	30
S3	Low	S3-1-1	19.1	14.5	31.9	0.87	99.3	10.9	1.6	26.1	67.8	6.1	41
	Middle	S3-1-4	18.9	14.2	33.0	0.91	98.3	13.1	2.0	17.2	79.9	2.9	46
	High	S3-1-6	18.9	14.7	28.9	0.84	91.0	16.0	4.5	19.4	75.0	5.6	39
S5	Low	S5-1-1	19.4	14.8	31.1	0.83	99.9	6.2	1.3	30	58.8	11.2	37
	Middle	S5-1-4	19.1	14.5	31.4	0.87	99.6	1.8	0.6	29.8	61.0	9.2	40
	High	S5-1-7	19.3	14.9	29.7	0.82	96.1	4.5	2.3	8.9	73.2	17.9	25
S6	Low	S6-1-1	19.3	14.7	30.5	0.84	98.3	3.9	0.5	34.7	64.3	1.0	54
	Middle	S6-1-3	18.7	14.2	31.4	0.90	94.5	6.4	1.2	40.6	51.8	7.6	37
	High	S6-1-5	19.2	15.0	28.2	0.81	92.1	7.7	1.5	21.9	72.4	5.7	40

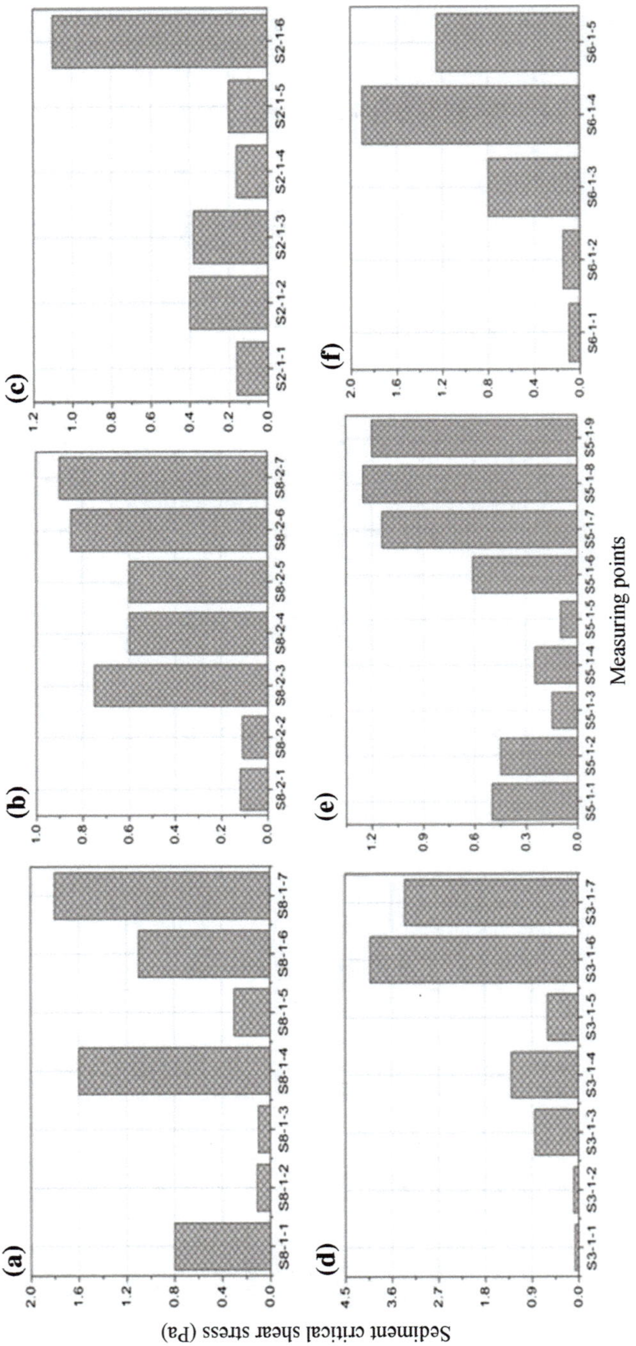

Fig. 4.8 Distribution of sediment critical shear stress levels perpendicular to the coastline at different study sites. **a** Measuring line 1 in S8, **b** measuring line 2 in S8, **c** measuring line in S2, **d** measuring line in S3, **e** measuring line in S5, and **f** measuring line in S6

findings on distribution characteristics of sediment critical shear stress in the middle tidal flat are consistent with those of Andersen et al. (2002), who show that biological activity and excretion can affect the physical properties of surface sediments and thus ocean dynamics.

(2) ***Sediment Properties in Different Sedimentary Areas***

In the modern Yellow River Delta, the wet bulk density ranges from 19.1 to 20.1 kN/m^3 with small deviations; the dry bulk density ranges from 14.5 to 15.8 kN/m^3 and is the highest to the north (S4) and east (S7); water content levels range from 26.6 to 33.0%, with high saturation degrees of over 90% and are highest in the northeast (S8); the pore ratio ranges from 0.72 to 0.90 and is the lowest to the north (S1 and S4) and the highest to the northeast (S8); and the shear strength ranges from 3.9 to 14.2 kPa and is low in both northern (S2) and eastern areas (S6).

In the modern Yellow River Delta, sediment sand content (2–0.075 mm) ranges from 2.0 to 44.3%, silt content (0.075–0.005 mm) ranges from 48.2 to 75.9%, clay content ranges from 1.0 to 37.4%, and mean grain size ranges from 8 to 54 m (Table 4.5). Sediment here can be divided into two categories: sandy fine sediment and silty fine sediment. At study sites S2, S3, S5, S6, S7, and S8, sediment sand content in the low tidal flat ranges from 25 to 50%, classifying it as sandy fine sediment. The clay content of sediment in the low tidal flat at study sites S6, S7, and S8 is less than 6%, classifying it as sandy fine silt. The clay content of sediment in the low tidal flat at study sites S2 and S3 ranges from 6 to 10%; therefore, it can be classified as sandy sediment. The sand content of sediment in the low tidal flat of study site S6 ranges from 10 to 15%, classifying it as sandy clay. The sand content of sediment in the low tidal flat at study sites S1 and S4 is less than 25%, classifying it as sandy fine sediment. The clay content of sediment in the low tidal flat at study site S1 ranges from 10 to 15%, and it can thus be classified as silty clay.

The sediment critical shear stress level ranges from 0.11 to 3.45 Pa across the entire modern Yellow River Delta, with significant variations found in different sedimentary areas (Fig. 4.9). Stress levels are lowest in the northern part of the delta (S1 and S2), which can be easily eroded. Previous field investigations show that erosion rates reach 56.6 cm/d in the Chezigou sea area in the northern section, and low sediment critical shear stress levels can be viewed as a main cause of high erosion rates. In other study sites presenting slight erosion patterns, there are no obvious variations in sediment critical shear stress levels. From relationships found between levels of sediment critical shear stress and states of sediment erosion and accretion in the study sites, it can be concluded that sediment critical shear stress plays a significant role in dynamic sediment processes in estuarine deltas. On the other hand, sediment critical shear stress is highly homogenized in each study site when comparing the highest, lowest, and middle values of sediment critical shear stress found. At least three measuring points at each measuring station were used, and the highest value of sediment critical shear stress can be 1.63–50.25 times larger than the lowest value found for the same measuring station. This variation was observed in the east (S1), and it was highest in the northeast (S3). From this, it can be concluded that the nonuniformity of erosional features can be significant.

Table 4.5 Sediment properties of different study sites

Study site	Wet bulk density/(kN/m^3)	Dry bulk density/(kN/m^3)	Water content/%	Pore ratio	Saturation degree/%	Shear strength/kPa	Penetration strength/N	Particle composition			
								Sand/%	Silt/%	Clay/%	Mean grain size/μm
S1	19.9	15.8	26.6	0.72	100	6.8	1.1	17.6	68.9	13.5	17
S2	19.3	14.8	30.5	0.83	99.9	4.0	0.8	44.3	48.2	7.5	48
S3	19.1	14.5	31.9	0.87	99.3	10.9	1.6	26.1	67.8	6.1	41
S4	20.1	15.8	26.9	0.72	100	10.2	1.9	2.0	60.6	37.4	8
S5	19.4	14.8	31.1	0.83	99.9	6.2	1.3	30	58.8	11.2	37
S6	19.3	14.7	30.5	0.84	98.3	3.9	0.5	34.7	64.3	1.0	54
S7	19.8	15.2	30.2	0.88	93.0	14.2	2.8	25.8	69.4	4.8	34
S8	19.3	14.5	33.0	0.90	99.4	4.0	1.4	34.1	61.5	4.4	46

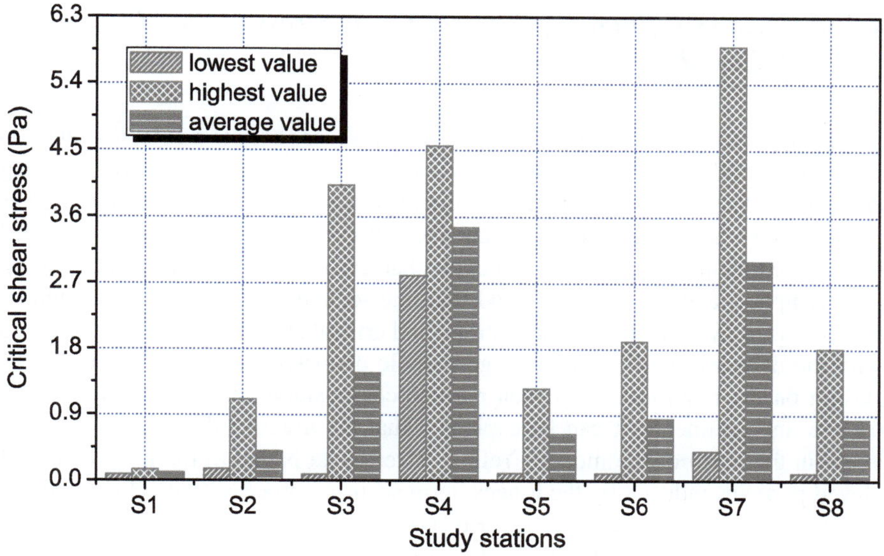

Fig. 4.9 Histogram diagram of sediment critical shear stress in the modern Yellow River Delta

4.3.3 Implications for Erosional Landforms of the Modern Yellow River Delta

The erosional landforms of tidal flats are intimately related to erosional characteristics of sediment in the modern Yellow River Delta. The tallest slope of the tidal flat in the northern part of the modern Yellow River Delta can reach 1/500, and the tallest slope in the middle area reaches 1/1,500. Our field investigations show that scarp erosional landforms exist between high and low tidal flats and that the elevations of high tidal flats exceed those of middle and low tidal flats in the northern part of the modern Yellow River Delta (Fig. 4.10a). Our investigation of sediment critical shear stress in the modern Yellow River Delta (S2, S3, and S7) shows that levels are clearly higher in the high tidal flat than in the low tidal flat. It can be speculated that the difference in sediment critical shear stress levels between the high, middle, and low tidal flats can be viewed as central to the formation of scarp erosional landforms in the modern Yellow River Delta. Other typical landforms widely found across the modern Yellow River Delta are collapses and scallops with diameters of several meters, a few tens of meters, and hundreds of meters (Prior et al. 1986; Keller et al. 1990). From our investigation, we found that many collapses and scallops of varying shapes and sizes exist in the low tidal flat of the delta, but few are found in the high tidal flat (Fig. 4.11b). It also can be speculated that the nonuniform spatial distribution of sediment critical shear stress plays a significant role in the formation and development of collapses and scallops in the modern Yellow River Delta.

4.3.4 Factors Influencing Critical Shear Stress of the Modern Yellow River Delta

Previous literature has demonstrated that critical shear stress is a complex function of shear strength, clay content, structure, and other geotechnical properties (Zheng et al. 2014). Particle composition, determining sediment types, can be considered as the important factor influencing critical shear stress. Aberle et al. (2004) revealed that fine sediment with high sand content is characterized by low critical shear stress, and Kamphuis and Hall (1983) found that fine sediment with high clay content is characterized by high critical shear stress, and critical shear stress can be increased with the decrease in sediment-average particle diameter (Grabowski et al. 2011). Basing on the relationship between particle composition and critical shear stress for the fine sediments, it can be explained that the lowest value of critical shear stress in the north of the modern Yellow River Delta is due to the low content of clayed particles induced by the longest term scouring process. Critical shear stress in the east part of the modern Yellow River delta is in the highest level since the fine sediment is composed by the highest clay content and sand content, respectively, for the newly deposited area (Table 4.5). In terms of the sedimentary area average level, particle composition significantly influences the critical shear stress, and the sediment particle composition is reworked by currents and waves in the long-term periods. Sediment in different sedimentary lobes deposited in different periods can be characterized with different sediment particle composition. That is to say, sediment critical shear stress is closely related with the deposition period of the sedimentary areas.

The field observation showed that abundant benthic organisms spread over the middle tidal flat compared to the high and low tidal flats (Fig. 4.10, Jia et al. 2011). Compared to the effect of sediment physical properties to critical shear stress, biological process can be considered as an important factor for the heterogeneous distribution of critical shear stress in middle tidal flat. Benthos community structure and its benthic effect can play a significant role in the formation, development, and stability of the sedimentary characteristics of estuarine tidal flat (Li et al. 2017). Bioturbation and fecal material can change sediments through physics and chemicals and then affect the topographic features and the chemical cementation between sediment particles. In the middle tidal flats in different sedimentary areas of the modern Yellow River Delta, benthos of different type and density play a different role to the fine sediment with different particle composition, then affect sediment critical shear stress in different degrees and mechanisms resulting in the heterogenetic distribution of critical shear stress on different spatial scales.

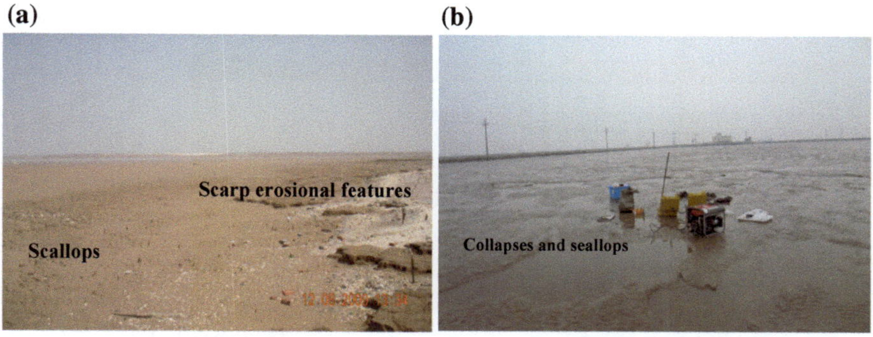

Fig. 4.10 Photos of a typical intertidal landform in the modern Yellow River Delta. **a** High and middle tidal flats in the north section and **b** low tidal flat in the eastern section

4.3.5 Comparisons with Critical Shear Stress from Other Estuarine Deltas

The sediment critical shear stress value is an important parameter for predicting sediment erosion processes under hydrodynamic conditions in sediment dynamic studies, and several researchers have promoted the measurement of sediment critical shear stress in different estuarine deltas (Schünemann and Kühl 1993; Widdows et al. 1998; Houwing 1999; Tolhurst et al. 1999; Paterson et al. 2000; Watts et al. 2003; Bale et al. 2006; Tolhusrt et al. 2006; Amos et al. 2010; Andersen et al. 2010; Meng et al. 2012). Measured values of sediment critical shear stress vary across different measuring devices. In this study, sediment critical shear stress levels measured using a circulating flume and cohesive strength applied to different estuarine deltas were collected and compared to sediment critical shear stress levels in the modern Yellow River Delta.

In comparing the highest and lowest values of sediment critical shear stress in different areas, it is evident that nonuniform spatial distributions of sediment critical shear stress are found around the world to varying degrees. The difference between the highest and lowest sediment critical shear stress levels tested through circulating flumes can reach 1.70 Pa while a value of 6.88 Pa was calculated using a CSM positioned in the modern Yellow River Delta (Fig. 4.11). On the other hand, sediment critical shear stress levels are known to vary significantly in different estuarine deltas around the world. From data obtained from the circulating flume experiment, it can be concluded that sediment critical shear stress levels in the modern Yellow River Delta are relatively low compared to those of other estuarine deltas around the world, as can be determined by comparing the lowest sediment critical shear stress levels found using a CSM to values found for other estuarine deltas.

From data for the lowest and highest values of sediment critical shear stress for the Yellow River Delta and for other estuarine deltas, it is evident that the nonuniformity of the spatial distribution of sediment critical shear stress in the modern Yellow

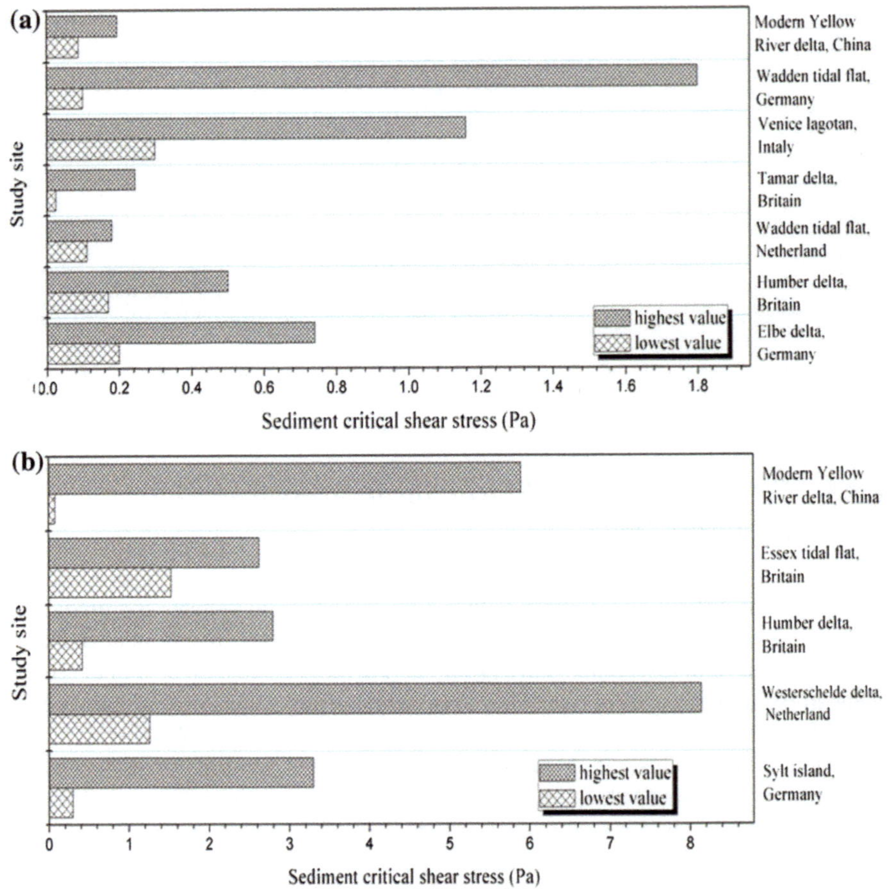

Fig. 4.11 Sediment critical shear stress levels of different estuarine deltas around the world. **a** Circulating flume experiment and **b** cohesive sediment strength meter tests

River Delta is more significant because the difference between the highest and lowest values of sediment critical shear stress is larger than that of other estuarine deltas. As the lowest levels of critical shear stress generally appear in the low tidal flat and as the highest critical shear stress levels generally appear in the high tidal flat, it can be inferred that erosion occurs more easily in the low tidal flat of the modern Yellow River Delta but hardly occurs in the high tidal flat. In addition, eroded and resuspended sediment in the low tidal flat produced through hydrodynamic forces in estuarine areas form a major component of sediment removed close to the shore.

4.3.6 Summary

The data show that the erosion thresholds vary at different sites along the seashore of the Yellow River delta. For the same intertidal area, the heterogeneous erodibility of the surficial sediment under hydrodynamic forces may lead to the formation of scour pits on the tidal flat. The factors influencing surficial sediment erodibility are complicated due to physical and biological reworking after the sediment deposited. The critical erosion shear stress of the surficial sediment can not clearly be related to single physical-mechanical sediment parameter except shear strength. The geotechnical shear strength is positively correlated to the critical erosion shear stress. The surficial sediment is more easily eroded at crab burrowing sites than in the nearby smoothed areas. It is possibly related to sediment bottom roughness.

Field measurements of sediment critical shear stress and geotechnical properties perpendicular to the coastline show that critical shear stress levels are high (ranging from 1.1 to 4.02 Pa) in the high tidal flat where erosion rarely occurs, and much lower (ranging from 0.08 to 0.80 Pa) in the low tidal flat where it erodes easily. The nonuniform spatial distribution of sediment critical shear stress plays a significant role in the formation of intertidal erosional landforms and in local changes. The nonuniform spatial distribution of sediment critical shear stress is a common pattern found in estuarine deltas around the world. Levels of nonuniformity were found to be significant in the modern Yellow River Delta. Sediment critical shear stress levels were found to be low, and sediment was easily eroded under marine hydrodynamic conditions.

References

Aberle J, Nikora V, Walters R (2004) Effects of bed material properties on cohesive sediment erosion. Mar Geol 207:83–93

Amos CL, Feeney T, Sutherland TF et al (1997) The stability of fine-grained sediments from the Fraser River delta. Estuar Coast Shelf Sci 45(4):507–524

Amos CL, Umgiesser G, Ferrarin C et al (2010) The erosion rates of cohesive sediments in Venice lagoon, Italy. Cont Shelf Res 30:859–870

Andersen TJ, Fredsoe J, Pejrup M (2007) In situ estimation of erosion and deposition thresholds by acoustic doppler velocimeter (ADV). Estuar Coast Shelf Sci 75:327–336

Andersen TJ, Lanuru M, Van Bermem C et al (2010) Erodibility of a mixed mudflat dominated by microphytobenthos and Cerastoderma edule, East Frisian Wadden Sea, Germany. Estuar Coast Shelf Sci 87:197–206

Bale AJ, Widdows J, Harris CB et al (2006) Measurements of the critical erosion threshold of surface sediments along the Tamar Estuary using a mini-annular flume. Cont Shelf Res 26:1206–1216

Black K, Cramp A (1995) A device to examine the in situ response of intertidal cohesive sediment deposits to fluid shear. Cont Shelf Res 15(15):1945–1954

Black KS, Tolhurst TJ, Paterson DM et al (2002) Working with natural cohesive sediments. J Hydraul Eng 128(2):2–8

British Standards Institution (1990) Bs 1377: British standards methods of test for soils for civil engineering purposes. B.S.I, London

Cardoso AH, Graf WH, Gust G (1990) Uniform flow in a smooth open channel. J Hydraul Res 27(5):603–616
Eckman JE, Nowell ARM, Jumars PA (1981) Sediment destabilization by animal tubes. J Mar Res 39(2):361–374
Einsele G, Overbeck R, Schwarz HU et al (1974) Mass physical properties, sliding and erodibility of experimentally deposited and differently consolidated clayey muds. Sedimentology 21:339–372
Endler R (2009) Sediment physical properties of the DYNAS study area. J Marine Syst 75:317–329
Fukuda MK, Lick W (1980) The entrainment of cohesive sediments in freshwater. J Geophys Res 85(C5):2813–2824
Gage JD (1977) Structure of the abyssal macrobenthic community in the Rockall Trough. In: Keegan BF, Ceidigh PO, Boaden PJS (eds) Biology of Benthic Organisms. Pergamon Press, Oxford, pp 247–260
Grabowski RC, Droppo IG, Wharton G (2011) Erodibility of cohesive sediment: the importance of sediment properties. Earth Sci Rev 105:101–120
Hjulstrom F (1939) Transportation of detritus by moving water. In: Trask PD (ed) Recent marine sediments. American Association of Petroleum Geologists, Tulsa, pp 5–31
Houwing E-J (1999) Determination of the critical erosion threshold of cohesive sediments on intertidal mudflats along the Dutch Wadden Sea coast. Estuar Coast Shelf Sci 49:545–555
Jepsen R, Roberts J, Lick W (1997) Effects of bulk density on sediment erosion rates. Water Air Soil Pollut 99(1–4):21–31
Jia Y, Liu X, Shan H et al (2011) The effects of hydrodynamic conditions on geotechnical strength of the sediment in Yellow River Delta. Int J Sedim Res 26:318–330
Jia Y, Zhang L, Zheng J et al (2014) Effects of wave-induced seabed liquefaction on sediment re-suspension in the Yellow River Delta. Ocean Eng 89:146–156
Kamphuis JW, Hall KR (1983) Cohesive material erosion by unidirectional current. J Hydraul Eng 109(1):49–61
Keller GH, Zhen J, Yang Z-S (1990) Mass physical properties of Huanghe delta and Southern Bohai Sea near-surface deposits, China. Mar Geotechnol 9:207–225
Li G, Zhuang K, Wei H (2000) Sedimentation in the Yellow River Delta. Part III. Seabed erosion and diapirism in the abandoned subaqueous delta lobe. Mar Geol 168:129–144
Liu X, Jia Y, Zheng J et al (2017) An experimental investigation of wave-induced sediment responses in a natural silty seabed: new insights into seabed stratification. Sedimentology 64:508–529
Liu X, Jia Y, Zheng J et al (2013a) Experimental evidence of wave-induced inhomogeneity in the strength of silty seabed sediments: Yellow River Delta, China. Ocean Eng 59:120–128
Liu X, Jia Y, Zheng J et al (2013b) Field and laboratory resistivity monitoring of sediment consolidation in China's Yellow River estuary. Eng Geol 164:77–85
Li B, Cozzoli F, Soissons LM, Bouma TJ et al (2017) Effects of bioturbation on the erodibility of cohesive versus non-cohesive sediments along a current-velocity gradient: a case study on cockles. J Exp Mar Biol Ecol 496:84–90
Meng X, Jia Y, Shan H (2012) An experimental study on erodibility of intertidal sediments in the Yellow River Delta. Int J Sedim Res 27:240–249
Meng X, Jia Y, Song J et al (2010) Study of erodibility change of Yellow River sediment into sea in process of consolidation. Rock Soil Mech 31(12):3809–3815 (in Chinese)
Mitchener H, Torfs H (1996) Erosion of mud/sand mixtures. Coast Eng 29:1–25
Paterson DM, Tolhurst TJ, Kelly JA et al (2000) Variations in sediment properties, Skeffling mudflat, Humber Estuary, UK. Cont Shelf Res 20:1373–1396
Postma H (1967) Sediment transport and sedimentation in the estuarine environment. In: Lauff GH (ed) Estuaries. American Association for the Advancement of Science, Washington, pp 158–179
Prior DB, Yang ZS, Bornhold BD et al (1986) Active slope failure, sediment collapse, and silt flows on the modern subaqueous Huanghe (Yellow River) Delta. Geo-Marine Lett 6:85–95
Roberts J, Jepsen R, Gotthard D et al (1998) Effects of particle size and bulk density on erosion of quartz particles. J Hydraul Eng 124(12):1261–1268

Scoffin TP (1970) The trapping and binding of subtidal carbonate sediments by marine vegetation in Bimini Lagoon, Bahamas. J Sediment Petrol 40:249–273

Schünemann M, Kühl H (1993) Experimental investigations of the erosional behavior of naturally formed mud from the Elbe estuary and adjacent Wadden Sea, Germany. In: Mehta AJ (ed) Nearshore and estuarine cohesive sediment transport, American Geophysical Union. D. C, Washington

Shi L, Li J, Ying M et al (2006) Erosion experiments on natural sediment from the Modern Yellow river delta. Ocean Eng 24(1):46–54 (in Chinese)

Teisson C, Ockenden M, Le Hir P et al (1993) Cohesive sediment transport processes. Coast Eng 21:129–162

Torfs H (1997) Erosion of mixed cohesive/non-cohesive sediments in uniform flow. In: Burt N, Parker R, Watts J (eds) Cohesive sediments. Wiley, Chichester, pp 245–252

Tolhurst TJ, Black KS, Shayler SA et al (1999) Measuring the in situ erosion shear stress of intertidal sediments with the Cohesive Strength Meter (CSM). Estuar Coast Shelf Sci 49:281–294

Tolhusrt TJ, Defew EC, de Brouwer JFC et al (2006) Small-scale temporal and spatial variability in the erosion threshold and properties of cohesive intertidal sediments. Cont Shelf Res 26:351–362

Watts CW, Tolhurst TJ, Black KS et al (2003) In situ measurements of erosion shear stress and geotechnical shear strength of the intertidal sediments of the experimental managed realignment scheme at Tollesbury, Essex, UK. Estuar Coast Shelf Sci 58:611–620

Widdows J, Brinsley MD, Salkeld PN et al (1998) Use of annular flumes to determine the influence of current velocity and bivalves on material flux at the sediment-water interface. Estuaries 51:552–559

Winterwerp JC, van Kesteren WGM, van Prooijen B et al (2012) A conceptual framework for shear-flow induced erosion of soft cohesive sediment beds. J Geophys Res 117:113–128

Zang Q (1996) Nearshore sediment along the Yellow River delta. Ocean Press, Beijing (in Chinese)

Zheng J, Jia Y, Liu X (2013) Experimental study of the variation of sediment erodibility under wave-loading conditions. Ocean Eng 68:14–26

Zheng J, Jia Y, Liu X (2014) Sediment characteristics as a function of variable hydrodynamics in a tidal flat of the Yellow River Delta, China. J Offshore Mech Arct Eng 136:1–9

Zheng J, Shan H, Jia Y et al (2011) Field tests and observation of wave-loading influence on erodibility of silty sediments in the Huanghe (Yellow River) estuary, China. J Coast Res 27:706–717

Chapter 5
Sediment Resuspension Process in the Modern Yellow River Delta

5.1 Overview

Sediment resuspension in the continental portions of the river and in the subaqueous delta and coastal areas are significant to physical, chemical, and biogeochemical aspects of hydrogeology (Green and Coco 2014). Human activities and natural processes have roles in regulating the sediment resuspension process (Ferré et al. 2008), but natural factor influences operate on a larger scale (Schoellhamer 2002; Zhu et al. 2017). Waves and currents are the primary causes that lead to sediment resuspension. Sediment resuspension affects sediment redistribution and influences regional particulate matter budgets and sediment movement to deeper ocean areas (Ferré et al. 2008).

The Yellow River is the second largest river in the world in terms of sediment load and contributes approximately 6% to the total estimated global river sediment flux entering the ocean (Liu et al. 2012; Wang et al. 2010). Apart from its lower reaches, large quantities of suspended sediments are deposited in coastal areas, Bohai Sea, and the North Yellow Sea (Zhang et al. 1990). The flow path in the Yellow River tail has experienced significant shifting over the last one hundred and fifty years. The dynamic marine processes in the region such as waves, tides and currents have interacted with the sediments through re-consolidation, resuspension and transportation and reshaped the subaqueous delta, which contains high silt content and experiences rapid deposition. It has been reported that waves played a crucial role in sediment transformation.

In the process of wave-induced seabed sediment suspension, one of main concerns has been the effects of wave shear stress on the onset of the movements of the silty seabed surface sediments (Wright et al. 2001; Wang 2003; Hoque et al. 2010). Among these studies, Wright et al. (2001) reported that a significant amount of the seabed sediment resuspension during storm events was attributed to friction shearing stress increases at the seabed surface, caused by the elevated currents and wave orbital speed. Through field observations, Wang (2003) found that suspended sediment concentrations (SSC) suddenly increased when wave shear stress exceeded

Y. Jia et al., *Wave-Forced Sediment Erosion and Resuspension in the Yellow River Delta*, Springer Oceanography, https://doi.org/10.1007/978-981-13-7032-8_5

the sediment critical shear stress. Hoque et al. (2010) reported that SSC and sediment fluxes were substantially higher in spring tidal cycles compared to neap tidal cycles. Kong and Zhu (2008) reported from their observations on wave flume experiments, that SSC increased with water depth under wave action, and that the increases were greater under regular and irregular wave action than those under the current. Later, Pang et al. (2011) further established a linear relationship between the vertical concentration and relative water depth.

Historically, sediment resuspension mechanisms are either studied using controlled laboratory experiments or field observations and both the approaches have advantages. For example, in situ data is closer to the real law of the nature while the controlled indoor tests are more reliable for establishing quantitative relationship between detected parameters. Therefore, in this chapter, the process of sediment erosion and resuspension was studied through in situ observations and wave flume experiments.

5.2 In Situ Observations of on Sediment Resuspension Under Ocean Dynamics

Most previous studies on the influence of waves and currents on sediment resuspension were conducted using numerical modeling, flume experiments, remote sensing, and short-term continuous observations conducted onboard boats (Podsechin et al. 2006; Rajaee et al. 2009; Yang et al. 2007) due to the difficulty of long-term continuous observations near the seabed. Continuous in situ observation is costly and presents technical difficulties. Although remote sensing can cover a large area, it can only determine suspended sediment concentrations near the surface. Little is known about the detailed dynamic properties of suspended sediment concentrations near the seabed caused by waves and currents. Therefore, research using in situ long-term continuous observation is needed to understand suspended sediment dynamics.

In this book, a submarine site was selected to deploy a tripod that operated in the Yellow River delta for more than 120d to measure the time series of waves, currents, suspended sediment concentrations, temperature, and water depth. We found the dominated role of wind waves in causing resuspension. More specifically, our flume experiments preliminarily found the important role of wave-induced seabed liquefaction in contributing to resuspension, in addition to the commonly known source of coming from seabed surface.

5.2.1 Methodology

(1) ***Study Area***

The Shandong peninsula is bordered by the Liaodong peninsula and the Chinese mainland, The Bohai Sea is a semi-enclosed coastal sea in the West Pacific Ocean. With a mean water depth of 18 m, the shallow areas (<10 m) of the Bohai Sea account for about 26% of the total area (Guo et al. 2016). Since the present Yellow River first drained into the Bohai Sea in 1855, its river course has shifted 11 times, forming a delta of 5400 km^2 (Saito et al. 2000). The latest large change occurred in 1976 when the river mouth channel migrated ~50 km southeastward from Diaokou to Qingshuigou (Kong et al. 2015).

Our study site was located next to the Diaokou channel in the Yellow River delta and the submarine in situ tripod was deployed at 6–7 m water depth and about 50 m to the south of a subaqueous depression (Fig. 5.1). The study area elevation ranges from high in the north and low in the south, with a mean slope of 7–8°. In surface sediments, the water content is 27.4–32.3% and the density is 19.4–20.1 kN/m^3.

Sediment grain size is an important characteristic of sediment particles because it is directly related to the critical shear stresses affecting their entrainment, transport, and deposition (Knapen et al. 2007). Grain size analysis therefore provides information on sediment resuspension (Blott and Pye 2001). Sediment grain size is essential for sediment classification and calculation of critical bottom shear stress. One of the more common means to present and compare grain size is to plot sand, silt, and clay fraction data on a ternary diagram (Shepard 1954; Folk 1966). The grain size of sediment in the study site was evaluated by the Shepard (1954) classification scheme, which is simple and facilitates rapid sediment classification and sample comparison.

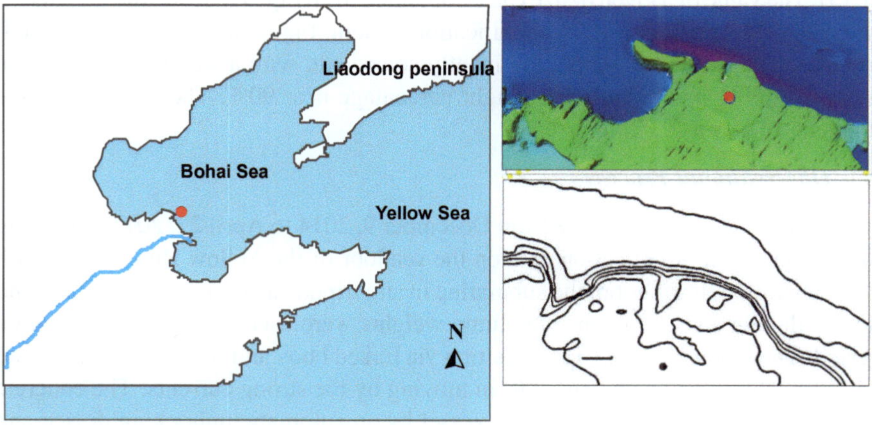

Fig. 5.1 Map of different scales showing the study area (left) and bathymetric chart (right). The mooring station is marked by a red solid dot

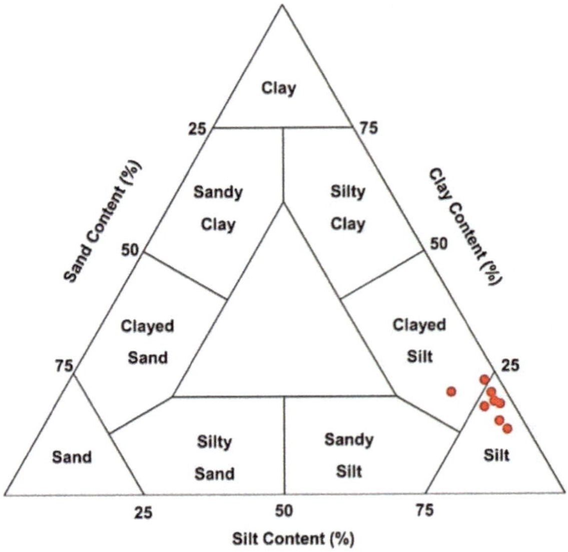

Fig. 5.2 Equilateral triangular diagram of the Shepard (1954) classification scheme. The samples are marked by solid red dots

Figure 5.2 shows that sediments in the studied site are mainly silty, with six of eight samples located in the silt fraction and two samples in the clayed silt fraction.

Apart from sediment classification for qualitative analysis, the weight percentages of sediment in various size classes for quantitative analysis are essential in calculating critical bottom shear stress. One of the methods used to display particle size distributions is d-size. The percentage below the given diameter is the number expressed after the "d". For example, the d_{50}, known as the median particle diameter, is the diameter at which 50% of the soil mass is finer than this size (Pholkern et al. 2015). The value of a given d-size can be obtained from Fig. 5.3 by the Massachusetts Institute of Technology (MIT) classification system. Figure 5.3 shows that the sediments at the study site are predominantly coarse silt, with a weight percentage of more than 50% and cumulative weight percentage near 90%. The median particle diameter (d_{50}) was 0.023 mm.

(2) ***Observational Process***

In situ observations were made from December 9, 2014 to April 22, 2015, providing more than 130 d of collection data on the seafloor of the Yellow River delta. With modularization installation, the submarine in situ tripod can easily be assembled and dismantled on site. Two concrete clump weights were used during the observation period. They indicated the tripod position via linked buoy and submersible buoy and kept the submarine in situ tripod from moving by the strong currents. The concrete clump weights were released on the vessel by an automatic unhooking device and the submarine in situ tripod was unhooked underwater by an aquanaut after it had stably landed on the seafloor. After unhooking the tripod, the aquanaut sampled the surface sediment for laboratory testing. With the aid of GPS localization and buoys,

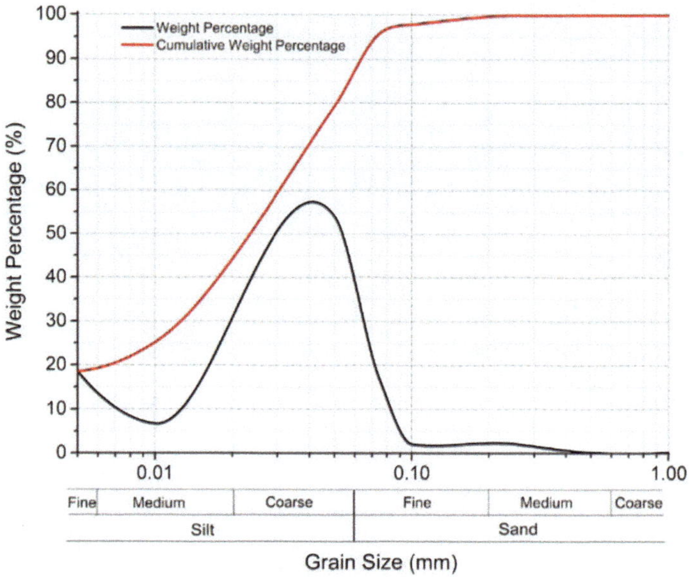

Fig. 5.3 Particle size distribution curve of the sediment samples in the study site using the Massachusetts Institute of Technology (MIT) classification system

the submarine in situ tripod was easily located for recovery on April 22, 2015. Layout of the in situ observation and implementation processes is shown in Fig. 5.4.

(3) ***Instrumentation***

The tripod was custom made of stainless steel. It was 1 m high and the side lengths were 2 m (Fig. 5.5). In order to maintain the tripod where it was deployed, steel clump weights were attached to each base of the tripod. A base plate with a big surface area was positioned beneath the steel clump weights to prevent the tripod from sinking. The main observation instruments installed on the tripod were a wave-tide recorder (RBR virtuoso Model, RBR Ltd. Canada), electromagnetic current meter (Infinity-EM Model, JFE Advantech Co., Ltd. Japan) and CTD + Tu (XR-620 Model, RBR Ltd. Canada). Wave parameters, such as significant wave height and significant wave period, were obtained by RBR virtuoso with the resolution of 0.05%. The velocity and direction of the currents were recorded by INFINITY-EM with a velocity resolution of 0.02 cm/s and a 0.01° direction resolution. Conductivity, temperature, depth, and turbidity were stored by XR-620 with the resolution of 0.003 mS/cm, 0.002 °C, 0.05%, 2% respectively. Parameter settings for each instrument were different and details are shown in Table 5.1.

The grain size characteristics of surface sediment samples were determined by sieve analysis according to the American Society for Testing and Materials. The suspended sediments concentration was obtained from the turbidity data by regression analysis. The turbidity of the suspension liquid, made of surface sediment samples

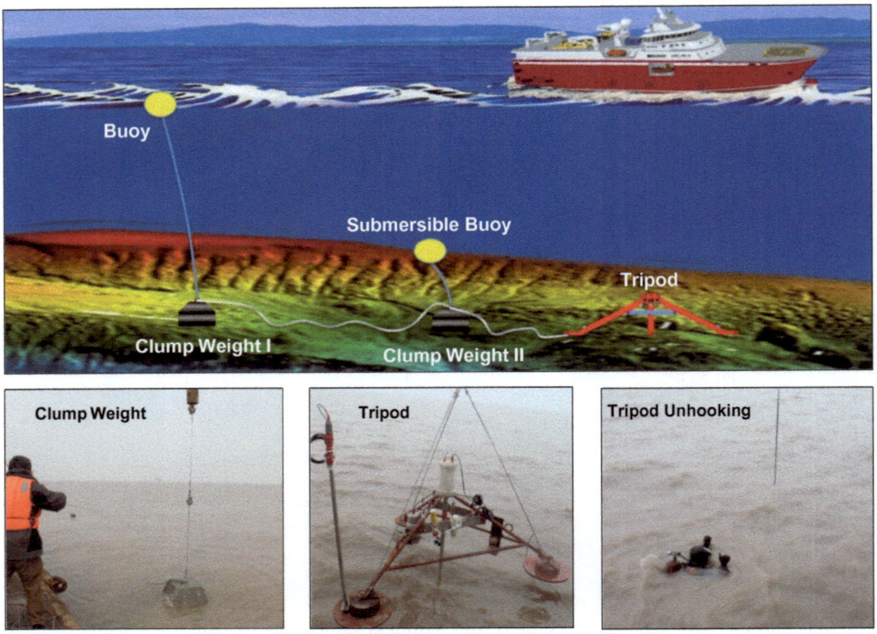

Fig. 5.4 Schematic diagram of in situ observation layout and some implementation process photos. The tripod being deployed at the Yellow River delta from the submarine at 15:35 on December 9, 2014

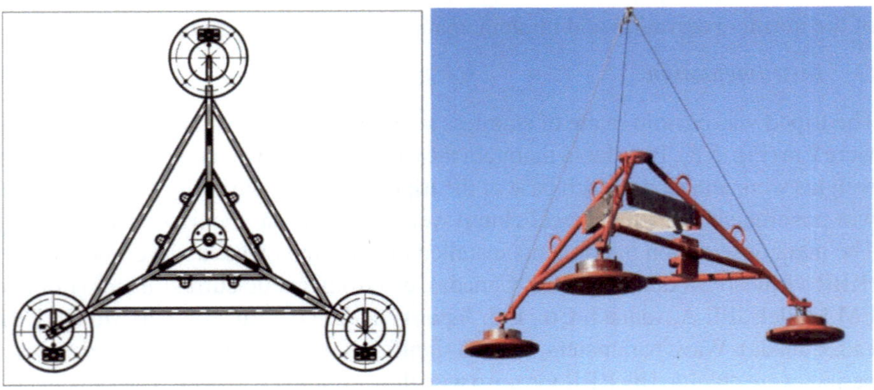

Fig. 5.5 The submarine in situ tripod. The left part is the vertical view of its blueprint and the right part is a photo of the finished tripod

Table 5.1 Parameter setting details for observation instruments

Instrument	Frequency (Hz)	Samples	Period (min)
RBR virtuoso	2	2048	60
Infinity-EM	1	20	30
XR-620	1	1	10

and seawater, was measured by a portable turbidity meter (HACH 2100Q Model, Hach Company, USA). With the sediment sample mass, suspension liquid volume, and turbidity data, the relationship between suspended sediments concentration and turbidity could be established. To enhance the comparability of the laboratory test, the sediment samples and water samples were taken from the study site.

(4) ***Calibration of Sensors***

Turbidity is an optical property of seawater that may be used to measure the relative water clarity. Since turbidity is mainly due to sediments in suspension, it is also used to estimate the suspended sediment concentration. Water color, temperature, and other factors may bias the results, but their effects are minor (Pavanelli and Bigi 2005). Therefore, we derived the suspended sediment concentration from turbidity in this study. In order to increase the comparability of the laboratory test, the sediment samples and water samples were taken from the study site. Even so, it is not possible to create identical water conditions in the laboratory. For example, the grain size distributions of the surface sediment and suspended sediment were different. These factors however do not affect the overall study results.

During the observation period, the turbidity in the study site mainly varied between 200 NTU to 800 NTU, with the highest turbidity reading of 920 NTU made on December 20, 2014. The relationship between suspended sediment concentration and turbidity has been studied by other researchers (Minella et al 2008; Pavanelli and Bigi 2005). Their results indicate that the relationship is linear or approximately linear. But, the turbidity in other study areas was not as high as in the present study. More importantly, the relationship between suspended sediment concentration and turbidity can also be influenced by the organic matter content (Slaets et al. 2014). Thus, their relational expression is not appropriate for the suspended sediment at the study site. To reduce the error of the relationship between suspended sediment concentration and turbidity, the turbidity range of the suspension liquid, made of surface sediment samples and seawater, was maintained as similar to the field turbidity as possible. Figure 5.6 presents the simple linear relationship between the suspended sediments concentration and turbidity with the Eq. 5.1.

$$y = 0.0048x \tag{5.1}$$

where y is the suspended sediments concentration in g/l, x is the turbidity in NTU. According to Eq. 5.1, the suspended sediment concentration at the study site, ~50 km away from the present Yellow River mouth, mainly ranges from 1 to 4 g/l.

5.2.2 Results

Although the study tripod was recovered on April 22, 2015, not all the observation instruments were still operating and collecting data up to that day. The electromagnetic current meter ceased operation on February 8. The collected data are reor-

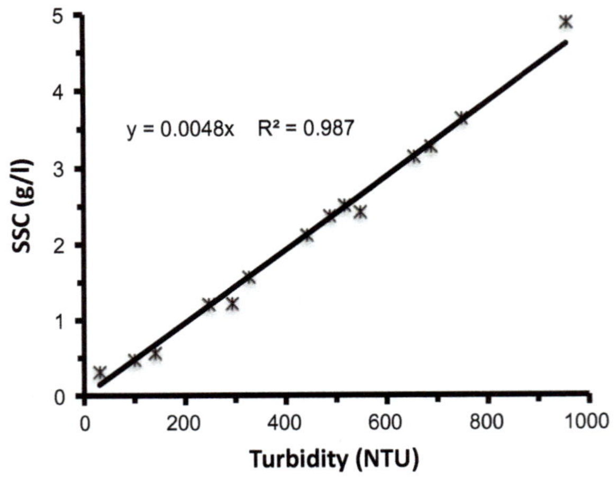

Fig. 5.6 Relationship between suspended sediment concentration (SSC) and turbidity determined from samples taken at the study site on the initial study day

ganized in an hourly time series in Fig. 5.7, showing the observation results during the time span when all the instruments were working. During the 60 d, the tide levels ranged from 4.92 m (on January 1) to 7.50 m (on January 21 and February 7), with the mean tide level of 6.38 m. The greatest current velocity was 0.84 m/s (on February 7), with a mean velocity of 0.22 m/s. The suspended sediment concentration calculated by Eq. 5.1 varied from 1.10 g/l (on January 18) to 4.41 g/l (on December 20), with a

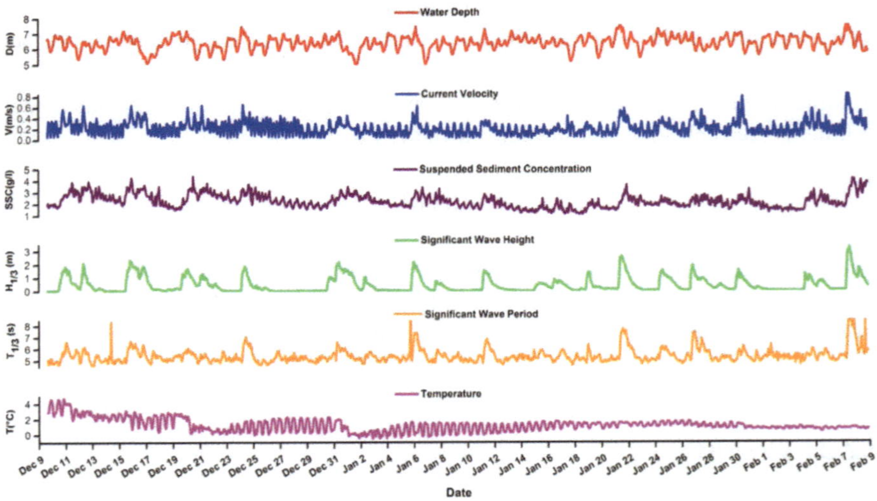

Fig. 5.7 Observation results in the study site from 16:00 December 9, 2014 to 21:00 February 8, 2015

mean concentration of 2.24 g/l. This is higher than that found in the waters of other deltas. More than 10 wave events with a significant wave height above 1 m were seen. The biggest significant wave height measured up to 3.30 m (on February 7). Based on the freezing point of seawater in this area (−1.7 °C) (Zheng et al. 2015), the seawater during the study period did not freeze. The lowest temperature recorded was –0.47 °C on January 2 and highest temperature was 4.79 °C on December 10.

It is difficult to maintain continuous in situ observations for more than one year due to limited battery supply, sensor drift, and other technical and monetary limitations (Tréhu 2016). Long-term continuous in situ observation might be easier to achieve by cable observations.

5.2.3 *Effects of Waves on Sediment Resuspension in the Yellow River Delta*

Although it is difficult to quantitatively determine the contribution of wave action to suspended sediment concentration, linear regression analysis can provide some insights. The daily mean values of the suspended sediment concentration and significant wave height were studied using linear regression analysis (Fig. 5.8). The relationship between the two variables is linear with a coefficient of 0.64, demonstrating wave height enhancing the concentration of suspended sediments. Despite the differences in correlation coefficients found in other areas (Bian et al. 2012; Houwing et al. 2002; Xie et al. 2010), the two variables have a positive relation. The intercept value of the linear equation indicated that the mean suspended sediment concentration during calm waves can reach 1.89 g/l.

Figure 5.7 shows the overall and general observations, with little detail. To illustrate the specific effects of waves on suspended sediment concentration, a plot of significant wave height and suspended sediment concentration from December 13 to December 26 is shown in Fig. 5.9. Figures 5.7 and 5.9 demonstrate that the suspended sediment concentration increases up to ~4 g/l during wave events. On December 15, wave heights up to 2.36 m and concentrations of 4.27 g/l were recorded. However,

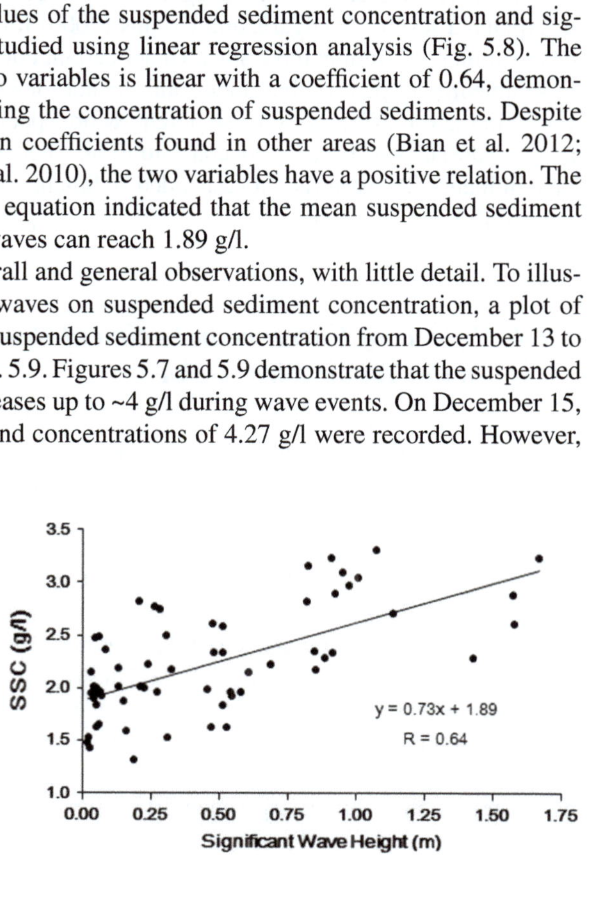

Fig. 5.8 Relationship between the daily mean suspended sediment concentration and significant wave height

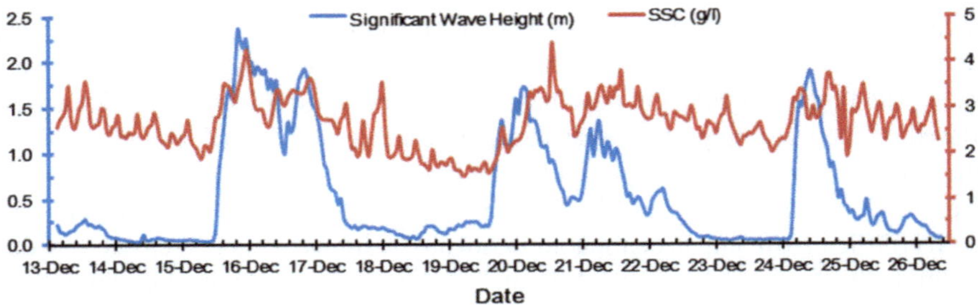

Fig. 5.9 The plot of significant wave height and suspended sediment concentration from Dec. 13 to Dec. 26

during calm waves, the suspended sediment concentration oscillated with almost no change in significant wave height such as on the December 23 (dotted box A in Fig. 5.9) and December 14 (dotted box B in Fig. 5.9). Surprisingly, the suspended sediment concentration and significant wave height had a negative correlation (dotted box C in Fig. 5.9). Kana (1978) reported that suspended sediment concentration decreases with increasing wave height and considered this to be unexpected, and unexplained results. Voulgaris and Collins (2000) found a similar result based on observation of 17 min in three sites. Under certain conditions waves can play a role in sediment resuspension but under other conditions the suspended sediment concentration is not affected by wave height. More specifically, only large waves can increase the suspended sediment concentration, whereas small waves have little positive influence. A sharp rise in suspended sediment concentration is attributed only to the episodic high waves.

5.2.4 Effects of Currents on Sediment Resuspension in the Yellow River Delta

The chart of daily mean suspended sediment concentration in the study site as a function of daily mean current velocity indicates that the relationship between the two quantities is linear, with higher current velocities increasing sediment concentration (See Fig. 5.10). The positive relation is clearly visible although some scatter is present. Besides, the correlation coefficient (0.68) in Fig. 5.10 is a little more than that in Fig. 5.8. Houwing et al. (2002) did not find a significant correlation between the mean current velocity and suspended sediment concentration, which may have been due to the low current velocity (<0.05 m/s) during the study. The intercept value of the linear equation in Fig. 5.10 showed that even if the seawater is motionless, the mean sediment concentration can reach 1.19 g/l. However, the seawater in the study site moves due to everyday tidal currents and other factors. The sediment concentration therefore is larger than 1.19 g/l in most cases.

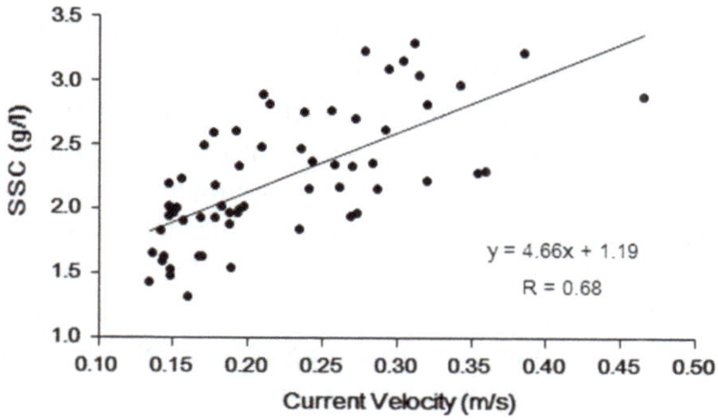

Fig. 5.10 Relationship between daily mean suspended sediment concentration and current velocity

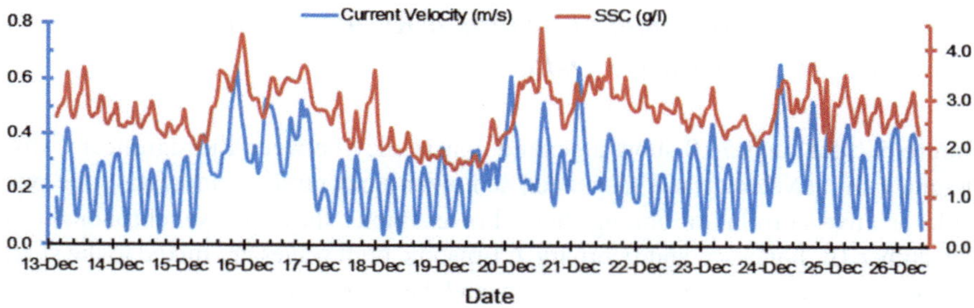

Fig. 5.11 Plot of current velocity and suspended sediment concentration from 13 December to 26 December

There is a slight lag of sharp rise in suspended sediment concentration caused by the episodic high waves in Fig. 5.9. However, there was no lag of suspended sediment concentration in Fig. 5.11. The traditional model of "wave-lifting-sand, current-transporting-sand" can account for the phenomenon. Episodic high waves can lift sediments into water but there must be an action time before the sediments are suspended. Besides, wave-induced liquefaction can enhance sediment resuspension. The pore pressure needs time to accumulate before sediment liquefaction. Hence, there is a lag of suspended sediment concentration in Fig. 5.9. Once the sediments have been suspended into seawater, small disturbances, such as current changes, can affect the suspended sediment concentration. Thus, there is no lag of suspended sediment concentration. Both human activities and natural processes play a role in suspended sediment concentration. Therefore, the correlation coefficient is low when calculated between a single factor and suspended sediment concentration.

Apart from sharp rise in suspended sediment concentration that is attributed to episodic high waves, there are more detailed changes that are not caused by the waves. One such change is the decreasing trend in periodic fluctuations after the episodic high waves. Settling of the suspended sediments can account for the decreasing trend, due to the diminishing suspended sediment in the seawater (Hill et al. 2000; Xia et al. 2004). The dynamic descending rate of suspended sediment concentration caused by currents can be determined as 0.37 g/l per day, namely the absolute value of the slope of the black dotted line in Fig. 5.11. The periodic fluctuation may be related to periodic hydrodynamic factors, such as tidal currents. To examine the specific effect of currents on suspended sediment concentration, more details are shown in Fig. 5.11. Figure 5.11 shows that the sediment concentration and current velocity fluctuate four times a day especially during calm waves, which accords with the irregular semidiurnal tide (Fagherazzi and Priestas 2010). The dynamic changes in sediment concentration and current velocity are consistent. We conclude that suspended sediment settling and currents are responsible for the decreasing trend with periodic fluctuation in suspended sediment concentration.

5.2.5 Conceptual Model of Sediment Resuspension in the Yellow River Delta

Figure 5.7, 5.8, and 5.10 show a general consistency between the significant wave height, current velocity and suspended sediment concentration, indicating the possible positive correlation among them. For example, they rose synchronously on December 16, January 6 and February 7. Episodic high waves are linked to the sharp rise in suspended sediment concentration, and the periodic fluctuation is attributed to the quotidian currents. Both the significant wave height and current velocity are positively correlated with suspended sediment concentration but which of these factors has the greatest influence? The static and dynamic models of suspended sediment concentration caused by waves and currents may help answer this question.

(1) ***Static Model of Sediment Resuspension Process***

The static model of suspended sediment concentration influenced by waves and currents is a time-independent view of the system. We cannot compare the values of waves and currents directly, because significant wave height and current velocity have difficult dimensions, the former in L and the latter in LT^{-1}. Multiple linear regression analysis can address this problem (Zhu et al. 2015). Multiple linear regressions of the hourly mean values of significant wave height, current velocity, and suspended sediment concentration are shown in Eq. 5.2 and Fig. 5.12, without consideration of sediment settling velocity and other control factors. The correlation coefficient of Eq. 5.2 is 0.57; the coefficient (0.73) of current velocity (V) is 1.55 times greater than (0.47) of significant wave height ($H_{1/3}$), indicating that the current velocity may have a greater influence regardless of episodic high waves enhancing sediment concentration. In Fig. 5.12, the black dots show the observation values averaged every hour, and the 3D curved surface fit in Fig. 5.12a indicates the function image of Eq. 5.2; its top view is shown in Fig. 5.12b. Different colors correspond to the value of suspended sediment concentration (see color bar in Fig. 5.12). The current velocity and significant wave height can influence the coefficient. Equation 5.2 uses

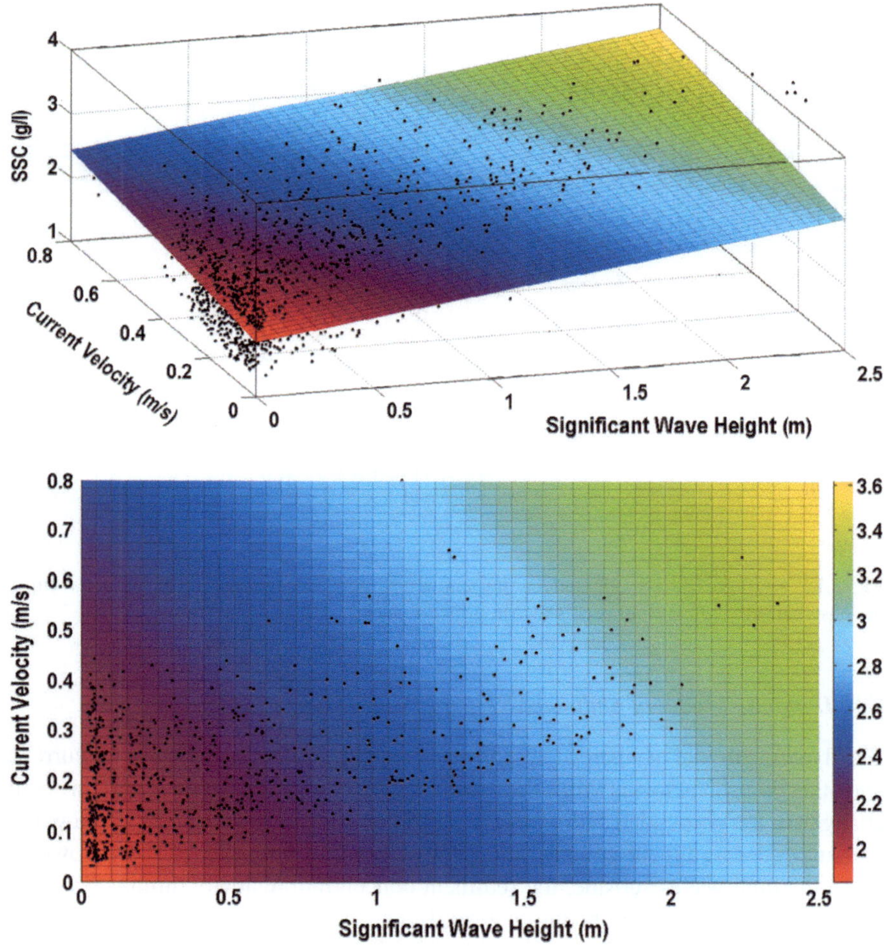

Fig. 5.12 Multiple linear regression of the hourly mean values of significant wave height, current velocity, and suspended sediment concentration

the International System of Units (SI) while the largest current velocity was less than 1 m/s. This indicates that the unit of current velocity in Eq. 5.2 is inappropriate. If we use cm/s as the unit, the coefficient of current velocity becomes 0.0073 which is much less than the coefficient of significant wave height. In this way, the significant wave height carries a greater weight. Though, the static model is approximate, it provides a simple way to analyze the influence of waves and currents on the suspended sediment concentration.

$$SSC = 0.47 * H_{1/3} + 0.73 * V + 1.85 \tag{5.2}$$

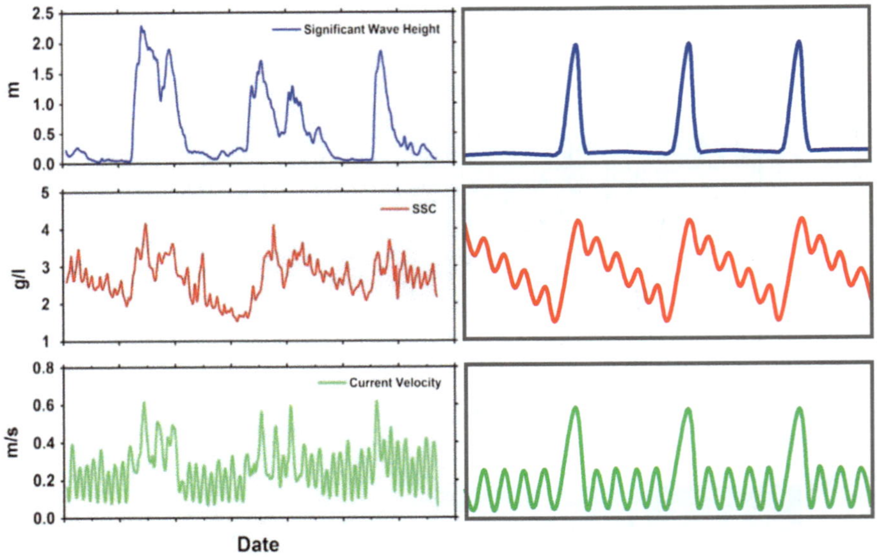

Fig. 5.13 Superposed sawtooth model of suspended sediment concentrations. The left part includes the observations and the right part is the simplified model

(2) *Dynamic Model of Sediment Resuspension Process*

The dynamic model of suspended sediment concentration caused by waves, currents and other control factors is a time independent view of the system and shows, in detail, changes in the suspended sediment concentration. Both Figs. 5.7 and 5.9 show that the suspended sediment concentration rises sharply during high waves. However, when the high waves subside, the sediment concentration slowly diminishes rather than abruptly dropping to the initial level. Hence, the curves of suspended sediment concentration represent a sawtooth shape. In this slowly diminishing process, the sediment concentration fluctuates with the current velocity. The curves of suspended sediment concentration therefore show another sawtooth shape which is smaller than the one induced by high waves. The curves of suspended sediment concentration show a superposed sawtooth model (red line in Fig. 5.13); the big sawtooth shape is mainly induced by large episodic waves and the small sawtooth shape is caused by the periodic currents. The number of small sawtooth shapes depends on the time interval between the next high waves and the characteristics of the tides. In the areas with semidiurnal tides, the tides fluctuate four times a day while in the diurnal tide areas, they fluctuate twice daily.

Although the significant wave height and current velocity have different dimensions, the maximum wave-induced shear stress and current-induced shear stress have the same in $MT^{-2}L^{-1}$. Therefore, we can compare the two different shear stresses to find possible explanations for the superposed sawtooth model of suspended sediment concentration. Figure 5.14 shows that, the maximum wave-induced shear stress is usually close to zero while during episodic waves it can reach 3.50 Pa. Compared

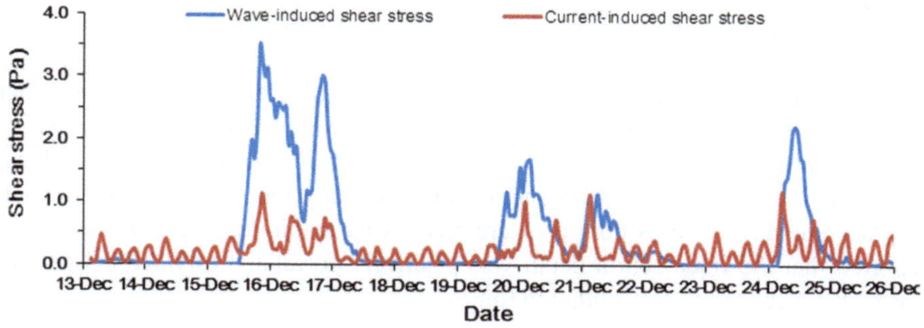

Fig. 5.14 Maximum wave-induced shear stress and current-induced shear stress

with maximum wave-induced shear stress, the current-induced shear stress constantly changes but not on such a large scale. It generally fluctuates four times daily in the areas with irregular semi-diurnal tides and varies within 0.4 Pa during calm waves. Though comparatively low compared to maximum wave-induced shear stress, the current-induced shear stress is more than the critical shear stress that must be exceeded before sediment begins to move (Ma et al. 2015). The shear stress results help explain why episodic waves are responsible for the big sawtooth shape and currents account for the small sawtooth shape. Both the wave energy and maximum wave-induced shear stress are proportional to the wave height squared. Similarly, both current energy and current-induced shear stress are proportional to the current velocity squared. Therefore, the shear stress can be regarded as a reflection of energy. The superposed sawtooth model of suspended sediment concentration is the external manifestation of wave energy and current energy.

5.3 Laboratory Experiment on Sediment Resuspension Under Ocean Dynamics

The sediment of the Yellow River Delta can be rapidly consolidated and an overconsolidated hard shell may be formed within the surface sediment 1–2 days after deposition, which led to the shear strength of the seabed sediments generally increasing to several kilopascals. Using numerical simulations, Gao and Jia (2003) concluded that the magnitudes of the shear stresses under the combined wave and current loading were only several pascals at the soil bed surface. From a comparison of the two values, it seemed difficult for the wave/current to erode and resuspend the seabed sediments. However, the stiff sediment may be quickly liquified resulting in its strength to rapidly decreasing, even under slight vibrations.

The phenomenon of the wave-induced accumulated pore pressure and the resultant soil bed liquefaction has been studied by numerous researchers. Foda and Tzang (1994) found that silt bed was liquified after a sudden increase in pore pressure under wave action. Using flume experiments, De Wit and Kranenburg (1997) established the pore pressure threshold at which soft soil liquefaction will occur. Through one-dimensional experimental studies, Zen and Yamazaki (1990) and suggested that the mechanism of the wave-induced seabed liquefaction was dominated by excess pore pressure redistributions, which were caused by wave damping and phase lags. Sumer et al. (2004, 2006, 2012) concluded that the silty sediment is liquified when the "accumulated" excess pore pressure reaches its maximum value. Feng (1992) reported that the liquefaction depth was positively correlated to wave height and reversely correlated to consolidation time.

Among the aforementioned studies, the characteristics of wave-induced sediment resuspensions, and the mechanism of liquefaction were particularly acknowledged. However, only a few studies have focused on the internal responses of sediment under wave action with regard to its role in sediment resuspension. Wave-induced pore pressure accumulation in the liquified cohesive sediments was found to enhance bed erosion and suspension (Maa and Mehta 1987; Aldridge and Rees 1997). Fluidisation and its subsequent upward fluid propagation caused the cohesive sediment particles to separate, and essentially affected the fluvial erosional strength of sediments. Thus, liquefaction plays an important role in resuspension. Foda and Tzang (1994) initially observed that the phenomenon of sediment transportation iand suspension was much more spectacular over fluidised beds than in unfluidised beds under the action of non-breaking waves. Tzang (1998) observed in a laboratory flume that the silt seabed formed a liquid surface layer after a short time and produced suspended materials. More recently, Tzang and Ou (2006) and Tzang et al. (2009) further explored some parameters inside fluidisation beds for the SSC and mechanism of sediment suspension, and found that the SSC occurs several wave cycles after the occurrence of the fluidisation response.

Therefore, the effect of wave-induced liquefaction in silt sediment on the resuspension feature and component is critical to understanding the mechanism of sediment resuspension. However, to date this aspect has not been investigated experimentally. The objectives of the work in this section are to use fine-grained sediment from the Yellow River Delta, exploring the relationships between wave-induced residual liquefaction in silt sediment and the seabed sediment resuspension, and quantifying the contribution of silt bed fluidised responses to the quantity of resuspended sediment.

5.3.1 Methodology

(1) *Experiment Flume and Sediments*

The tests were conducted in a wave flume at the Geotechnical Laboratory, Ocean University of China. The experiment setup is depicted in Fig. 5.15. The flume was 14 m (L) × 0.7 m (H) × 0.5 m (W) in size, equipped with a piston-type wave generator at one end and a 1:4 dissipating gravel beach at the other. The dissipation system is one of the most important parts of the wave flume used in this experiment, and the new

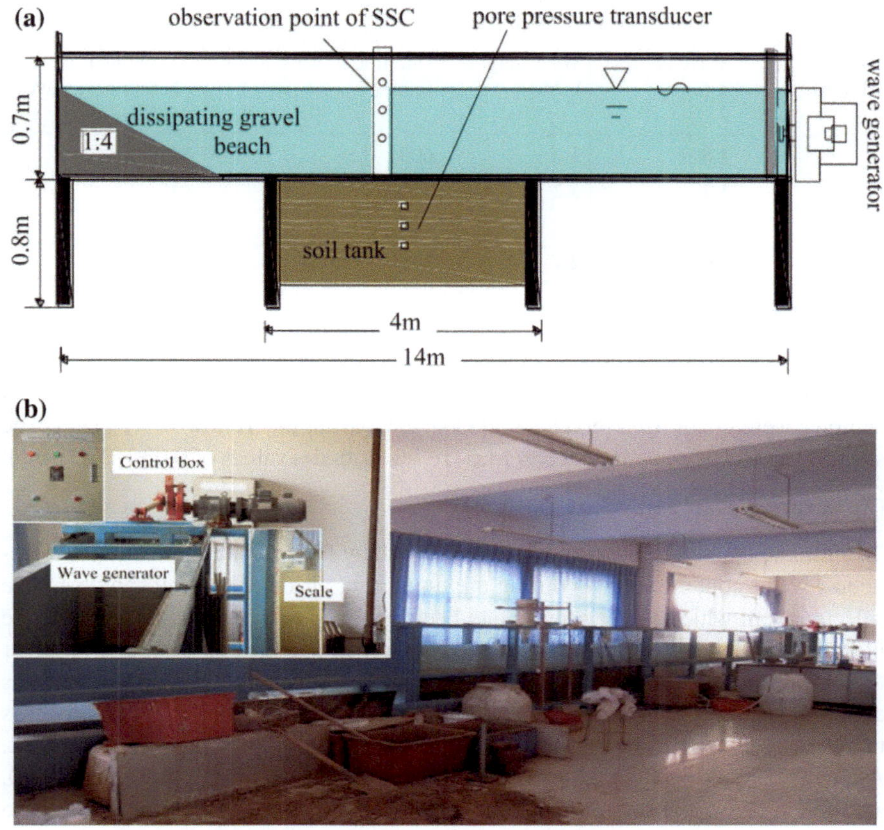

Fig. 5.15 **a** Sketch of the wave flume and **b** photograph of the wave flume during experiments

generated monochromatic waves can be considered regular wave group parading in the direct of paddle wave-maker when the front waves reach the dissipation system and disappear during the test running-period. As shown in Fig. 5.15, three pore pressure transducers (20 mm in diameter and 60 mm in length) were deployed at different depths along the central site of the soil tank. Before the pore pressure transducers were embedded in the soil bed they were soaked in water for 24 h with continuous shaking to ensure gas removal. A TURB335IR-type nephelometer (WTW-Munich, Germany) and XR-420 turbidimeter (RBR Ltd.-Ottawa, Canada) were used to measure the turbidity of the water during the flume test. The relationship between turbidity and the SSC was predetermined by an indoor test. The turbidity values measured during the flume tests were then converted to SSC, which was used to quantify the sediment resuspension.

Table 5.2 Wave conditions in the experiments

Wave height H (cm)	Wave period T (s)	Wave length L (m)	Duration of wave loads (min)	Standing time before next wave function (h)
5	1.8	3.2	600	12
7	1.7	3.2	570	12
10	1.7	3.2	440	12
18	1.015	2.4	525	*

Note "a" stands for data not available

Artificially produced seawater with salinity of 35% was used in the tests. The soil samples were taken from the Diaokou Course Coast at the Yellow River Delta. The silt soil had a clay content between 13.6–16.7% with d_{50} values of 0.036–0.042 mm.

(2) ***Experimental Procedure***

To simulate the natural water content of the Yellow River Delta sediment (homogeneous, water content at 25%) during the flume test, air-dried silt soil samples were selected from which larger gravels had been removed. Weighed soil samples were placed in a blender and the standard seawater was added proportionally in order to achieve the desired water content. The soil was stirred while the seawater was added to ensure the mixture developed into a homogeneous slurry. The slurry was then slowly poured along the flume wall into the soil tank up to about 50 cm thick. After the soil bed settled, three pore pressure transducers were installed at 20, 30, and 40 cm below the soil surface and 24 h was allowed for the soil to resettle. The standard seawater was then gradually added into the wave flume up to a depth of 40 cm above the soil surface. The soil bed was left to consolidate under the still water pressure for 24 h before the flume test started.

A wave loading was imposed using the wave generator mounted at the end of the flume. Four designed wave conditions (with a wave height of 5 cm, 7 cm, 10 cm and 18 cm respectively) were tested and the associated wave parameters were summarized in Table 5.2. The waves have been run at least 440 s, and then break for 12 h under each wave conditions. Wave flume experiments were conducted several turns for each case to study sediment dynamics under waves by our research group, and similar experiment phenomena on the resuspension process studied in this paper was found in all test runs. In this section we conducted the wave flume experiment focusing on the relationship between pore water pressure and sediment resuspension.

During the experiments, the TURB335IR-type nephelometer was used to measure the turbidity of the water at depths of 15, 25 and 35 cm in predetermined time intervals. Due to the rapid onset of the sediment suspension at the beginning, measurements were taken every 5 min, increasing to 15 min and then 30 min as the turbidity values became stable. To continuously measure the changes of suspended sediment concentrations near the surface of the soil bed under different hydrodynamic conditions, the XR-420 turbidimeter was placed at a depth of 40 cm from the water surface. After each of the wave action process, the silt layer thickness and the change of the soil bed elevation was measured until the soil bed became steady.

(3) ***Data Processing Method***

Parameterisation of the onset of liquefaction

The effective stress was found to decrease when pore pressure built up in the saturated silt during cyclic wave loading for sediments with poor permeability. Furthermore, as the total stress remained constant during the wave action process, the excess pore pressure increased continually (Foda and Tzang 1994; Sumer et al. 2006, 2012). If excess pore pressure is defined as Δu, liquefaction then occurs when Δu reaches the maximum excess pore pressure Δu max (McDougal et al. 1989). That is, the quantity Δu max may be set equal to the initial mean normal effective stress σ_0 (Sumer et al. 2006). This consideration for the experimental conditions in the original tests was similar to Sumer et al. (2006), can be expressed as

$$\Delta u_{\max} = \sigma_0' = \gamma' d(1 + 2K_0)/3 \tag{5.3}$$

where K_0 is the ratio between the horizontal and vertical effective stresses, i.e., $K_0 = \sigma_h'/\sigma_v'$, which is not a constant value for silt, so $K_0 = 0.5$ is used as an average value. Substituting $K_0 = 0.5$ into Eq. (5.3), $\Delta u_{\max} = \sigma_0' = 2\gamma' d/3$ was obtained, where d is the thickness of the overlaying soil; and $\gamma' = \gamma_s - \gamma_w$ is the submerged unit weight, which is assumed to be constant throughout the whole process, 7.9 kN/m^3 in this experiment.

According to Eq. (5.3), when the wave-generated maximum pore pressure $\Delta u_{\max}$ equals the effective self-weight stress $(2\gamma' d/3)$ of the overlaying soil $(2\gamma' d/3)$, the soil at the corresponding depth is in a critical state of liquefaction. As mentioned by Sumer et al. (2012), when the accumulated period-averaged pore pressure reaches the initial mean normal effective stress at the corresponding depth, it will liquefy. When $\Delta u < 2\gamma' d/3$, the soil at the depth remains stable, and the degree of stability can be expressed by the ratio of $3\Delta u/2\gamma' d$. On the other hand, when $\Delta u > 2\gamma' d/3$, liquefaction will occur.

To make the statistical results of soil liquefaction more intuitive, we define the liquefaction index Y as

$$Y = 3\Delta u/2\gamma' d \times 100 \tag{5.4}$$

where Y varies between 0 and 100, and the greater Y, the soil is closer to the liquefied state, and thus more prone to movement and collapse. The strength of the soil bed decreases gradually during the process of liquefaction. Meanwhile, the pore water in the soil bed will discharge with the upwelling fine-grained sediments to the soil bed surface. The sediments generated due to liquefaction and seepage to sediment surface becomes available for resuspension under the action of waves and currents.

Quantification of the resuspended sediment

The quantity of soil bed resuspended sediment is the amount of sediment transferred from underneath the soil bed into the water. The quantity resuspended sediment (M) per basal area water column was adopted as an index representing the variability in suspended sediment, which can be expressed as

$$\mathrm{M} = \sum_{i=1}^{n} h_i C_i,\ \mathrm{n} = 5 \tag{5.5}$$

where M denotes the quantity of resuspended sediment within a unit basal area water column (kg/m^2); i is the partitioning layer number in the water column; and h_i represents the thickness of each layer. For this study we divided the water column into five layers with their respective thicknesses as $h_1 = 0.05$ m, $h_2 = 0.1$ m, $h_3 = 0.1$ m, $h_4 = 0.1$ m, and $h_5 = 0.05$ m, because the turbidity has greater variation at the boundaries (near the bottom and water surfaces) the two boundary layers were designed smaller in order to catch the variations. C_i is the SSC of each layer (g/L), converted from the real-time measurement of flume water turbidity.

5.3.2 Results

(1) ***Accumulative Pore Pressure***

The wave-induced pore pressure responses to waves at 20, 30, and 40 cm below the bed surface are presented in Fig. 5.16. Figure 5.16a was composed by four sub-graphs which were changes of pore pressures over time under waves with the wave height of 5, 7, 10, 18 cm labeled above each sub-graph respectively. Figure 5.16b–e were obtained from the sub-graphs with the wave height of 5, 7, 10, 18 cm in Fig. 5.16a correspondingly changing x-axis from time to wave cycles. Before wave loading, the water level was still (the initial pore pressure for 10 cm and 18 cm wave height scenarios were not measured) and soil pore pressure remained constant. The pore pressure started to fluctuate after wave loading was exerted. The greatest fluctuations were found at 30 cm depth, compared to those at 20 and 40 cm.

The pore pressure remained stable in the initial period under various wave heights (7, 10, and 18 cm), then a sharp increase was observed after a certain number of wave cycles, especially at shallower depths. The number of wave cycles needed to reach the pore pressure peak were about 360, 3060 and 11,400, respectively (as shown in Fig. 5.16c–e). After the peak was reached, the pore pressure then gradually decreased to the initial value and became steady.

The statistical results of the cumulative pore pressure changes are summarized in Table 5.3. where T_0 is the start time of wave loading, n is the number of wave cycles

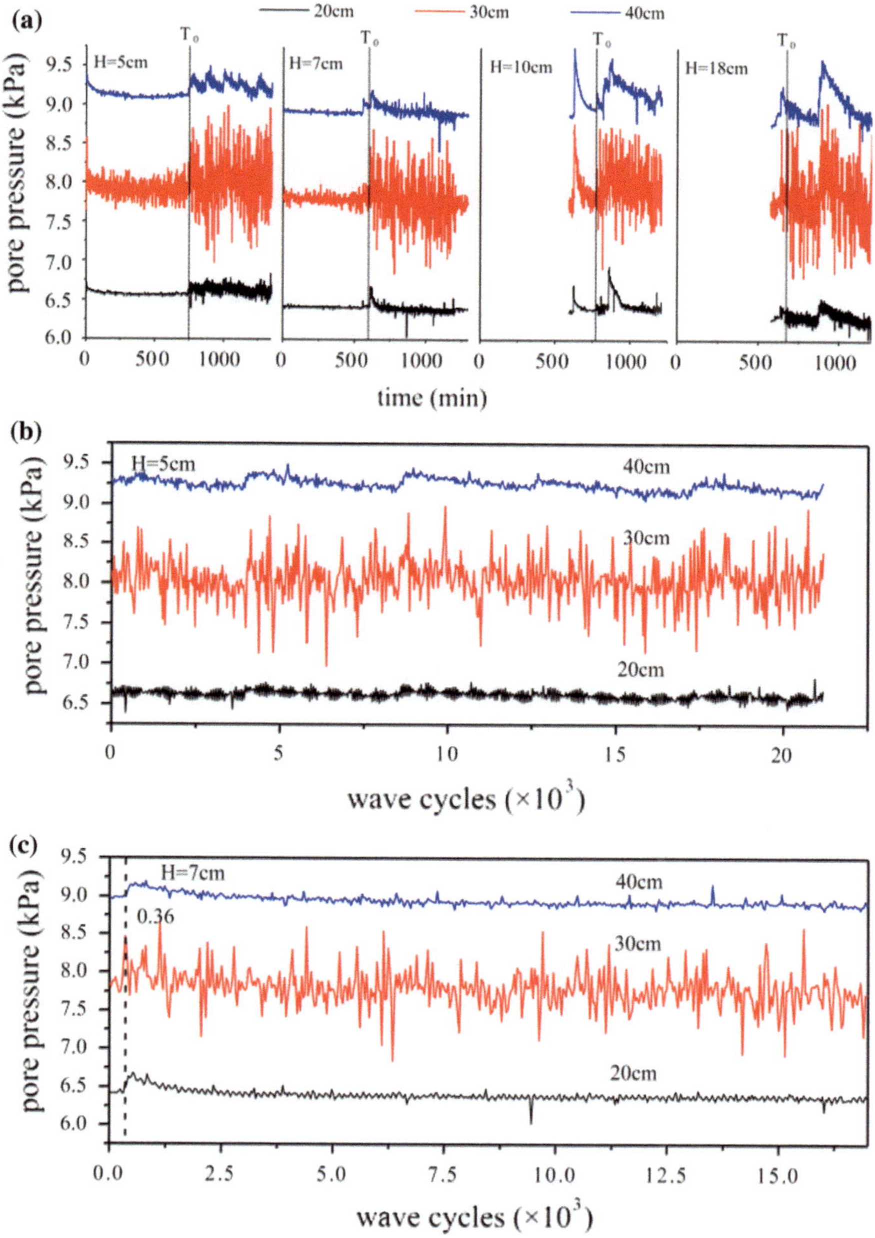

Fig. 5.16 Pore pressure responses to different wave heights at various observation depths (20, 30 and 40 cm) below the sediment surface. **a** Pore pressure changes over time for all scenarios; **b** pore pressure responses to 5 cm wave; **c** pore pressure responses to 7 cm wave; **d** pore pressure responses to 10 cm wave; and **e** pore pressure responses to 18 cm wave

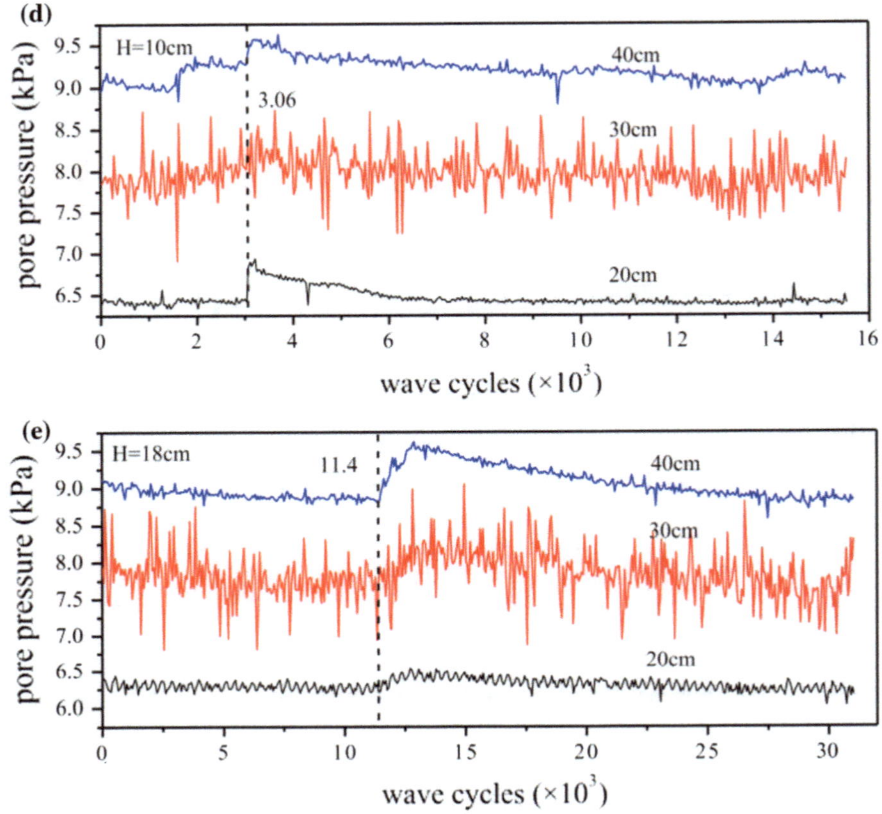

Fig. 5.16 (continued)

Table 5.3 Statistical analysis of pore pressure at different wave heights

H (cm)	T_0 (min)	n	$\overline{U_1}$	$\overline{U_2}$	$\Delta\overline{U}$ (*kpa*)
5	750	*	*	*	*
7	600	360	7.77	8.19	0.42
10	770	3060	7.93	8.43	0.50
18	675	11,400	7.56	8.40	0.84

as the pore pressure changes, U_1 is the average values before pore pressure change, U_2 is the average values after pore pressure change, ΔU is the average variation in magnitude of pore pressure, and "n" stands for data not available. To measure pore pressure changes under different waves, an average variation magnitude, ΔU was defined as the difference between an initial stable value and the maximum pore pressure at each depth under different wave heights. From Table 5.3 and Fig. 5.16, it appeared that ΔU increased with wave height, from 0.42 kPa at 7 cm wave height to 0.84 kPa at 18 cm wave height.

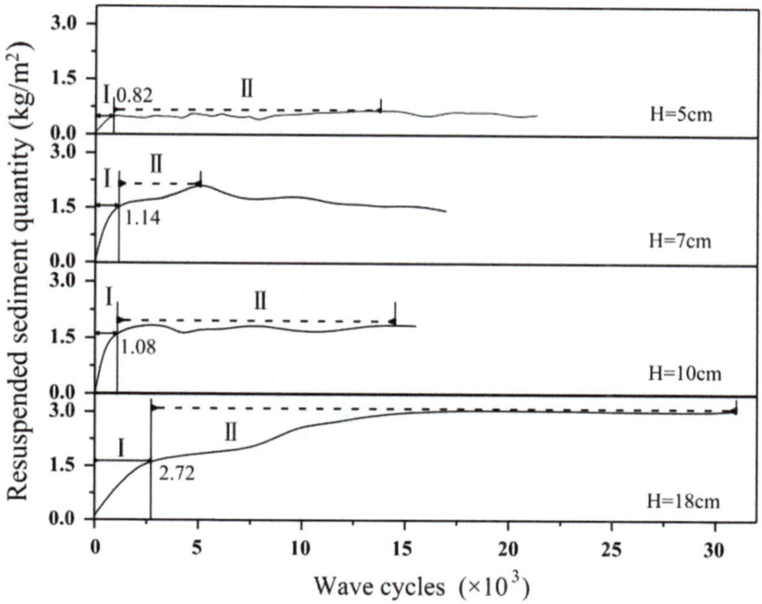

Fig. 5.17 Distributions of resuspended sediment quantity versus wave cycle for various wave heights

(2) ***Sediment Resuspension Quantity***

Initiation time of the surface sediment movement under wave action was short in the tests. The suspended sediment then spread quickly into the water near the bed surface, and reached a relatively steady state after a while (Fig. 5.17). It formed a high concentration bed load layer (about 10 cm high) above the bed surface after the onset of soil bed sediment movement, and its existence was sustainable during the entire wave process. There was only wave circulatory orbital velocity in the flume and no current velocity. The bed load layer did not have absolute movement, only reciprocate cyclical movement. Figure 5.17 presents the quantity of resuspended sediment with wave cycles under various wave height scenarios. The quantity of resuspended sediment under wave action may be characterized by two stages, the initial rapid increase stage (I) and the following slow increase stage (II). The corresponding resuspended sediment quantity may be called the initial increment (I) and the later resuspension increment (II). The two resuspension stages, and resuspended sediment quantity variations and the associated parameters are summarized in Table 5.4, and illustrated in Fig. 5.17.

Even though the two-stage resuspension appeared to be similar for different wave action scenarios, the rates of sediment resuspension were significantly different under various wave heights. Under a 5 cm wave height, the resuspension rate was 1.99×10^{-2} kg m^{-2} min^{-1}in stage I. The resuspension rate was higher at 4.80×10^{-2} kg m^{-2} min^{-1} for the 7 cm wave height scenario. The 10 cm wave scenario has the

Table 5.4 Summary of the quantity of resuspended sediment variations under different wave loadings

H(cm)	$M_1\,(\mathrm{kg \cdot m^{-2}})$	$M_2\,(\mathrm{kg \cdot m^{-2}})$	n_{max}	$M_{max}\,(\mathrm{kg \cdot m^{-2}})$	$v_1\,(\mathrm{kg \cdot m^{-2} min^{-1}})$	$v_2\,(\mathrm{kg \cdot m^{-2} min^{-1}})$	M_1/M_{max} (%)	M_2/M_{max} (%)
5	0.49	0.18	1280	0.67	1.99×10^{-2}	4.98×10^{-4}	73.1	26.9
7	1.55	0.55	5080	2.10	4.80×10^{-2}	49.3×10^{-4}	73.8	26.2
10	1.61	0.34	13,590	1.95	5.26×10^{-2}	9.56×10^{-4}	82.6	17.4
18	1.70	1.37	31,000	3.07	3.69×10^{-2}	28.6×10^{-4}	54.4	44.6

Note M_1 and v_1 are the initial resuspended increment and resuspension rate in stage I, respectively, M_2 and v_2 are the later resuspended increment and resuspension rate in stage II, respectively, M_{max} is the maximum quantity of resuspended sediment, n_{max} is the occurrence wave cycles of maximum quantity of resuspended sediment, and M_1/M_{max} and M_2/M_{max} are the suspended sediment increment proportion in two stages maximum resuspension rate compared to the three other scenarios. The total amount of resuspended sediment showed a close positive correlation with wave height. The maximum resuspension sediment quantity increased from about 0.67 kg m^{-2} at 5 cm wave height to 3.07 kg m^{-2} at 18 cm wave height, which means an increase of 2.40 kg m^{-2}.

5.3.3 *Pore Pressure Accumulation and Seabed Liquefaction Process*

(1) ***Liquefaction Index***

When the accumulative pore pressure reaches its maximum following Eq. 5.4, the maximum liquefaction indexes Y_{max} (Eq. 5.5) were calculated for each observation depth using the maximum pore pressure value. Abnormal pore pressure values were removed in advance. The Y_{max} and its average value Y_{max} under the action of the same wave height are tabulated in Table 5.5, where d is the thickness of soil, u_{max} is the maximum accumulative pore pressure, Δu_{max} is the maximum excess pore water pressure, $\gamma' d$ is the effective self-weight stress of overlaying soil, Y_{max} is the maximum liquefaction index, and Y max is the average maximum liquefaction index.

The liquefaction index ranged from 69.0 to 116.4%. For sandy sediment, when Y comes to 100%, it can be considered as fully liquefied, but for silty sediments, cohesive strength exists between particles and the constitutive properties affects the sediment liquefaction. The unique phenomenon for silty sediments is that it may be not fully liquefied even excess pore water pressure exceeds effective stress, i.e., Y is over 100%.

Table 5.5 Summary of experimental statistical results of pore pressure

H(cm)	d(cm)	$u_{max}\,(kpa)$	$\Delta u_{max}\,(kpa)$	$\gamma' d\,(kpa)$	Y_{max}	$\overline{Y}_{max}$
5	20	6.71	0.53	1.58	50.25	60.10
	30	8.41	1.20	2.37	75.90	
	40	9.38	1.14	3.16	54.15	
7	20	6.68	0.50	1.58	47.55	62.30
	30	8.69	1.48	2.37	93.75	
	40	9.20	0.96	3.16	45.60	
10	20	6.86	0.68	1.58	64.50	64.75
	30	8.30	1.09	2.37	69.00	
	40	9.52	1.28	3.16	60.75	
18	20	6.54	0.36	1.58	34.20	72.05
	30	9.05	1.84	2.37	116.4	
	40	9.62	1.38	3.16	65.55	

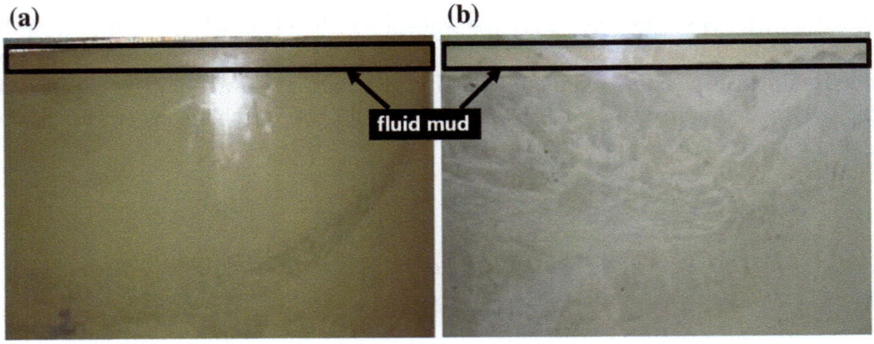

Fig. 5.18 Liquefaction interface and sediment migration phenomenon in soil bed

Liu and Jeng (2007) found that the soil responses and the amplitude of oscillatory pore pressure at the soil bed were greater for longer wave period cases than that of shorter period cases, based on random wave loadings by semi-analytical solution. Waves with longer periods induced a smaller number of waves, and less load actions on the soil due to its low wave frequency. Similarly, in this study, with the gradual increase in wave height from 5 to 18 cm, the average maximum liquefaction index increased from 60.1 to 72.0, which demonstrated that an increase in wave energy was more conducive to the occurrence of liquefaction.

It can be observed from Table 5.5 that almost all maximum liquefaction indexes did not reach 100 (complete soil liquefaction) under the different wave heights in the tests. However, the phenomenon of liquefaction and concurrently upward transport of soil particles, were indeed observed in the experiments. As shown in Fig. 5.18, a series of interlayers formed on the initially homogeneous soil bed under wave action, displayed alternatively with light-colored fine-grained materials and dark coarse particles. The findings in the wave flume experiments for the silty sediment fetched from the modern Yellow River delta in this paper indeed have a difference to the case of surface fluidization since no arc motion was found for the sediment bed at spatial or temporal multi-scale in Tzang (1992, 1998) but found in this paper indicating the effect of pore water accumulation on sediment stability under waves was different. However, no fully sediment liquefaction like mud flow except the surface redeposit sediment was found. In this flume experiment, the skeleton changed with the fine particles upward transport within sediment bed but did not completely damaged like sandy sediment under wave loadings, sediment strength still existed. The driving force promoted the inner fine particle upward transport was essencely the excess pore water pressure gradient, that was seepage gradient force, and the higher wave intensity could induce more excess pore water pressure accumulation and then lead to more internal sediment suspension and more complete bedding structure.

Fine-grained materials had undergone an obvious upward migration and the fine-grained content decreased at the liquefaction edge of the liquefied soil (Fig. 5.18). The liquefaction zone was then relatively weakened because of the reduction in the internal cohesion force. Therefore, wave-induced pulsating channeled seepage became more likely to take place within the soil bed. This could be the reason why liquefaction failure occurs again under the action of similar or higher hydrodynamic conditions, similar to the findings of the submarine landslide "reactivation" phenomenon by storm waves (Prior et al. 1989).

The liquefaction process of the silty sediment under wave loadings observed in this experiment can be attributed to the following reasons. First, the flume experiment was conducted under undrained loading conditions, when the excess pore pressure increased and the effective stress decreased to a certain value, the physical and mechanical properties of soil were subsequently changed. Although sometimes overlooked, the shear stresses caused by stream wise variations in wave-induced pressures seem to play an important role in the liquefaction process in cohesive sediments (Suhayda 1984). Therefore, the combined effect of wave shear stress and excess pore pressure, accelerate the trend of complete soil liquefaction

In addition, wave forces would reduce the bed strength and the effective viscosity of the fluid mud (Yamamoto 1982; Van Kessel and Kranenburg 1998), which may lead to the reduction in soil shear strength. This would result in an enhancement of soil bed liquefaction when the maximum liquefaction index had not reached 100. Thus, we concluded that wave-induced sediment liquefaction is a gradual process, which may occur even though he wave-generated maximum pore pressure (Δu_{max}) was less than the effective self-weight stress of the overlaying soil ($\gamma' d$) in the corresponding depth of soil.

(2) ***Liquefaction Thickness***

The thickness of the fluidised layer d_f, referring to the distance from the sediment surface to the deepest layer of liquefaction, is a very important parameter to evaluate the wave-induced liquefaction behavior of a seabed. The above analysis showed that part of the resuspended sediment into the water were from inside the soil bed, alternatively d_f also represented the resuspend sediment source depth. The following formulation was proposed to calculate the thickness of the fluidised layer d_f for an experimental study of silt (Foda and Tzang 1994) and sand (Tzang and Ou 2006; Tzang et al. 2009)

$$d_f = \Delta u_{max}/(1 - n)(\rho_s - \rho_w)g \tag{5.6}$$

where ρ_s and ρ_ω denote the solid particles and pore water of soil, respectively, n is the soil porosity, and g is the gravitational acceleration.

The thickness of the fluidised soil layer was calculated using Eq. 5.6 (where n was 0.54 and g is 9.81 m/s^2), which was based on the assumption of no vertical transportation of fine particles within soil bed. The calculated fully fluidized thickness was less than the measured fluidized thickness because fine particles within soil bed can transport to the bed surface, and made contribution to the fluid mud thickness. The depth profiles of the measured and calculated thicknesses of the fluidised soil layer under various wave height scenarios are presented in Fig. 5.23 (except the 5 cm scenario, which was not fluidised).

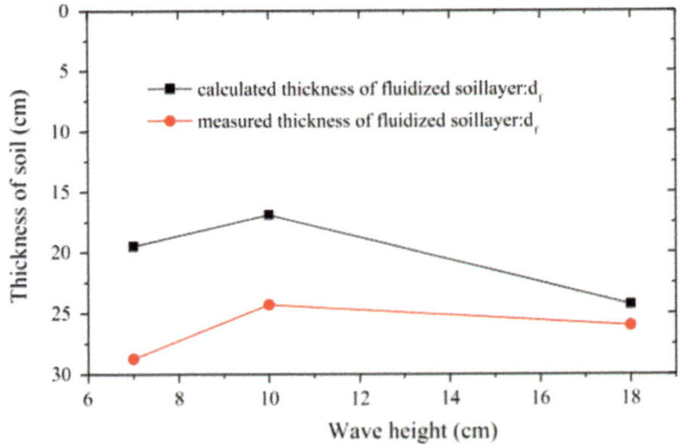

Fig. 5.19 Measured and calculated thicknesses of fluidised soil layer versus wave height

As shown in Fig. 5.19, Eq. 5.6 under estimated all fluidised soil layer thicknesses. However, the overall development trends were correctly predicted. As shown in a parametric study conducted by Liu et al. (2009), a number of wave and soil characteristic parameters contributed significantly to liquefaction depth. The soil sample for this experiment was silt, the liquefaction phenomenon was observed even though pore pressure accumulation was less than effective. That is Δu_{max} was relatively small, which led to the underproduction of the fluidised soil layer thickness. To adjust the underestimation, an empirical equation was developed for predicting the liquefaction thickness for the Yellow River Delta silt soil as follows:

$$d_f = -1.393d_f^2 = 74.43d_f - 969.1 \tag{5.7}$$

5.3.4 *Quantitative Contribution of Liquefaction to Sediment Resuspension*

Tzang et al. (2009) reported that there is a phase lag (about 7–26 s or 5–17 wave cycles) between the occurrence time of suspended sediment concentration and the excess pore pressure events. Our results indicated that, as shown in Fig. 5.16 and 5.17, during the initial period of wave action the pore pressure had a gradual increase overall, and then the quantity of resuspended sediment gradually increased correspondingly. The pore pressure first reached its maximum after a certain number of wave cycles, then the concentration of resuspended sediment reached its maximum with a time lag following the peak of the pore pressure. Despite the time lag, the pore pressure and resuspended sediment quantity demonstrated consistent correlations.

On the other hand, only drainage (swell) and surface erosion will occur when the flow-induced stresses exceed the critical shear stress for erosion, which relates to the true cohesion, or drained strength of the sediment at the bed surface (Winterwerp et al. 2012). The pore pressure variations were not obvious and soil has not lique-

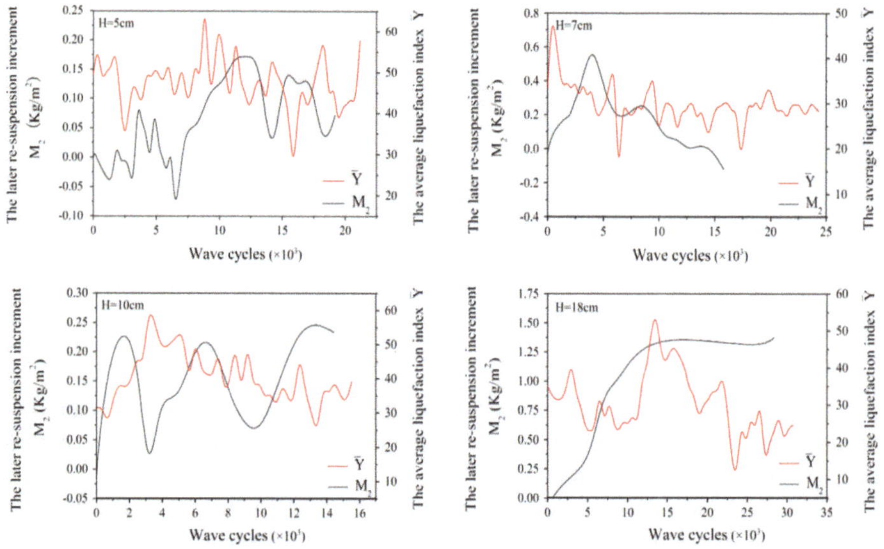

Fig. 5.20 Wave cycle-series of later resuspension increment (M_2) and average liquefaction index (Y)

fied during this stage, the quantity of initial suspended sediment was mainly from the resuspension of fine particulate matter on the seabed surface. However, after a period of wave action, both the pore pressure and the quantity of resuspension sediment present significant accumulation (Figs. 5.16 and 5.17). Pore water pressure development or wave-induced bed liquefaction can significantly affect the sediment transported volume and the fine sediment budget (Clukey et al. 1985; Lambrechts et al. 2010). Thus the resuspension increment at the later stage was also closely related to the variations in pore pressure, and the growth of sediment resuspension occurred after a certain building-up of excess pore pressure through our experiment.

Changes in M_2 and Y with the wave cycle under various wave heights are shown in Fig. 5.20. The Y value demonstrated a rapid growth at a certain time under different wave action, indicating that the degree of soil liquefaction increased. M_2, on the other hand, changed irregularly at the beginning and became relatively steady, then significantly increased to the maximum after Y reached its peak. It appeared that variations in M_2 were closely associated with Y, which suggested that the wave-induced soil bed liquefaction may contribute to the stage II resuspended sediment in the water.

In this study, we defined "liquefaction index" to describe the liquefaction degree of seabed sediment. The "liquefaction" mentioned in this paper did not indicate soil bed was fully fluidized considered as the traditional view, but referred to the pore water pressure accumulated and sediment strength reduced. Therefore, the mechanism of internal sediment resuspension proposed by Tzang (1998) can be accordance with the findings in our work. The variation trend of M_2 with Y shown in Fig. 5.20 confirmed

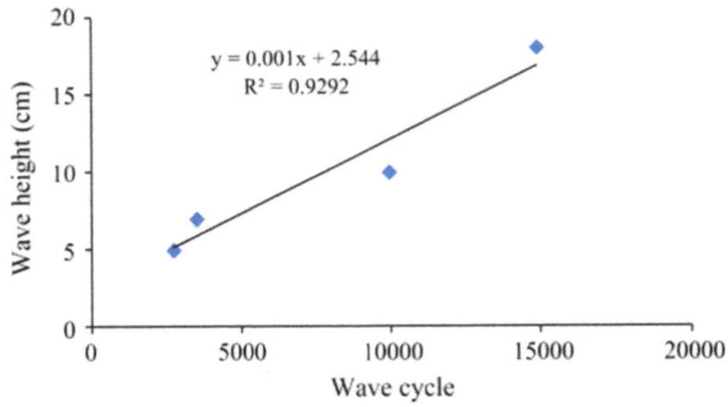

Fig. 5.21 Phase lag (in terms of wave cycle) of later resuspension increment (M_2) to average liquefaction index (Y) under different wave conditions

the assumption by Foda and Tzang (1994) that sediment suspensions should be closely dependent on bed responses and their patterns could be substantially changed in cases of soil bed fluidisation.

Figure 5.21 shows the relation between the phase lags (in terms of wave cycle) of the average liquefaction index and the peak of stage II resuspended sediment under different wave heights. The lagging time was the time difference between the peak of Y and the peak of M_2. Under wave action of 5–18 cm wave heights, the lagging time was about 2730, 3500, 9920, and 14,850 wave cycles, respectively. Figure 5.21 demonstrated a typical linear regression relationship between lagging time and wave height derived from our experiments. The lagging time linearly increased with wave height with a correlation coefficient of 0.93. A similar phenomenon was reported by others. Vincent and Hanes (2002) found that suspended sand concentrations lagged the forcing waves and the lag increased with distance from the seabed surface. They also quantified the lag of the suspended sediment in relation to regular waves and wave groups. Dohmen-Janssen and Hanes (2005) proposed that the suspended sediment concentrations varied on the time scale of the wave group, with a time delay relative to the peak wave within the wave group.

The contribution of wave-induced liquefaction to the quantity of resuspended sediment was quantified by analyzing the relationship between the later resuspension increment (M_2) and the average maximum liquefaction index (Y_{max}), as shown in Fig. 5.22. A linear correlation with a correlation coefficient was 0.87. This demonstrated that a greater liquefaction index predicts a higher quantity of resuspended sediment generated through liquefaction seepage.

The silt sediments from the Yellow River Estuary have demonstrated fast initiation and resuspension characteristics, and moderate wave energy could cause bottom silt resuspension (Wright et al. 1986). Therefore, the initial resuspended increment generated in stage, as shown in Fig. 5.17, was the resuspended fine-grained material from the soil bed surface when the wave-induced shear stress reached the critical shear stress. The results of the particle size distribution analysis indicated that particles smaller than 0.005 mm in the soil were about 13–16% before wave loading, whereas this portion increased up to about 65–75% in the deposition layer (Fig. 5.18) after wave loading ceased. In addition, based on the observed reciprocating arc-shaped

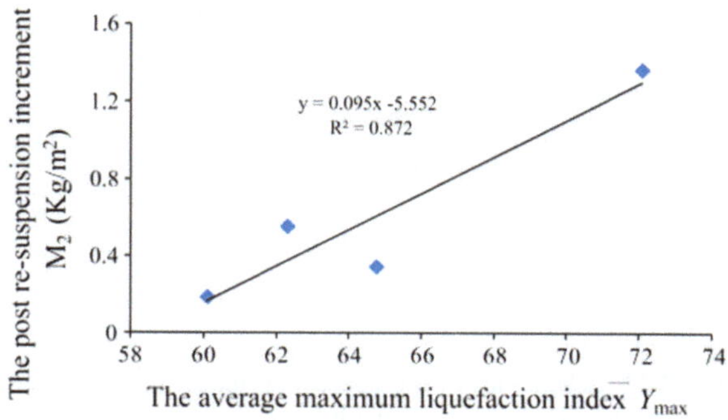

Fig. 5.22 Relation curve between the later resuspension increment (M_2) and the average maximum liquefaction index (Y_{max})

oscillation motion of part of the soil bed during the experiment and continuously increase of the deposition layer thickness h_d (Table 5.6) at different wave heights after the cessation of wave loading, this demonstrated that there may be a continuous upward transportation of fine particles within the soil bed during cyclic wave loading.

The initial state of the soil deposition was homogeneous before wave actions in this wave flume experiment. After several rounds of wave action, the homogeneous soil bed turned to nonhomogeneous, and this was the result of the vertical transportation of fine particle within the soil bed under waves. Because the total soil content was constant during the experiment, this result suggested that the later suspension increment in stage II might come from fine sediment migration into the water from the internal of soil bed through liquefaction and seepage. It can be explained using the conceptual model of soil fluidisation as shown in Fig. 5.23). Some of the fine sediment particles within the fluidised layer are extracted from the soil's solid skeleton and fed into the seepage flow, and then migrated or suspended along the oscillation interface, and on the other aspect, the inner particle upward transportation led to the nonuniform distribution of seabed silt and clay particles. Tzang (1992) reported that fine particles detached from the soil skeleton could be suspended in the pore water, and the suspended particles near the bed surface would then be brought out by seepage action from the pores into the water. Ni and Meng (2001) reported that as particle concentrations increased in the water, the interparticle forces became more significant, which enhanced the upward movement of particles and density segregation effects within the sediment layer, as observed in various laboratory and field studies (Best 1989; Druitt 1995). Finer grained particles were expected in the surface layers rather than in the bulk soils after the deposition processes. Furthermore, the resuspension of fine sediments proceeded at a higher rate during the early stage of wave loading, whereas, as mentioned by Park et al. (2001), the coarser particles were

Table 5.6 Average values of deposition layer thickness (h_d) variations under different wave loadings

Wave height H(cm)	5	7	10	18
h_d (cm)	*	1.37	1.47	1.62

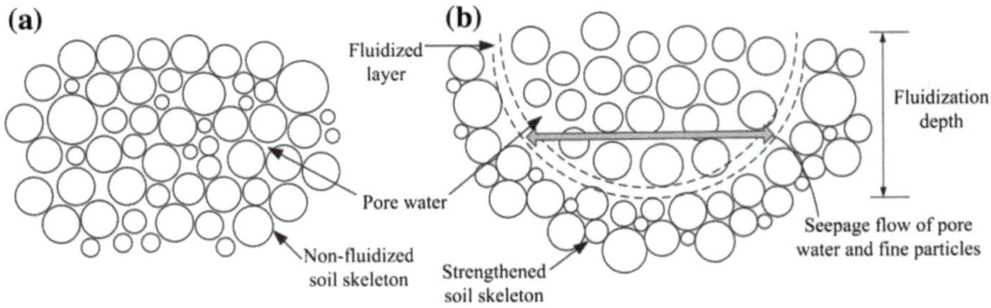

Fig. 5.23 Schematics of the conceptual model of soil fluidisation and the resulting arc-shaped oscillation in **a** pre-fluidised and **b** fluidised seabeds

resuspended from bottom sediments in later stage at a lower rate. Thus there was no linear correlation between wave height and resuspension rate, which was influenced by the sorting of soil particles.

As mentioned previously, the resuspension process in this experiment can be divided into two subphases according to the occurrence of fluidization. Sediment resuspension during the period before fluidization mainly came from the surface of the seabed, finally reached the first equilibrium state of the SSC. However, the oscillation of subsurface sediments after seabed fluidization led to the second increase in the SSC and finally reached the second equilibrium state of the SSC. According to the existing findings and experimental results, the contribution of seabed fluidization to sediment resuspension was successfully evaluated through comparing the stabilized resuspension loads before and after seabed fluidization, which was given as

$$C = (M_{\mathrm{a}} - M_{\mathrm{b}})/M_{\mathrm{a}}, \tag{5.8}$$

where C is the contribution of fluidization to resuspension; M_{b} (kg) is the first stabilized total suspended load before fluidization; and M_{a} (kg) is the final stabilized total suspended load after fluidization. The calculated results from the present study were shown in Table 5.7. Moreover, in order to make a more comprehensive analysis, previous reported results on this issue using both the same flume and sediments were also summarized in Table 5.7.

The experiments of Jia et al. (2014) and Guo et al. (2016) were both conducted in the same wave flume using similar Huanghe River silts. Different experimental parameters are also listed in Table 5.6. It is worth noting that Jia et al. (2014) applied four rounds of waves in sequence over the same seabed, which meant that the experimental condition for each round of the wave height was not single-variable. Therefore, we believe that the initial seabed conditions can only account for the sediment behaviors during the first wave round of wave height 5 cm. In this case, the results for wave height 7, 10, and 18 cm rounds were merely listed but not employed for discussion. In addition, Jia et al. (2014) divided the two substages according to their self-defined terms of the initial resuspended increment and the latter resuspended increment, whereas Guo et al. (2016) and the present study both divided the

Table 5.7 Contribution of seabed fluidization to sediment resuspension

Parameters	Tests						
	Jia et al. 2014	Jia et al. 2014	Present study	Jia et al. 2014	Present study	Guo et al. 2016	Jia et al. 2014
H_w/cm	5	7	10	10	15	15	18
D/cm	40	40	50	40	50	40	40
T_c/h	2	2	5	2	5	10	2
D_{50}/um	36–42	36–42	43	36–42	43	43.67	36–42
M_s	2.5/20	3.5/20	4/20	5/20	6/20	7.5/20	9/20
M_b [M_1]	0.49	1.55	3314	1.61	5716	2.17	1.70
M_a [M_2]	0.18	0.55	6975	0.34	16498	7.05	1.37
C (%)	26.9	26.2	52.5	17.4	65.3	76.4	44.6
Use/not	√	–	√	–	√	√	–

Notes The previously reported and present experimental results were summarized for a contrastive analysis. D is the water depth; T_c is the consolidation time; D_{50} is the median grain size; and M_s is the model scale (the ratio of wave height to water depth). In the experiment of Jia et al. (2014), M_1 (kg/m^2) is the initial resuspended increment; M_2 (kg/m^2) is the latter resuspended increment; and M_{max} is the maximum quantity of resuspended sediment; thus the contribution was estimated by $C = M_2/M_{max}$

two substages according to the two stabilized SSC levels. However, this made no difference because the influence of units was ruled out when the contribution was calculated. Besides, Guo et al. (2016) measured the SSC profiles under a single wave height of 15 cm using an instrument for profiling the suspension turbidity in the benthic boundary layer (Argus Surface Meter IV, Argus, Germany), instead of the traditional method of extracting water samples in the present study. However, this also made no difference as long as the accuracy of both measuring methods was assured. Therefore, we believe that the data from Jia et al. (2014) and Guo et al. (2016) are both comparable and convincing enough for reference.

The contrastive analysis results were shown in Fig. 5.24. It is clear that the contribution of fluidization under 10 cm waves (H_w/D = 4.0/20.0) reached 52.5% of the total suspension, while the contribution under 15 cm waves (H_w/D = 6.0/20.0) accounted for 66.8% in the present study. Such large percentages strongly demonstrated a non-ignorable effect of fluidization on sediment resuspension, seabed fluidization should double the resuspension load under certain conditions. Besides, the contribution seems to be positively correlated with the ratio of the wave height to the water depth. This inference is verified by the previous results because Guo et al. (2016) reported that the contribution exceeds 76.4% under 15 cm waves (H_w/D = 7.5/20.0) and Jia et al. (2014) indicated the contribution is approximately 26.9% under 5 cm waves (H_w/D = 2.5/20.0). The contribution of fluidization increases with the model scales at a decreasing rate, and the relationship well fits a logarithmic profile. Temporarily, we believe the decreasing rate is resulted from a synchronous

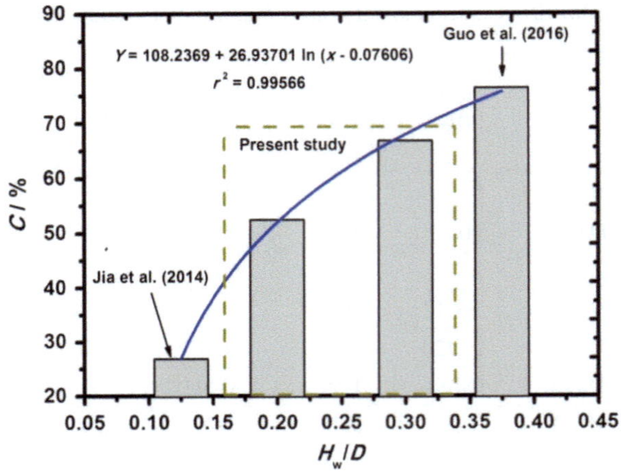

Fig. 5.24 The contribution of seabed fluidization to the total resuspension under various ratios of the wave height to the water depth

increase of resuspension flux due to an enhanced orbital shear erosion from the seabed surface.

5.3.5 Mechanisms of the Contribution of Liquefaction to Sediment Resuspension

Combined with the previous findings from the waves flume experiments, we further proposed two modes of wave-induced resuspension which are closely related to pore pressure responses. Namely transient pumping and residual pumping of sediments respectively.

Transient vertical seepage across water-seabed interface is well known for coastal engineers and researchers. As seabed is often permeable in coastal areas, seepage flow in the seabed might occur naturally due to a pressure gradient caused by the difference between the pressures on the seabed under wave crests and wave troughs. This difference will vary in time and space following the wave motion, thus inducing flow into the seabed as the wave crest passes and out of the seabed as the wave trough passes (Precht and Huettel 2004; Obhrai 2002; Myrhaug et al. 2014).

It was also reported that sediment resuspension under steady flow and wave motion is governed by forces on individual sediment grains (Putnam 1949; Madsen 1978). Should we assume the existence of a potential migratory soil particle within the shallow layer of seabed, this particle is exposed to seepage forces due to passage of overlying wave oscillations. For each oscillation, it is possible to compute a vertical component of wave-induced seepage force and to distinguish two special cases, namely, a vertical force acting upwards F_Z (+) under wave trough, which tends to suck the soil particle out of the seabed, and a vertical force acting downwards F_Z (–) under wave crest, which tends to press the soil particle down. This force is directly related to the stresses induced by bottom pressure fluctuations and always decreases

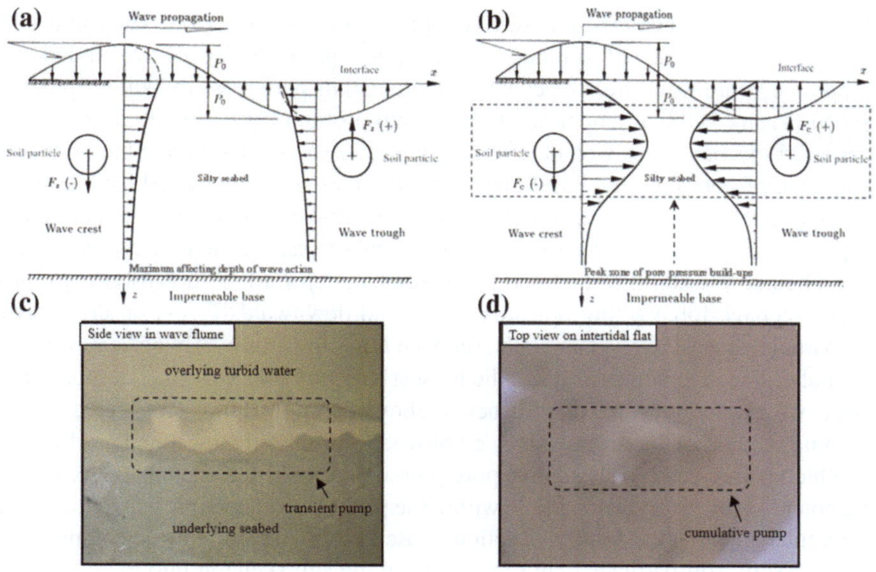

Fig. 5.25 **a** Schematic depth profile of bottom pressure fluctuations and mechanism of "transient pump". **b** Schematic depth profile of wave-induced cumulative pore pressure and mechanism of "cumulative pump". Note: arrows to the left and top both denote the direction of force acting upwards, whereas the arrows to the right and bottom both denote the direction of force acting downwards; length of arrows denotes the magnitude of force. **c** A typical photograph of "transient pump" in transparent wave flume. **d** "cumulative pump" on intertidal flat

within depth where wave could affect (Fig. 5.25a). Thus sediments within shallow layers are most obviously exposed to this effect. Seepage force acting on soil particles has its direction and magnitude that alternates cyclically with the same frequency as surface wave oscillations, increasing or reducing the stability of soil particles and determine their final destiny. Theoretically, many soil particles may separate from the soil skeleton into the pore fluid and be subsequently exposed to forces including gravity, buoyancy, and transient seepage force F_Z. Should the sum of buoyancy and transient seepage force exceed the self-gravity of the separated soil particle, it will moved upwards and might finally be pumped into the overlying water column and become resuspended materials. As this process is closely related to transient pore pressure, we define this process as "transient pump" which refers to the transient particle flow exchange across the water-sediment interface that occurs with the same frequency as the wave crest-trough alternation, only across the water-seabed interface or within sub-bottom shallow layers.

Cumulative pump is thought to be closely related to cumulative pore pressure (i.e. pore pressure build-ups) and cumulative seepage. Cumulative vertical seepage in seabed is also well known for coastal engineers and researchers. As we all know, bottom pressure fluctuations have a direct and continuous influence on the changes of pore pressure within a porous permeable seabed (Bennett et al. 1977;

1979). When the downward pressure is applied on the seabed, pore water in it will surely show a tendency of drainage. However, low permeability always hinders the compressed pore water from discharging in time, thus pore pressure builds up in the seabed (Tzang 1998; Tzang et al. 2009). Depth profile of wave-induced pore pressure build-ups have been well determined, typically shown in Fig. 5.25b, in which a peak zone of pore pressure build-ups exists. If we assume the seabed has the same degree of pore pressure build-ups along the horizontal direction, then the horizontal seepage gradient is negligible to make the vertical seepage gradient play a dominant role. Then pore water has the trend of the drainage upwards or downwards to thus induce seepage force acting upwards F_c (+) and downwards F_c (–) on the nearby soil particles, respectively. Generally, seepage flows towards overlying seabed is the optimal drainage direction as it has the loosest structure due to weaker compression of self-weight and repeated disturbance of above-mentioned transient seepage.

Same method of force analysis is employed to discuss the mechanism of origin (a) which is induced by cumulative pore pressure. Should we assume the existence of a potential migratory soil particle within the peak zone of pore pressure build-ups. Theoretically, under the combined action of oscillatory pore pressure and cumulative pore pressure, soil particles may separate from the soil skeleton into the pore fluid and be subsequently exposed to forces including gravity, buoyancy, and cumulative seepage force F_c. When the sum of buoyancy and cumulative seepage force exceed the self-gravity of the separated soil particle, it will be moved upwards by pore water flows and might be finally pumped into overlying water column and become resuspended materials, act as "mud chimney" or "mud volcano".

Compressible seabed skeleton and low permeability are both necessary for the formation of cumulative pore pressure. Silty seabed of the Yellow River delta meets both the requirements for it. However, when pore pressure build-ups are large enough to induce seabed fluidization or natural seabed bioturbation exists, vulnerable spot or paths could provide preferred channels (Nichols et al. 1994; Lambrechts et al. 2010) for the vertical delivery of soil particles. Therefore, the cyclic accumulation or dissipation of pore pressure build-ups may be act to induce seepage flows to transport fine-grained sediments upwards and finally be pumped onto the seabed surface. As this process is closely related to cumulative pore pressure (build-ups), we define this process as "cumulative pump" which refers to the upward migration of fine-grained sediment particles along with the seepage flows under the driving of pore pressure build-ups.

5.4 Summary

In this chapter, the process of sediment erosion and resuspension was studied through in situ observations and wave flume experiments. Sediments are more frequently resuspended by wind waves in the Yellow River Delta, the contribution of waves to sediment resuspension was not only wave orbital velocities, but also seabed liquefaction, and these lately resuspended sediments seems to come from the interior of the seabed, rather than seabed surface as the conventional opinion holds. Detailed conclusions can be summarized as follows: (1) The wave-induced pore pressure initially remained stable and then a significant increase (jump) was observed after a certain number of wave cycles had been applied to the soil. The pore pressure jump occurred at the same time for different observation depths as long as the wave height was maintained. However, this jump appeared earlier under smaller wave (lower wave height) action and appeared later under greater wave action (greater wave height). The jump was almost nonexistent for the 5 cm wave scenario. (2) The quantity of resuspended sediment over the fluidised bed was positively correlated to wave height (wave energy). The resuspended sediment consisted of materials from the seabed surface (the initial resuspended increment) and upward transported fine-grained materials from the inner seabed (generated by liquefaction seepage), and the increase in later resuspended sediment took a longer time in comparison to the initial resuspended sediment. (3) There was a strong linear correlation between the average value of the maximum liquefaction index (Y_{max}) and the later resuspension increment in stage II (M_2). It was observed that part of the soil bed was under reciprocating oscillation motion, and during the experiment the deposition layer thickness h_d continued to increase. Such an observation revealed that the quantity of later resuspended sediment was indeed generated by the upward transport of deep materials from the lagging-behind liquefaction, and the delay time increased with wave height. (4) A liquefaction layer prediction equation developed from other studies was used to predict the thickness of the fluidised soil layer, but was found to consistently under estimate the measurements. Potential reasons may be due to the greater porosity of the silty soil, which caused a lower pore pressure accumulation and shallower liquefaction layers.

References

Aldridge JN, Rees JM (1997) Interpreting observations of near-bed sediment concentration and estimation of "pick-up" function constants. In: Burt N, Parker J, Watts J (eds) Cohesive sediments. Wiley, Chichester, pp 289–303

Bennett RH (1977) Pore-water pressure measurements: Mississippi delta submarine sediments. Mar Georesources Geotechnol 2(1–4):177–189

Bennett RH, Faris JR (1979) Ambient and dynamic pore pressures in fine-grained submarine sediments: Mississippi delta. Appl Ocean Res 1(3):115–123

Best JL (1989) Fluidisation pipes in volcaniclastic mass flows, Volcan Hudson, southern Chile. Terra Nova 1:203–208

Bian S, Hu Z, Xue Z et al (2012) An observational study of the carrying capacity of suspended sediment during a storm event. Environ Monit Assess 184(10):6037–6044

Blott SJ, Pye K (2001) GRADISTAT: a grain size distribution and statistics package for the analysis of unconsolidated sediments. Earth Surf Process Landf 26(11):1237–1248

Clukey EC, Kulhawy FH, Liu PLF et al (1985) The impact of wave loads and pore-water pressure generation on initiation of sediment transport. Geo-Mar Lett 5:177–183

De Wit PJ, Kranenburg C (1997) The wave-induced liquefaction of cohesive sediment beds. Estuar Coast Shelf Sci 45:261–271

Dohmen-Janssen CM, Hanes DM (2005) Sheet flow and suspended sediment due to wave groups in a large wave flume. Cont Shelf Res 25(3):333–347
Druitt TH (1995) Settling behaviour of concentrated dispersions and some volcanological applications. J Volcanol Geotherm Res 65:27–39
Fagherazzi S, Priestas AM (2010) Sediments and water fluxes in a muddy coastline: interplay between waves and tidal channel hydrodynamics. Earth Surf Process Landf 35(3):284–293
Feng J (1992) Laboratory experiments on cohesive soil-bed fluidisation by water waves (Thesis, Report no. UFL/COEL-92/005). University of Florida, Gainesville, USA, p 108
Ferré B, Durrieu De Madron X, Estournel C et al (2008) Impact of natural (waves and currents) and anthropogenic (trawl) resuspension on the export of particulate matter to the open ocean. Cont Shelf Res 28(15):2071–2091
Foda MA, Tzang SY (1994) Resonant fluidisation of silty soil by water waves. J Geophys Res 99–10(20):463–475
Folk RL (1966) A review of grain-size parameters. Sedimentology 6(2):73–93
Gao S, Jia JJ (2003) Modeling suspended sediment distribution in continental shelf upwelling/downwelling settings. Geo-Mar Lett 22(4):218–226
Green MO, Coco G (2014) Review of wave-driven sediment resuspension and transport in estuaries. Rev Geophys 52(1):77–117
Guo W, Wu G, Liang B et al (2016) The influence of surface wave on water exchange in the Bohai Sea. Cont Shelf Res 118:128–142
Hill PS, Milligan TG, Geyer WR (2000) Controls on effective setting velocity in the Eel River flood plume. Cont Shelf Res 20:2095–2111
Hoque MA, Ahad BG, Saleh E (2010) Hydrodynamics and suspended sediment transport at tidal inlets of Salut Mengkabong Lagoon, Sabah, Malaysia. Int J Sediment Res 25(4):399–410
Houwing EJ, Nczos ICT, Kroon A et al (2002) Interaction of submerged vegetation, hydrodynamics and turbidity; analysis of field and laboratory studies. Proc Mar Sci 5:441–453
Jia YG, Zheng JW, Yue ZQ et al (2014) Tidal flat erosion of the Huanghe River Delta due to local changes in hydrodynamic conditions. Acta Oceanol Sin 33(7):116–124
Kana TW (1978) Surf zone measurements of suspended sediment. Coast Eng Proc 1(16):1725–1743
Knapen A, Poesen J, Govers G et al (2007) Resistance of soils to concentrated flow erosion: a review. Earth-Sci Rev 80(1–2):75–109
Kong D, Miao C, Borthwick AGL et al (2015) Evolution of the Yellow River delta and its relationship with runoff and sediment load from 1983 to 2011. J Hydrol 520:157–167
Kong YZ, Zhu CF (2008) Experimental study on vertical sediment mixing coefficient under waves. J East China Norm Univ (Nat Sci) 6:9–13
Lambrechts J, Humphrey C, Mckinna L et al (2010) Importance of wave-induced bed liquefaction in the fine sediment budget of Cleveland Bay, Great Barrier Reef. Estuar Coast Shelf Sci 89:154–162
Liu HJ, Jeng DS (2007) A semi-analytical solution for random wave-induced soil response and seabed liquefaction in marine sediments. Ocean Eng 34:1211–1224
Liu SM, Li LW, Zhang GL et al (2012) Impacts of human activities on nutrient transports in the Huanghe (Yellow River) estuary. J Hydrol 430–431:103–110
Liu Z, Jeng DS, Chan AH et al (2009) Wave-induced progressive liquefaction in a poro-elastoplastic seabed: a two-layered model. Int J Numer Anal Methods Geomech 33(5):591–610
Madsen OS (1978) Wave induced pore pressures and effective stresses in a porous bed. Geotechnique 28:377–393
Ma R, Cai C, Wang J et al (2015) Partial least squares regression for linking aggregate pore characteristics to the detachment of undisturbed soil by simulating concentrated flow in Ultisols (subtropical China). J Hydrol 524:44–52
Maa PY, Mehta AJ (1987) Mud erosion by waves: a laboratory study. Cont Shelf Res 7:1269–1284
McDougal WG, Tsai YT, Liu PL-F et al (1989) Wave-induced pore-water pressure accumulation in marine soils. J Offshore Mech Arct Eng 111:1–11
Minella JPG, Merten GH, Reichert JM et al (2008) Estimating suspended sediment concentrations from turbidity measurements and the calibration problem. Hydrol Process 22(12):1819–1830

Myrhaug D, Holmedal LE, Ong MC (2014) Seepage effects on bedload sediment transport rate by random waves. Ocean Eng 82:123–127
Ni JR, Meng XG (2001) Forces on particles and their effects on vertical sediment sorting in solid-liquid two- phase flows. Int J Sedim Res 16–2:128–138
Nichols RJ, Sparks RSJ, Wilson CJN (1994) Experimental studies of the fluidization of layered sediments and the formation of fluid escape structures. Sedimentology 41(2):233–253
Obhrai C, Nielsen P, Vincent CE (2002) Influence of infiltration on suspended sediment under waves. Coast Eng 45:111–123
Pang CG, Yu W, Yang Y et al (2011) An improved method for evaluating the seasonal variability of total suspended sediment flux field in the Yellow and East China Seas. Int J Sedim Res 26(1):1–14
Park YA, Choi JY, Gao S (2001) Spatial variation of suspended particulate matter in the Yellow Sea. Geo-Mar Lett 20(4):196–200
Pavanelli D, Bigi A (2005) Indirect methods to estimate suspended sediment concentration: reliability and relationship of turbidity and settleable solids. Biosyst Eng 90(1):75–83
Pholkern K, Srisuk K, Grischek T et al (2015) Riverbed clogging experiments at potential river bank filtration sites along the Ping River, Chiang Mai, Thailand. Environ Earth Sci 73(12):7699–7709
Podsechin V, Tejakusuma I, Schernewski G et al (2006) On parameters estimation in dynamic model of suspended sediments. J Hydrol 318(1–4):17–23
Precht E, Huettel M (2004) Rapid wave-driven advective pore water exchange in a permeable coastal sediment. J Sea Res 51:93–107
Prior DB, Suhayda JN, Lu NZ et al (1989) Storm wave reactivation of a submarine landslide. Nature 341(6237):47–50
Putnam JA (1949) Loss of wave energy due to percolation in a permeable sea bottom. EOS Trans Am Geophys Union 30:349–356
Rajaee T, Mirbagheri SA, Zounemat-Kermani M et al (2009) Daily suspended sediment concentration simulation using ANN and neuro-fuzzy models. Sci Total Environ 407(17):4916–4927
Saito Y, Wei H, Zhou Y et al (2000) Delta progradation and chenier formation in the Huanghe (Yellow River) delta, China. J Asian Earth Sci 18(4):489–497
Schoellhamer DH (2002) Comparison of the basin-scale effect of dredging operations and natural estuarine processes on suspended sediment concentration. Estuaries 25(3):488–495
Shepard FP (1954) Nomenclature based on sand–silt–clay rations. J Sediment Petrol 24:151–158
Slaets JIF, Schmitter P, Hilger T et al (2014) A turbidity-based method to continuously monitor sediment, carbon and nitrogen flows in mountainous watersheds. J Hydrol 513:45–57
Suhayda JN (1984) Interaction between surface waves and muddy bottom sediments. In: Mehta AJ (ed) Estuarine cohesive sediment dynamics. Springer, Berlin, pp 401–428
Sumer BM, Hatipoglu F, Fredsøe J (2004) The cycle of soil behaviour during wave liquefaction. In: Proceedings of the 29th international conference on coastal engineering, Lisbon, Portugal, 19–24 September
Sumer BM, Hatioglu F, Fredsøe J et al (2006) The sequence of sediment behaviour during wave-induced liquefaction. Sedimentology 53:611–629
Sumer BM, Kirca VSO, Fredsøe J (2012) Experimental validation of a mathematical model for seabed liquefaction under waves. Int J Offshore Polar Eng 22(2):1–9
Tréhu AM (2016) Measuring slow slip offshore. Science 352(6286):654–655
Tzang SY (1992) Water wave-induced soil fluidisation in a cohesionless finegrained seabed (PhD thesis). University of California, Berkeley, USA
Tzang SY (1998) Unfluidized soil responses of a silty seabed to monochromatic waves. Coast Eng 35:283–301
Tzang SY, Ou SH (2006) Laboratory flume studies on monochromatic wave-fine sandy bed interactions: part 1. Soil fluidisation. Coast Eng 53(11):965–982
Tzang SY, Ou SH, Hsu TW (2009) Laboratory flume studies on monochromatic wave-fine sandy bed interactions: part 2. Sediment suspensions. Coast Eng 56(3):230–243
Van Kessel T, Kranenburg C (1998) Wave-induced liquefaction and flow of subaqueous mud layers. Coast Eng 34:109–127

Vincent CE, Hanes DM (2002) The accumulation and decay of near-bed suspended sand concentration due to waves and wave groups. Cont Shelf Res 22(14):1987–2000

Voulgaris G, Collins MB (2000) Sediment resuspension on beaches: response to breaking waves. Mar Geol 167:167–187

Wang H, Bi N, Wang Y et al (2010) Tide-modulated hyperpycnal flows off the Huanghe (Yellow River) mouth, China. Earth Surf Process Landf 35(11):1315–1329

Wang YH (2003) The intertidal erosion rate of cohesive sediment: a case study from Long Island sound. Estuaine Coast Shelf Sci 56(5–6):891–896

Winterwerp JC, van Kesteren WGM, van Prooijen B et al (2012) A conceptual framework for shear-flow induced erosion of soft cohesive sediment beds. J Geophys Res 117:C10020

Wright LD, Yang ZS, Bornhold BD et al (1986) Hyperpycnal plumes and plume fronts over the Huanghe (Yellow River) delta front. Geo-Mar Lett 6(2):97–105

Wright LD, Friedrichs CT, Kim SC et al (2001) Effects of ambient currents and waves on gravity-driven sediment transport on continental shelves. Mar Geol 175:25–45

Xia XM, Li Y, Yang H et al (2004) Observations on the size and settling velocity distributions of suspended sediments in the Pearl River Estuary, China. Cont Shelf Res 24:1809–1826

Xie M, Zhang W, Guo W (2010) A validation concept for cohesive sediment transport model and application on Lianyungang Harbor, China. Coast Eng 57(6):585–596

Yamamoto T (1982) Experiments of wave-driven soil transport in clay beds. Geo-Mar Lett 2(3–4):205–208

Yang SL, Li P, Gao A et al (2007) Cyclical variability of suspended sediment concentration over a low-energy tidal flat in Jiaozhou Bay, China: effect of shoaling on wave impact. Geo-Mar Lett 27(5):345–353

Zhang J, Huang WW, Shi MC (1990) Huanghe (Yellow River) and its estuary: sediment origin, transport and deposition. J Hydrol 120(1–4):203–223

Zen K, Yamazaki H (1990) Mechanism of wave-induced liquefaction and densification in seabed. Soil Found 30(4):90–104

Zheng W, Ye T, You S et al (2015) The thermal performance of seawater-source heat pump systems in areas of severe cold during winter. Energy Convers Manag 90:166–174

Zhu C, Wang Q, Jia Y et al (2015) Numerical analysis on the ultimate bearing capacity and parameter selection of U-shaped confined concrete short column. Electron J Geotech Eng 20(26):13051–13062

Zhu C, Jia Y, Wang Z et al (2017) Dynamics of bottom boundary layers in the Yellow River subaqueous delta based on long-term in-situ observations. Acta Geol Sin-Engl Ed 91(1):369–370

Chapter 6
Wave-Induced Pore Pressure in Relation to Sediment Erosion and Resuspension in the Modern Yellow River Delta

6.1 Overview

When water waves propagate in the ocean, they generate significant dynamic pressures on the sea floor. This pressure field induces pore water pressure and effective stresses within the seabed. With excess pore pressure and diminishing vertical effective stress, part of the seabed may become unstable or even liquified. Once liquefaction occurs, the soil particles are likely to be carried away as a fluid by any prevailing bottom current or mass transport owing to the action of ocean waves.

Generally speaking, two mechanisms of the wave-induced soil response have been observed in the laboratory and field measurements, depending on the manner that the pore pressure is generated, as illustrated in Fig. 6.1 (Zen and Yamazaki 1990a, b). One is caused by the progressive nature of the excess pore pressure, which appears at the initial stage of cyclic loading. The other is generated by the oscillatory pore pressure, which is accompanied by the amplitude damping and phase lag in the pore pressure. This type of soil response appears periodically during a storm sequence. In most marine sediments, the wave-induced soil response is oscillatory in nature, except for some special cases of non-cohesive sediments with loose to medium density.

Since the water pressure on the seabed surface varies according to the propagation of waves, it implies that not only the hydrostatic pressure but also the pore pressure changes. Excess pore pressure is defined as the excess component of pressure beyond the hydrostatic pressure. Total stress equals the sum of excess pore water pressure and effective force. Therefore, when the excess pore water increases, vertical effective stress vanishes, the soil will be liquefied. Thus, the soil matrix loses its strength to carry any load and consequently causes seabed instability. However, the mechanisms of the wave-induced soil liquefaction in marine sediments have not been clearly addressed in geotechnical terms at the present stage. The liquefaction is also affected by the state of soil compaction, permeability, the wave-induced cyclic stress as well as the degree of drainage.

Y. Jia et al., *Wave-Forced Sediment Erosion and Resuspension in the Yellow River Delta*, Springer Oceanography, https://doi.org/10.1007/978-981-13-7032-8_6

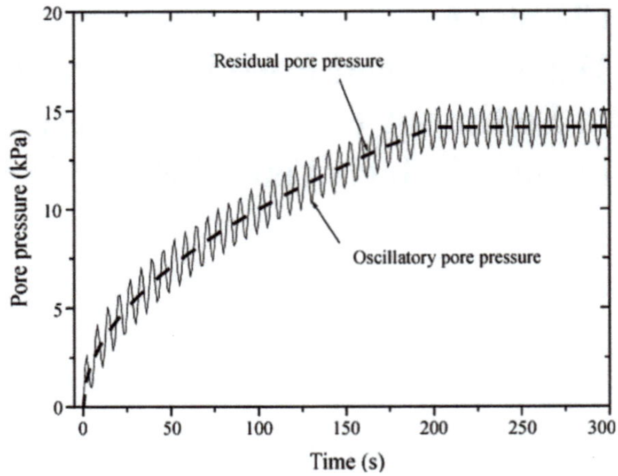

Fig. 6.1 Conceptual sketch of two different mechanisms of pore pressure (not in scale)

Mainstream wave-induced pore pressure models have been briefly reviewed in Sect. 1.3.2. However, to date, the pore pressure model has not been well established in the Yellow River delta area. Therefore, this section attempts to do some experiments on this issue and provide a research guide for coastal geotechnical engineers.

6.2 Dynamic Triaxial Test on the Pore Pressure Response Under Waves

6.2.1 Methodology

(1) ***Sample Collection and Regular Sediment Properties Tests***

The modern Yellow River Delta consists of 11 sedimentary lobes of different sedimentary ages (Chu et al. 2006). Sediment types on sedimentary lobes of different ages have different types, exhibiting different properties. In this study, six study areas (S3, S4, S5, S6, S7, and S8) deposited in fives different periods were selected (see more details about the study site in Fig. 2.1), and sediment samples were collected for dynamic triaxial simulation tests of pore water pressure and deformation processes under large-scale dynamic loading, to study the dynamic response process of deep sediments under extreme sea conditions. Among them, the study area S6 and S5 located in the same sedimentary age, and the sedimentary age is 1904–1929. However, there are different erosion conditions, S6 shows slight erosion, and S5 erosion is very serious. The rate of shoreline receding is as high as 56.6 cm/d (Yang et al. 2011). Therefore, two study areas S6 and S5 were selected on the sedimentary lobes of this sedimentary age, and one study area was identified on the lobes of other sedimentary ages.

(a) S3 (37°51.174'N,119°05.832'E)

(b) S4 (38°08.379'N, 118°42.634'E)

(c) S5 (38°06.129'N,118°27.872'E)

(d) S6 (38°08.051'N,118°12.771'E)

(e) S7 (37°42.137'N, 119°16.458'E)

(f) S8 (37°22.276'N, 118°56.869'E)

Fig. 6.2 Photos of sedimentary environment of sampling areas

In October 2008, the samples with 30 cm in depth were collected for dynamic triaxial tests from the natural sediments of the tidal flats of sub-delta lobe with different sedimentary periods in the modern Yellow River Delta. Six sample collection areas are located along the coast of the delta, and the photo of the sedimentary environment in each study area is shown in Fig. 6.2.

The natural samples were collected by PVC sampling tube, and the surface samples were taken back in sealed bags. The natural density and water content of tidal flats were directly tested by cutting ring method and drying method. The undrained shear strength and penetration strength of the surface sediments in the tidal flat are used respectively a pocket vane shear device with a range of 50 kPa and an electronic digital micro penetration instrument with a range of 20 N. The particle size analysis was carried out in the laboratory using the combined method of sieve and densitometer. The plasticity index was determined by the liquid-plastic limit combined tester, and the compression was tested by the KTG-ZY type triple medium pressure air pressure automatic consolidation instrument. The dry bulk density, pore ratio, and saturation of sediments were calculated by formulas 6.1, 6.2, and 6.3, respectively.

Calculation formula of dry density

$$\rho_d = \frac{\rho}{1 + w} \tag{6.1}$$

where ρ_d is the dry density, g/cm^3; ρ is the natural density, g/cm^3; w is the natural water content, %.

Calculation formula of pore ratio

$$e = \frac{\rho_s}{\rho_d} - 1 \tag{6.2}$$

where ***e*** is the pore ratio; ρ_s is the soil density, also known as proportion, the empirical value of the proportion of the silty soil in the Yellow River estuary is 2.71 g/cm^3.

Calculation formula of saturation

$$S_r = \frac{w \times \rho_s}{e \times \rho_w} \tag{6.3}$$

where S_r is the saturation, %; ρ_w is the pore fluid density, the seawater density is 1.025 g/cm^3.

The empirical value of the proportion of the silty soil in the Yellow River estuary is 2.71 g/cm^3. The conventional physical and mechanical properties and particle size composition characteristics of the six study areas are shown in Table 6.1.

(2) ***Dynamic Triaxial Testing System***

The dynamic triaxial test system used in this experiment is the W3ZB-20 microcomputer controlled hydraulic servo soil dynamic triaxial experiment instrument produced by China Xi'an Lichuang Measurement Instrument Co., Ltd. The instrument uses stress-controlled vibration, and the test results are recorded by photoelectric recorders, including dynamic stress, dynamic strain, and pore water pressure.

The main technical indicators of the vibration triaxial instrument: the axial excitation dynamic load is ±20 kN; the axial excitation static load is 0–20 kN; the accuracy is better than ±1% FS; the axial static deformation is controlled at 0.01–5 mm/min; deformation accuracy is better than ±1% FS; deformation range is better than ±20 mm; axial exciter stroke is 40 mm; displacement accuracy is better than ±1% FS; triaxial pressure chamber pressure is in the range of 0–1 MPa (Can be expanded according to user requirements); the sample size is φ 61.8 × 150 mm, φ 39.1 × 80 mm; the test waveform has several waveforms: sine wave, triangle wave, square

Table 6.1 Physical and mechanical properties and size composition of surface sediment in the research area

Test items	Research area					
	S3	S4	S5	S6	S7	S8
Natural density/(g/cm^3)	1.92	1.89	1.97	1.99	2.01	2.00
Dry density/(g/cm^3)	1.45	1.46	1.51	1.57	1.58	1.55
Pore ratio	0.87	0.86	0.79	0.73	0.72	0.75
Natural moisture content/%	32.52	29.58	29.88	26.60	26.90	28.30
Saturation/%	98.83	90.94	100.0	96.34	98.78	99.09
Plasticity index	10.4	7.3	5.3	7	11.2	6.7
Undrained shear strength/kPa	7.7	6.5	4.7	8.8	7.2	3.2
Penetration strength/N	0.2	0.5	0.8	1.6	1.9	0.4
Compression coefficient (100–200 kPa) (MPa^{-1})	0.151	0.153	0.120	0.169	0.149	0.128
Average particle size/mm	0.044	0.042	0.051	0.040	0.029	0.058
Sand content/%	3.4	5.8	20.8	8.6	1.4	25.4
Powder content/%	79.1	75.5	71.4	77.5	85.3	65.4
Clay content/%	17.5	18.7	7.8	13.9	13.3	9.2

wave, trapezoidal wave, oblique wave and random wave. This experiment uses sinusoidal waveform.

(3) ***Sample Preparation***

The dynamic triaxial test was carried out in August 2011. The natural sample is pushed out from the PVC sampling tube with a bulldozer, and the sample is cut. The cutting process is as follows. First, a wire saw or a ripper is used to take a soil column slightly larger than the specified size, and then it is placed between the upper and lower discs of the cutting disk. Holding the wire saw against the side plate, carefully cut from top to bottom, and turning the disc while cutting until the sample is cut to the specified diameter, then flatten the upper and lower ends. Then, the sample is loaded into a sample holder and placed in a vacuum saturator. After one hour of vacuum pumping, adding water to the vacuum saturator until water is over sample, then pumping vacuum for half an hour, and then placing the vacuum sample into the water for a night.

(4) ***Dynamic Triaxial Test***

The prepared and saturated sample is installed in a three-axis pressure chamber for loading and consolidation. In this experiment, isostatic consolidation is used, that is, the consolidation stress ratio is $K_c = 1.0$. During the loading process, the confining pressure and axial pressure were applied respectively. The experiment was set to 45 kPa, simulating the sediment stress environment at a depth of about 4 m. After loading and consolidation, consolidation process starts. This experiment uses drainage consolidation and the consolidation time is 2 h. As the loading and consolidation process was completed, a dynamic strength (liquefaction) test is performed. In this experiment, a consolidated undrained shear test (CU) was applied, and a cyclic dynamic load was applied to the sample. The dynamic load amplitude was set to 28 kPa, simulating the wave condition of 50-year reoccurrence (Chang 2009). The end condition of dynamic triaxial test is set to a total strain of 10%, a dynamic load cycle of 5 s, and a vibration frequency of 0.2 Hz.

6.2.2 Results

Temporal evolutions of pore pressure and axial deformation during the loading are obtained. The maximum pore pressure value in each cycle is averaged with the adjacent pore water pressure, and the averaged pore water pressure is plotted with vibration loading time. The difference between axial deformation and initial defor-

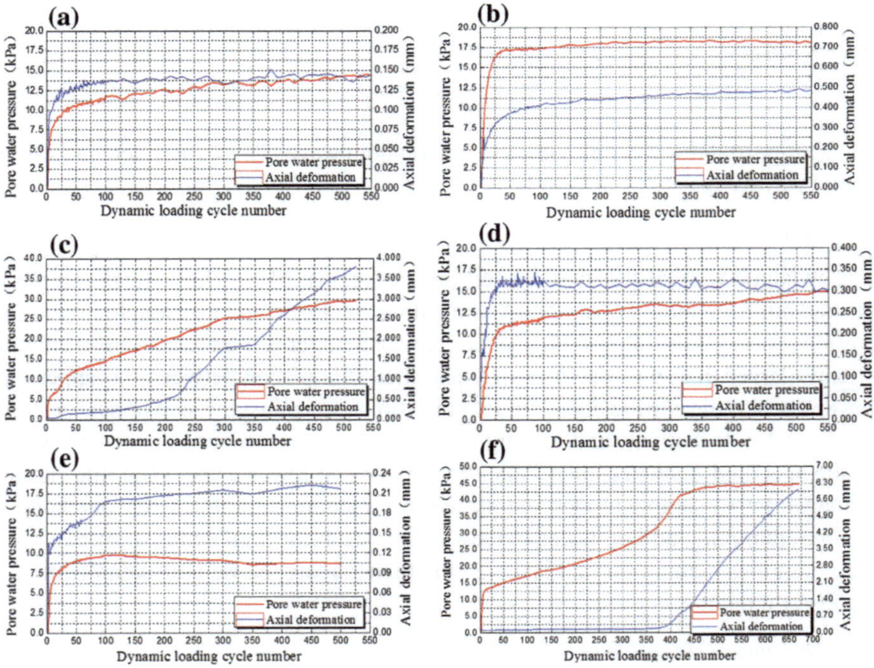

Fig. 6.3 Dynamic triaxial test results of the tidal flat undisturbed sediment in the modern Yellow River delta **a** S3 (sedimentary period 1947–1964); **b** S4 (sedimentary period 1964–1976); **c** S5 (sedimentary period 1904–1929); **d** S6 (sedimentary period 1904–1929); **e** S7 (sedimentary period 1976–1996); **f** S8 (sedimentary period 1929–1934)

Table 6.2 Pore pressure and axial deflection dynamic test results of sediment in the modern Yellow River delta

Station	Sedimentary age	u_{max}	N_s	L_{dmax}
S3	1947–1964	15	550	0.402
S4	1964–1976	18.50	225	0.509
S5	1904–1929	30.00	520	0.823
S6	1904–1929	15.00	530	0.414
S7	1976–1996	10.00	100	0.271
S8	1929–1934	45.00	600	0.995

mation value corresponding to the maximum pore water pressure value in each cycle is also calculated, and the curve of the axial deformation with the vibration time is also plotted, as shown in Fig. 6.3.

From the curves of pore water pressure and axial deformation with vibration times in Fig. 6.3, it can be determined that the maximum pore pressure (u_{max}) in the sample at each station corresponds to the maximum axial deformation amount (D_u) and the minimum dynamic load cycle (N_u), the maximum axial deformation (d_{max}) corresponds to the pore pressure value (u_d) and dynamic load cycle vibration (N_d), as shown in Table 6.2.

Pore water pressure (u_e) to the effective soil pressure (P_o), the larger the ratio, the higher the degree of liquefaction. In this test, the liquefaction degree (L_d) of the sample under different test conditions is the ratio of the cumulative pore water pressure (u) to the confining pressure (σ_3) of the sediment. The maximum liquefaction degree of the sample is the ratio of the maximum cumulative pore water pressure (u_{max}) to the confining pressure (σ_3).

6.2.3 *Dynamic Response Process of Pore Pressure in Dynamic Triaxial Test*

As shown in Fig. 6.3, S6 and S5 (deposited in 1904–1929), S4 (deposited in 1964–1976), and S3 (deposited in 1947–1964), the pore water pressure of the sediment increases rapidly in the early stage of dynamic load, and then changes slowly, finally tends to a stable state, showing three stages in the pore pressure curve. The pore pressure of S8 sediments (deposited in 1929–1934) shows four stages under dynamic load, which first rapidly rises, then changes relatively slowly, then rapidly rises, and finally tends to stabilize. Sediments deposited in the S7 (1976–1996) increase rapidly in the early stage of dynamic loading and then slowly increase. Finally, when the vibration number reaches 130, the pore water pressure begins to decrease slowly.

It can be seen that under the action of dynamic load, the pore pressure of the S5 sediments in the study area is rapidly increasing during the small deformation process after three vibrational loads, and the sediment deformation is very small in the range of vibrational 3–375, and the pore pressure accumulation rate is relatively slower than the initial stage, but it is significantly higher than the pore pressure accumulation rate in the second stage of pore water pressure in other research areas. When the dynamic load is more than 375, the accumulation rate increases, finally reaches the confining pressure and the sample is liquefied. The vibration cycles to reach total liquefaction are 650 and the dynamic load period is 5 s. Thus, it can be seen that the sediments under the confining pressure of 50 kPa can be completely liquefied after 54 min of 30 kPa dynamic load. At this time, sediment deformation increases linearly with the vibration times, sediment samples showed large deformation failure (Fig. 6.3d). This deformation model has also been found in the dynamic three-axis test of Xu (2006), but their pore pressure response mode is different. During the stage of slow deformation, their pore pressure showed a decreasing trend, no cumulative increase, but rapidly accumulated at the stage of large deformation period, and no complete liquefaction occurred, liquefaction degree was less than 0.5. Comparing with other study areas, only the pore pressure in at S8 reaches the confining pressure and complete liquefaction occurs, which indicates that for the sediments deposited from 1929 to 1934, the pore pressure accumulation rate is significantly higher than that of other study areas, 50-years reoccurrence storm waves could cause complete liquefaction of sediments within 4 m depth of the seabed.

Comparing the pore water pressure changes and deformation processes in the sediments of S6, S6, S4, S7, and S3 in the study area, it can be found that the pore water pressure in the sediments of the S5 increases rapidly to 4 kPa at the initial stage of dynamic load, and the deformation is very small at this stage. After the dynamic load vibration is greater than 20, the cumulative rate of pore pressure decreases, but still within the ascending phase. During this process, the deformation rate of the sample increases slightly, but the deformation is still small. When the vibration is greater than 250, pore pressure cumulatively increased to 23 kPa, and the liquefaction degree is 0.51, the sample deformation increased suddenly, and deformation rate also increased greatly. The above results show that the sediments deposited in the period of 1904–1929, which were not affected by human factors, could not fully liquefied with depth of 4 m under big wave condition, but the liquefaction degree could reach 0.51 when the wave loading continued for 21 min, when sediments began to enter the rapid deformation stage, and failure occurred.

The sediments deposited in the S6 research area, although deposited during 1904–1929, are significantly different from the study area S2 due to human factors such as dams (as shown in Table 6.1). The pore pressure accumulation pattern and deformation process are different from those in S5 study area, but very similar to that of S4 (1964–1976) and S3 (1947–1964). In the initial stage of dynamic loading, when vibration times have not exceeded 30, the pore pressure and deformation increase rapidly, after which enter a very slow rise phase. In the range of 700 vibration times, the pore pressure accumulation is not high, the maximum liquefaction degree never exceeded 0.5, and the sediment sample deformation never exceeded 0.5 mm. The maximum liquefaction degree of S7 (1976–1996) sediments is only 0.271. The above results show that the sediments from S6, S4, S3, and S7, which were affected by human factors, big wave conditions could make the pore pressure accumulation occur for the sediments with depth of 4 m, but the degree of accumulation is not high and not easy to liquefy.

6.2.4 *Pore Pressure Accumulation Model in Sediments of the Yellow River Delta*

Based on the analysis of the accumulation law of pore water pressure in different sedimentary ages of modern Yellow River Delta under dynamic load, it can be seen that under the extreme sea conditions in 50 years, the pore water pressure of sediments with depth of 4 m shows a rapid growth-slow growth-basically stable change process. Zeng et al. (2008) obtained the pore pressure accumulation model for sediments which can completely liquefy under dynamic loads by normalized curve fitting of the dynamic pore water pressure through dynamic three-axis tests. However, in our experiment, sediments never reached complete liquefaction, thus a new applicable model is needed.

The dynamic pore water pressure (u) and dynamic vibration load times (N) were normalized respectively. The relation function between the ratio of dynamic pore pressure to confining pressure (u/δ_c), the ratio of the total vibration times to vibration times when pore pressure stabilized (N/N_s) were established. As shown in Fig. 6.4, the pore pressure accumulation model under 50-year reoccurrence period for sediments from different sedimentary years in modern Yellow River Delta is obtained. The model is in the form of a logarithmic curve $u/\delta_c = a + b(\ln N/N_s)$, where a and b are test parameters, which are determined by sediment type and physical properties. a, b, c and its standard error (SE) and the correlation coefficient (R^2) are listed in Table 6.3.

It can be seen from Table 6.3 that the pore water pressure accumulation model in the newest and older study areas in the southern and eastern part of the modern Yellow River Delta has larger errors than the measured points, and the pore pressure model is less practical. For the sediments of other sedimentary age, the logarithmic curve model is more practical, in which the values of a and b are in the range of 0.34–0.63, 0.04–0.24 respectively.

6.2.5 *Influence Factors for Sediment Liquefaction in the Yellow River Delta*

In this book, the maximum liquefaction degree is used as an index to characterize the sensitivity of liquefaction. The larger the maximum liquefaction degree, the easier the sediment is to liquefy. On the contrary, sediments are more difficult to liquefy. Figure 6.5 shows the relation curves of maximum liquefaction degree and corresponding sediment density, dry density, water content, void ratio, plasticity index, averaged particle size, clay content, sand content, silt content, and sedimentary age of sediment within depth of 4 m in the modern Yellow River Delta under the 50-years reoccurrence period.

The influence of density and water content on the liquefaction of sediments under dynamic loading is not significant. The relationship between them shows irregular

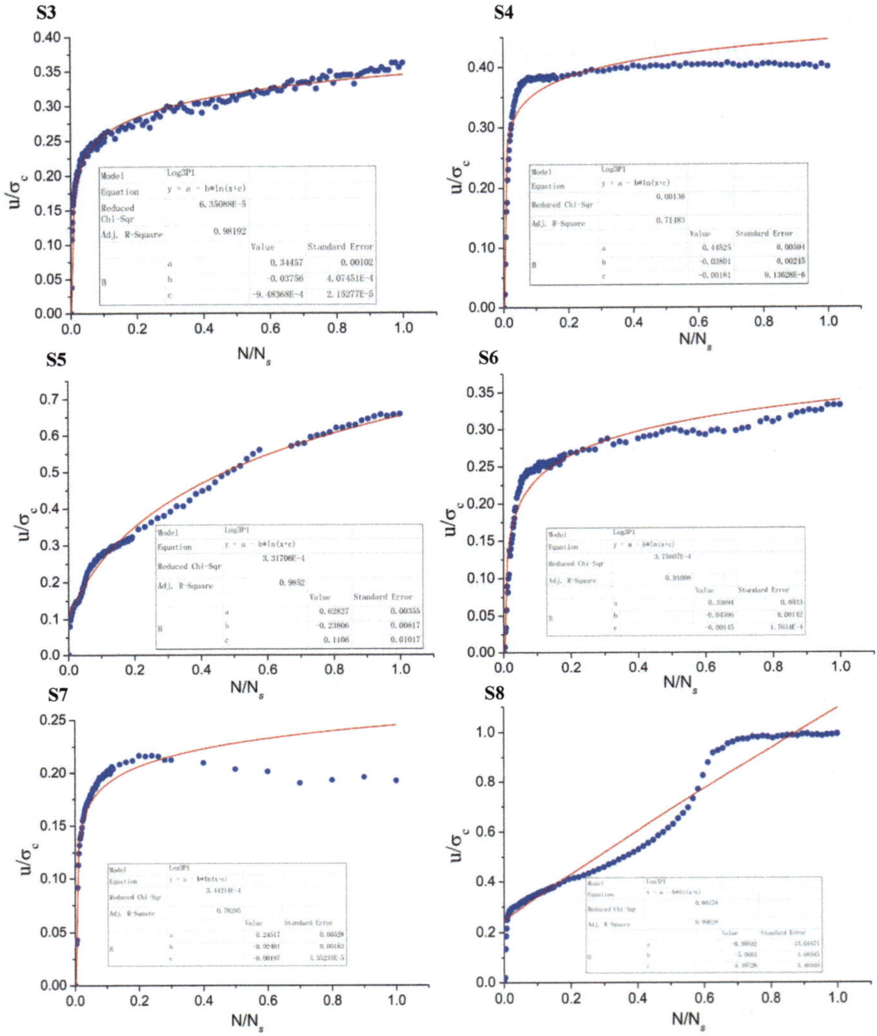

Fig. 6.4 The mode of pore water pressure accumulation for the sediments deposited in different ages in the modern Yellow River delta (The blue dots represent the experimental results; the red solid lines represent the fitted model)

laws, and other physical properties of sediments show obvious influence (Fig. 6.5). With the increase of dry density, the liquefaction resistance of sediment is smaller. When the value reaches about 1.55 g/cm^3, the sediment is most easily to liquefy. Then, with the increase of dry density, the liquefaction resistance of sediment decreases gradually. The influence of the pore ratio and plasticity index on the liquefaction resistance of the sediment is similar to the dry density. When the void ratio is about 0.75, the liquefaction resistance is the smallest. When the plasticity index is about

Table 6.3 Parameters of pore water accumulative model

Research area	Sedimentary age	a	b	SE	R^2
S3	1947–1964	0.34	0.04	0.01	0.98
S4	1964–1976	0.45	0.04	0.02	0.71
S5	1904–1929	0.63	0.24	0.003	0.99
S6	1904–1929	0.34	0.05	0.003	0.91
S7	1976–1996	0.25	0.02	0.02	0.78
S8	1929–1934	−8.39	5.07	13.01	0.96

8.3, the liquefaction resistance of the sediment is the smallest. The effects of average particle size, sand content and silt content on the liquefaction resistance of sediments show very obvious monotonous increasing and decreasing laws. The overall trend is that with the increase of average particle size and sand content, the liquefaction resistance of sediments decreases gradually. With the increase of silt content, the liquefaction resistance of sediments increases gradually.

The effect of clay content on the liquefaction resistance of sediment is obviously two stages, with the limit of 13%, the content of clay particles is higher than this value, the liquefaction resistance of sediment is very high, the content of clay particles is lower than this value, and the liquefaction resistance of sediment is very low. Liu et al. (2013) and Xu et al. (2012) carried out dynamic triaxial tests with four different clay contents: 3%, 9%, 15% and 21%. It is found that 9% is the turning point of pore water pressure development, and the cumulative pattern of pore water pressure is different before and after the turning point. It can be seen from their pore pressure accumulation curve that the liquefaction resistance of soil sample with clay content over 9% is larger than that with less than 9%, which is similar to the findings in this book. Li et al. (2012) found that the wave energy attenuation in soil bed with high clay content (10%) is higher than that of the soil bed with lower clay content (5%), so when the clay content is high, sediment is not easy to liquefy.

The influence of sedimentary history on the liquefaction resistance of sediments is also very significant. Excluding the interference of human factors, it can be seen from Fig. 4.8 that the anti-liquefaction ability of sediments in old sedimentary age is significantly lower than that of sediments in new sedimentary age. The mechanism can be analyzed from the physical characteristics of sediments of different sedimentary histories. From the analysis of the factors affecting the anti-liquefaction ability of sediments above, it can be seen that the particle size component has the most prominent influence on the anti-liquefaction ability of sediments. Combining with the particle size composition of sediments of different ages, it can be seen that the clay content in sediments of new sedimentary age is significantly higher than that is of old sedimentary age, so the longer the sediment deposition time is, the easier to liquefy, the "roughening" process of sediments in the sedimentary history (Jia et al. 2011) plays an important role in the influence of sedimentary history on the liquefaction resistance of sediments.

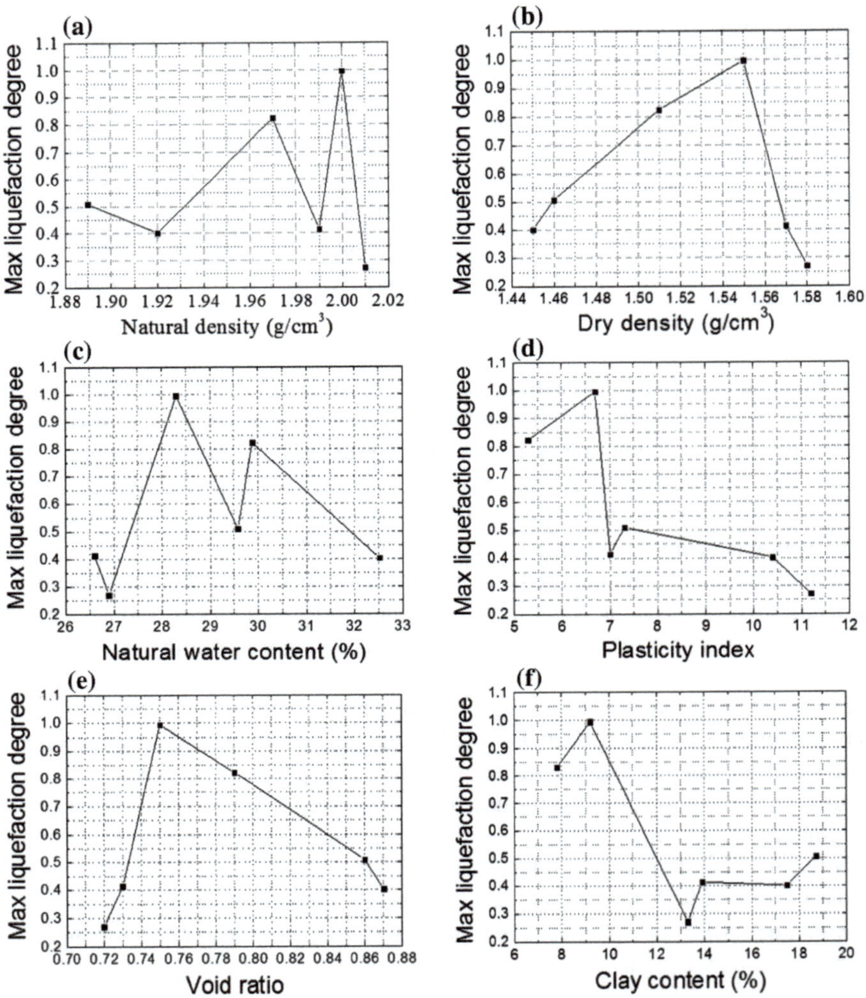

Fig. 6.5 Relationship between maximum liquefaction degree and the influence factor of the modern Yellow River delta deep sediment under extreme sea conditions. **a** Maximum liquefaction degree versus natural density. **b** Maximum liquefaction degree versus dry density. **c** Maximum liquefaction degree versus natural water content. **d** Maximum liquefaction degree versus plasticity index. **e** Max. liquefaction degree versus void ratio. **f** Max. liquefaction degree versus clay content. **g** Maximum liquefaction degree versus sand content. **h** Maximum liquefaction degree versus silt content. **i** Maximum liquefaction degree versus mean particle size. **j** Maximum liquefaction degree versus sedimentary periods

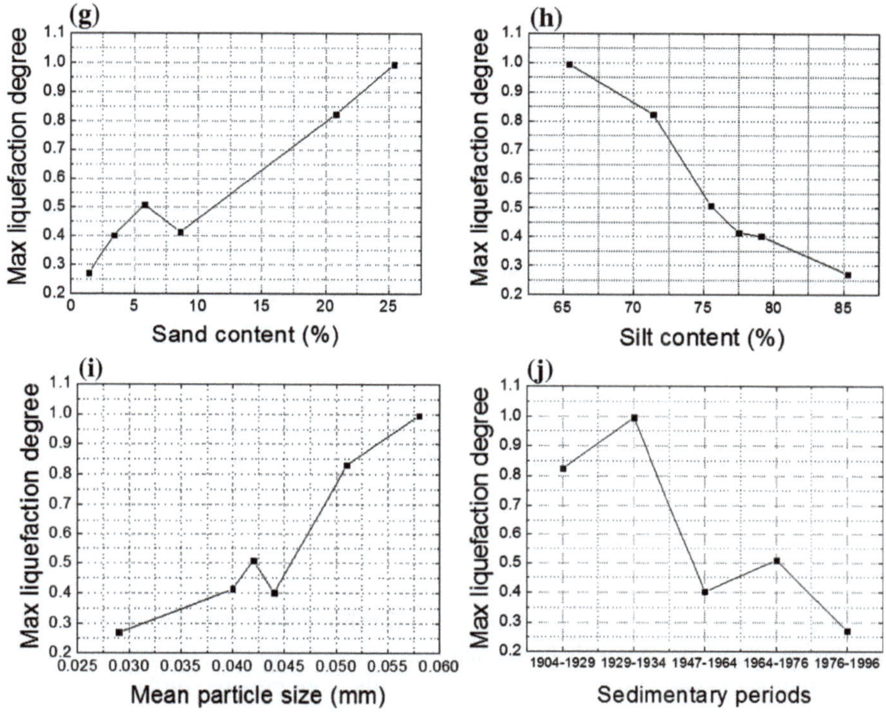

Fig. 6.5 (continued)

6.3 Field Experiments on Pore Pressure Response Under Waves

6.3.1 Methodology

(1) ***Field Test Area***

Field artificial vibration test and regular geotechnical property tests were carried out in November 2011 near the six test stations (S3, S4, S5, S6, S7, S8) that collected the original sediments of the tidal flat in October 2008. And two test stations added later. In order to discuss clearly, the test stations were remarked as C1, C2, C3, C4, C5, C6, C7, and C8.

(2) ***Experimental Process***

The wave experiments were conducted with a self-designed wave-making device to apply a cyclic wave load on the natural seabed (Fig. 6.6). Prior to applying the wave loading, pore pressure probe is first buried in the underlying hole: insert the

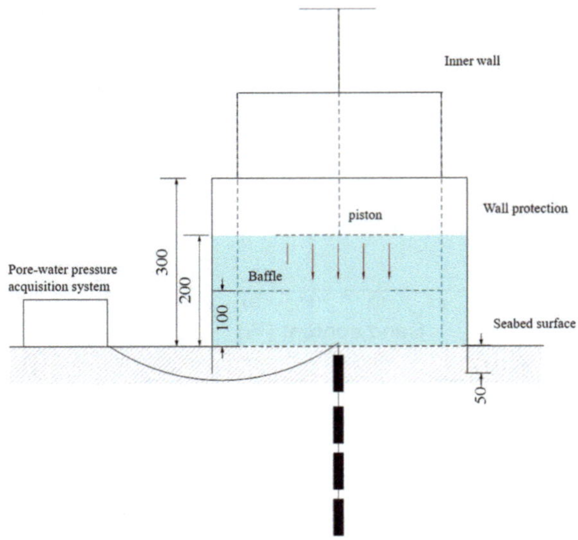

Fig. 6.6 Sketch map of the field wave simulator and experimental layout

graduated penetration probe to the designed depth together with the pore pressure sensor and data line. After that, the data line connected to the pore pressure probe was fixed on the surface of the sediment. After the pore pressure probe is arranged, deploy the wave-making device into the tidal flat sediments, put nearby seawater into the device to a predetermined height, which, in this study, was 20 cm. Then put into the wave barrel. After the wave-making device is arranged, it will be stationary until the pore pressure is dissipated and stable, start making waves. The waves simulated in the experiment were 5 s in period and 4 kPa in pressure, which was equivalent to real wave height of 0.8 m, on the condition that the ratio of water depth to wave length was very small.

Pore water pressure test adopts the pore pressure automatic acquisition system developed by Nanjing Institute of Water Science, the entire system consists of pore water pressure probe, a multichannel data acquisition system, a communication cable and a computer. The probe is cylindrical, 60 mm in height and 20 mm in diameter. The outer casing is made of stainless steel, which has good corrosion resistance and sealing, also has good contact with surrounding sediments. The probe converts the pressure signal into electrical signal that is automatically saved by the acquisition system in the computer. Before the pore pressure is collected, the pore pressure probe is calibrated in the laboratory to obtain the relationship between the electrical signal and the pore pressure value. Working power for the field test is supplied by small generators.

Test method of sediment penetration strength, shear strength, bulk density, water content, particle size composition test and calculation method of dry density, void ratio and saturation are same as that in Sect. 6.2. Empirical value of specific gravity in the Yellow River Delta is 2.71 g/cm^3. Table 6.4 shows the regular physical/mechanical properties and particle size composition in the seven study areas.

Table 6.4 Physical and mechanical properties and size composition of surficial sediment on the tidal flat

Test items	Test area							
	C1	C2	C3	C4	C5	C6	C7	C8
Natural density/(g/cm^3)	1.97	1.95	1.89	2.01	1.93	1.92	1.98	1.90
Dry density/(g/cm^3)	1.55	1.51	1.47	1.58	1.49	1.50	1.56	1.46
Void ratio	0.75	0.79	0.84	0.72	0.82	0.81	0.74	0.86
Natural water content/%	26.7	28.8	28.9	27.2	29.7	28.2	27	29.8
Saturation/%	94.1	96.7	91.0	99.9	96.1	92.1	98.7	91.6
Undrained shear behavior/kPa	4	4.5	13.5	3.5	3.9	8.7	9.8	6.5
Penetration strength/N	4.5	4.1	1.6	0	1.6	3.2	3.5	2.9
Average particle size/mm	0.035	0.039	0.04	0.008	0.038	0.024	0.016	0.023
Sand content/%	33.7	33.8	29.8	2.0	19.4	15.6	3.2	12.9
Silt content/%	53	56.9	61	60.6	74.4	77.5	63.2	82.6
Clay content/%	13.3	9.3	9.2	37.4	6.2	6.9	33.6	4.5

6.3.2 *Results*

The pore water pressure acquisition system collects the measured voltage value 3 times per second, and collects the voltage value 15 times in a hydrodynamic pressure cycle. Select the maximum and the minimum voltage value in each cycle, and determine the averaged value as the typical voltage value in the cycle. Brought the typical voltage value to the probe calibration curve to get the pore water pressure value in the cycle. Therefore, the variation of pore water pressure at different depths of tidal flat sediments during the dynamic loading in eight typical study areas of the modern Yellow River Delta is shown in Fig. 6.7.

6.3.3 *Dynamic Response Process of Pore Pressure in Field Experiment*

From the results of the laboratory dynamic triaxial shear test in Sect. 6.2 of this chapter, it can be found that the pore pressure response modes of sediments in different sedimentary ages under large dynamic loads can be different. Due to the difference of location and human factors, the pore pressure response process in same sedimentary ages is not the same as well. It can be found from the experiment conducted in eight locations from the modern Yellow River Delta, pore pressure response in different areas is different, there are also significant differences in the response of pore water pressure to wave loads at different sediment depths.

Under the protection of artificial dams, C1 (1904–1929) is currently in a slightly erosive state, in this area, dynamic load of 0.4 kPa in amplitude and 5 s in period for 25 cycles can lead to an increase in pore pressure in the surface depth of 15 cm.

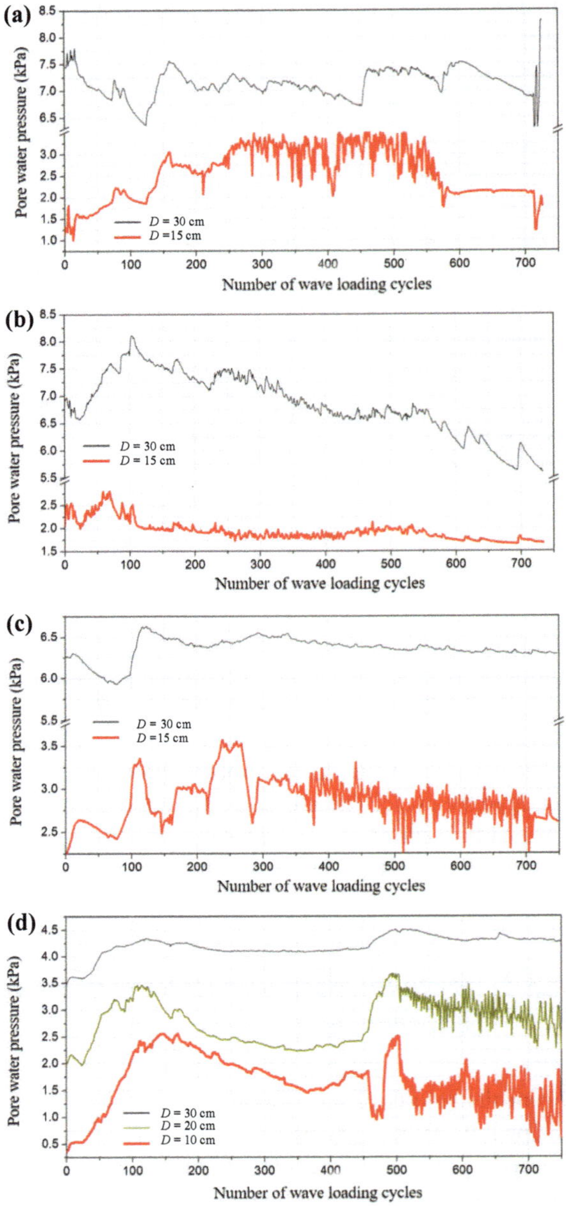

Fig. 6.7 Pore water pressure variation during the dynamic loading on surficial sediment in the Yellow River delta **a** C1 (sedimentary age 1904–1929); **b** C2 (sedimentary age 1904–1929); **c** C3 (sedimentary age 1964–1976); **d** C4 (sedimentary age 1976–1996); **e** C5 (sedimentary age 1929–1934); **f** C6 (sedimentary age 1947–1964); **g** C7 (sedimentary age 1964–1976); **h** C8 (sedimentary age 1964–1976)

Fig. 6.7 (continued)

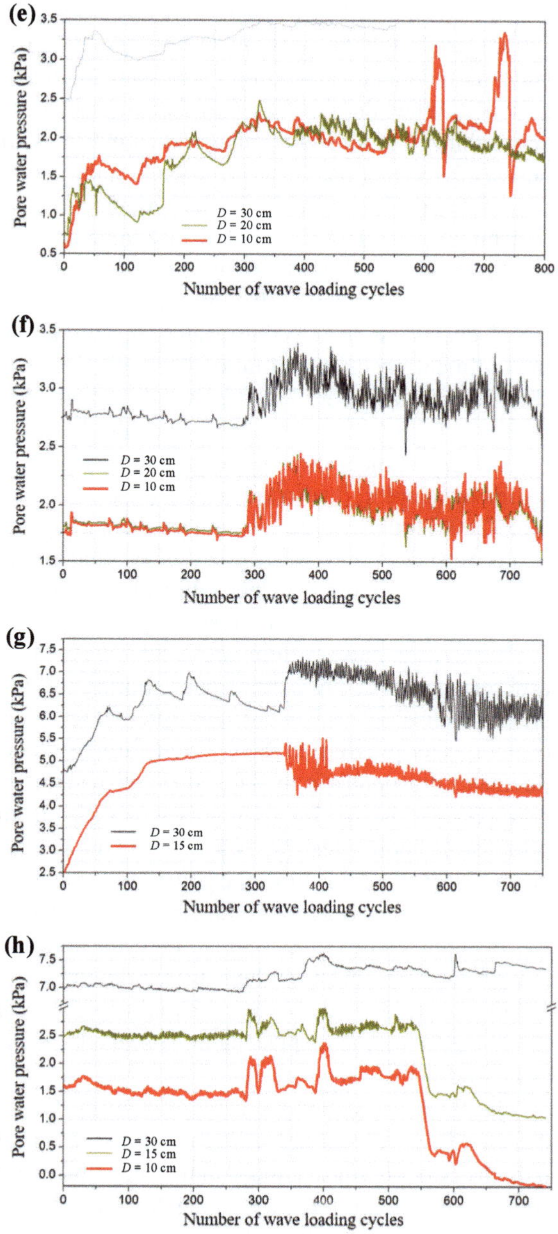

When the vibration times reached 250, the pore pressure fluctuated drastically. Chang (2012) also found this phenomenon in the vibration test of remolded soil in the field, and believed that this kind of pore pressure fluctuations indicates that the sediment is liquefied. When the vibration times are more than 550, pore water pressure decreases and stabilizes to a certain value, indicating the sediments are consolidating after a period of liquefaction. The pore pressure at the depth of 30 cm fluctuated at a different degree, but there was no obvious accumulation of pore pressure, indicating that the dynamic load of 0.4 kPa did not cause liquefaction at 30 cm depth.

However, C2 (1904–1929) is less affected by human factors and is currently in a state of severe erosion, dynamic load of 0.4 kPa in amplitude and 5 s in period for 25 cycles can lead to pore water pressure accumulation within 30 cm depth of the sediment, when the vibration times reached 100, the pore water pressure drops sharply and dissipates. The final steady values of pore pressure after dissipation at depths of 15–30 cm are both lower than those at the initial stage of dynamic loading, indicating that the sediment consolidation degree is higher than that of the original sediment samples, due to the action of dynamic loading. The vibration test of Shan et al. (2006) found that the loss and recovery of seabed strength is closely related to the two processes of pore water pressure accumulation and dissipation during dynamic loading. Unsimilar to C1, the sediment at C2 did not show significant fluctuations throughout the pore pressure response.

The C7, C8, and C3 study areas, all deposited in 1964–1976, showed significant differences in pore pressure response during dynamic loading. Both the study areas are located inside the dam but have different sediment types. The cumulative increase of pore water pressure in the C7 study area is obvious, and the pore water pressure fluctuates drastically in the later stage of dynamic load. The cumulative increase of pore pressure in the S8 study area is not significant. During the dynamic loading process, the pore pressure at a depth of more than 20 cm has a large amplitude protrusion, at the same time, the pore pressure at 30 cm depth increased cumulatively. At the late stage of dynamic loading, pore pressure decreased significantly within 20 cm depth. At this time, sediment may be deformed and destroyed. This changing law of pore pressure was the same as the field investigation results of Prior et al. (1989) during the storm waves. Simultaneously with the change of pore pressure, there is also a large displacement of seabed, indicating the seabed is unstable and destroyed at that time.

The tidal flat sediments deposited in the C3 (1964–1976) showed a dynamic increase in pore pressure at the depth of 15 cm at vibration times of 250. After that, the pore pressure reduced and violent fluctuations occurred. The dynamic load showed very small influence on the depth of 30 cm.

Within 30 cm depth of the tidal flat sediments of C4 (1976–1996), pore pressure increased in different degrees at the initial stage of dynamic loading. Shallower sensors showed more significant pore pressure accumulation. After which the pore pressure showed a downward trend. When the vibration times were more than 450, pore pressure increased suddenly, and then violent fluctuation occurred, but no significant fluctuation occurred at the depth of 30 cm. Under the dynamic load of 4 kPa, the pore pressure in this area experienced increase first, then decrease, again rise, and then a stage of violent fluctuations, which is completely consistent with the research of Chang (2009).

In the tidal flat sediments of C5 (1929–1934), the dynamic load of 4 kPa immediately caused the accumulation of pore pressure within 30 cm depth, followed by a downward trend, and finally violent fluctuations. Vibration times to the maximum pore pressure and the beginning of severe fluctuations at various depths are also different. Deeper sediments showed shorter vibration times and smaller amplitude.

For sediments of C6 (1947–1964), there was not significant cumulative increase of pore water pressure at the beginning of the dynamic load, but when the vibration times reached 300, pore water pressure fluctuated drastically, finally began to decrease slowly after accumulating to a certain value during the fluctuation.

6.3.4 Influencing Factors on the Liquefaction Characteristics of Sediments

The maximum liquefaction degree (the ratio of maximum cumulative excess pore pressure to overlying effective stress) of sediments at depths of 10, 15, 20, and 30 cm in different study areas under dynamic pressure amplitude of 4 kPa is plotted as a histogram (Fig. 6.8). The liquefaction capacity of sediments in the modern Yellow River delta and the influence of sedimentary history on the liquefaction characteristics of shallow sediments were analyzed. It can be seen from Fig. 6.9 that in the modern Yellow River Delta region, large wind waves can cause pore water pressure accumulation in the depth of 30 cm of sediment, and the accumulation of pore pressure at different depths is different. The shallower the depth, the higher the pore pressure accumulation. The maximum liquefaction degree of sediments at 10 cm depth is 0.7–2.2, the maximum liquefaction degree of sediments at 15 cm depth is 0.5–2.0, the maximum liquefaction degree of sediments at 20 cm depth is 0.3–0.9, and the maximum liquefaction degree of sediments at 30 cm depth is 0.1–0.9. At 20 cm depth, the accumulation of pore water pressure in shallow sediments is significantly higher than that in the lower 20 cm depth, indicating that the effect of heavy wind load on the upper sediment at 20 cm depth is significantly higher than that in the lower part. The above data also show that the highest value of excess pore water pressure accumulation ($\Delta\mu_{max}$) in surface sediments can reach more than twice the overburden effective stress ($\gamma' d$), and the sediments still have not reached full liquefaction, and there is a certain strength, which is completely different from the liquefaction characteristics of sand. The typical liquefaction properties of silty sediments are closely related to their structure (Chang and Jia 2012). It can be seen from this that the structural continuous relay of silty soil particles has a significant influence on its liquefaction characteristics.

Comparing the maximum liquefaction of sediments in different sedimentary years, it can be found that for shallow sediments of 20 cm, the accumulation of excess pore water pressure in the C1, C5, C7, C4 research area during the process of heavy wind

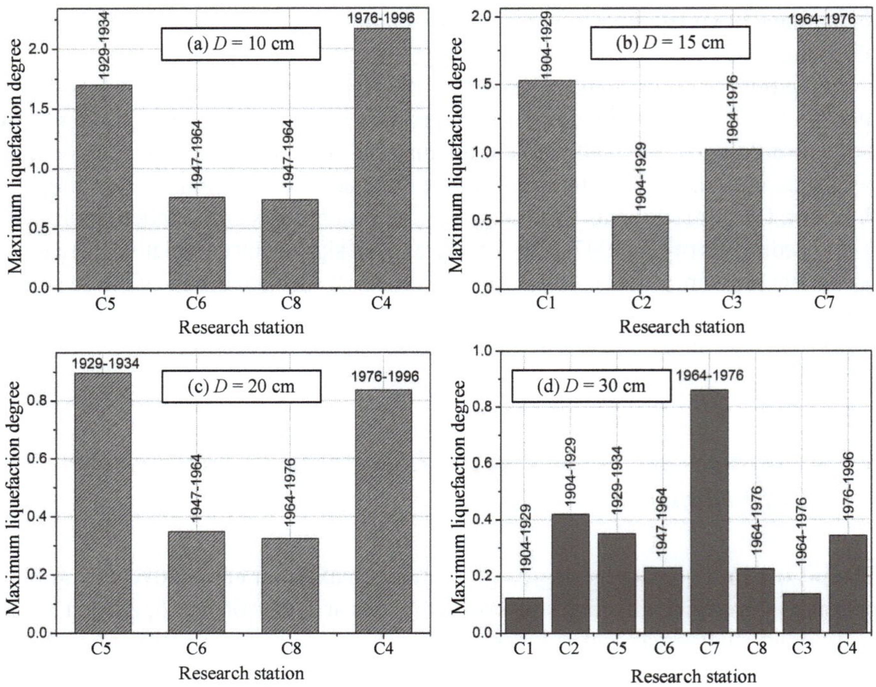

Fig. 6.8 Histogram of sediment max. liquefaction degree at different depths under rough sea conditions

and waves is very high, indicating that the sediments in the four study areas are prone to liquefaction. For sediments below 30 cm depth, in the C7 research area, the accumulation of excess pore water pressure was the highest, and the accumulation of excess pore water pressure in sediments in other sedimentary years was very low, and the maximum liquefaction degree was less than 0.5. It indicates that the sediments of C7 in the study area are highly liable to liquefy under the conditions of heavy wind and sea, and the liquefaction depth is high.

Sediment properties and particle size composition characteristics can influence the liquefaction of seabed under wave cyclic loading. Based on the analysis of Fig. 6.8, it can be seen that the difference in pore water pressure accumulation of shallow sediments of 20 cm in the seabed is small, indicating that the characteristics of sediments in this depth range are very similar. The sediment properties and particle size composition tests carried out in this section are based on sediment samples in the depth range of 10 cm. According to the above speculation, it can be considered that the sediment at 15 cm depth of the seabed has little difference in the nature and grain size composition of the sediment at a depth of 10 cm. Therefore, a scatter plot is drawn of the maximum liquefaction degree generated during the dynamic loading of 10–15 cm depth corresponding to different sediment characteristics in each study area (Fig. 6.9). The effects of sediment properties and particle size composition on the degree of liquefaction of shallow sediments in the modern Yellow River delta were analyzed.

It can be seen from Fig. 6.9 that the degree of liquefaction of sediments with a density less than 1.98 g/cm^3 is generally lower than that of dense sediments; The degree of liquefaction of sediments with a dry density of less than 1.51 g/cm^3 is generally lower than that of dry sediments; degree of liquefaction of sediments with a water content higher than 28.0% is generally lower than that with low water content; The degree of liquefaction of sediments with a void ratio higher than 0.78 is generally lower than that with low void ratio; the degree of liquefaction of sediments with an average particle diameter larger than 0.022 mm is generally lower than that with large average particle size; The degree of liquefaction of sediments with sand content greater than 10% is generally lower than that with high sand content; The degree of liquefaction of sediments with clay content greater than 30% is generally higher than that with low clay content; The degree of liquefaction of sediments with a particle content of about 63% is high.

Based on the analysis of the data points, it can be seen that the degree of liquefaction of the sediment does not show a clear single correspondence with the physical properties and particle size composition. The effects of the properties and particle size composition characteristics of the shallow sediments on the difficulty of liquefaction under the condition of large wind waves are significantly different from those of the deep sediments and particle size composition characteristics in Sect. 6.2. Under the extreme sea conditions, the difficulty of liquefaction of deep sediments under the dynamic load of storm waves and the dry density, void ratio, average particle size, silt content and sand content of sediments are in a certain monotonous correspondence.

6.3.5 Granulometric Composition Variation in Sediments

In the selected eight study areas, surface sediments were sampled before and after the wave load, and the particle size distribution test was carried out in the laboratory. Samples C1 and C7 were lost during transport. Therefore, only the particle size composition data of surface sediments in the other six study areas were measured, and shown in Fig. 6.10 and Table 6.5.

According to the grain size accumulation curve of surface sediments before and after dynamic loading in six study areas, we found fine particle content of surface sediments after dynamic loading is higher than that before dynamic loading, while the coarse particle content is lower. There is a phenomenon of upwelling of fines in the dynamic load zone when the on-site dynamic load is over. This typical wave dynamic response process has also been found in the vibration test of remolded soil (Chang 2009). It can be inferred that the increased fine particles in the surface layer are derived from the lower deposits. The test results of different research areas in the modern Yellow River Delta indicate that the upward transport of fine sediments under dynamic loading is common in the modern Yellow River Delta.

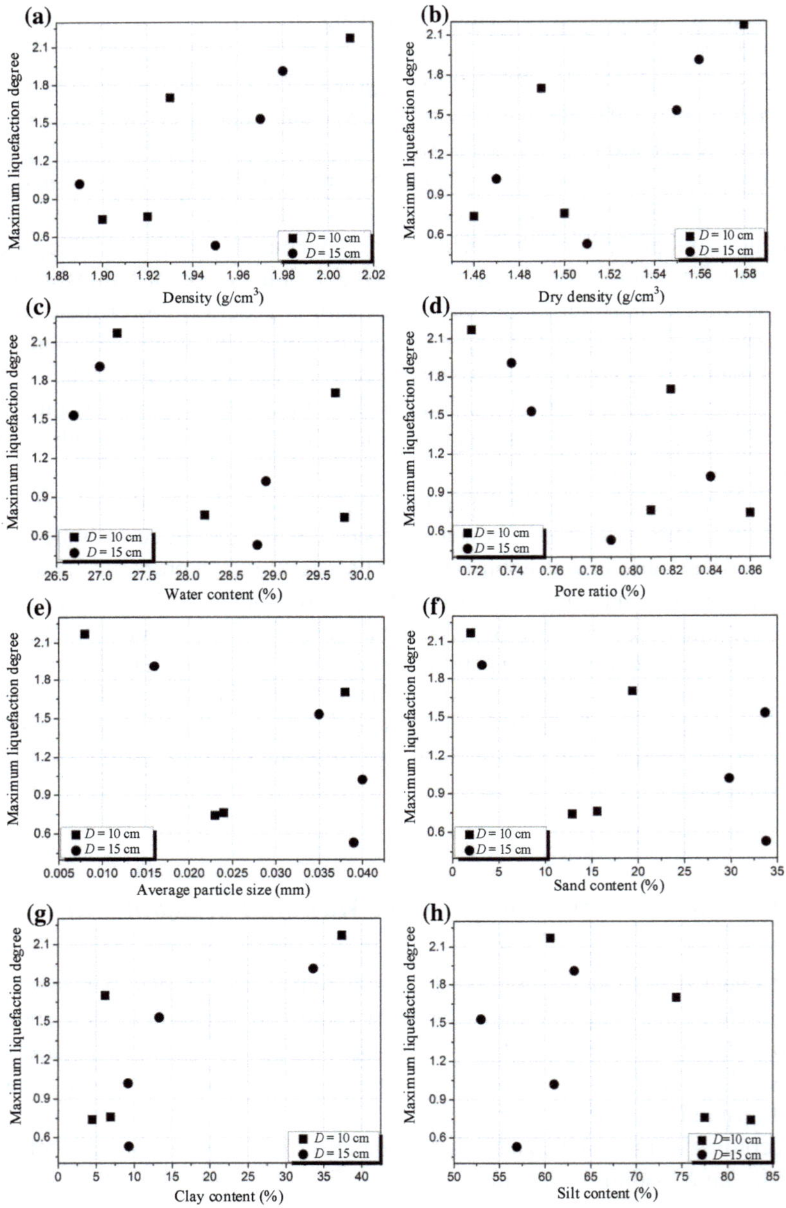

Fig. 6.9 Scatter diagram of max. liquefaction degree and influence factor of surficial sediment in the modern Yellow River delta under rough sea conditions. **a** Maximum liquefaction degree versus density. **b** Maximum liquefaction degree versus dry density. **c** Maximum liquefaction degree versus water content. **d** Maximum liquefaction degree versus void ratio. **e** Max. liquefaction degree versus average particle size. **f** Max. liquefaction degree versus sand content. **g** Maximum liquefaction degree versus clay content. **h** Maximum liquefaction degree versus silt content

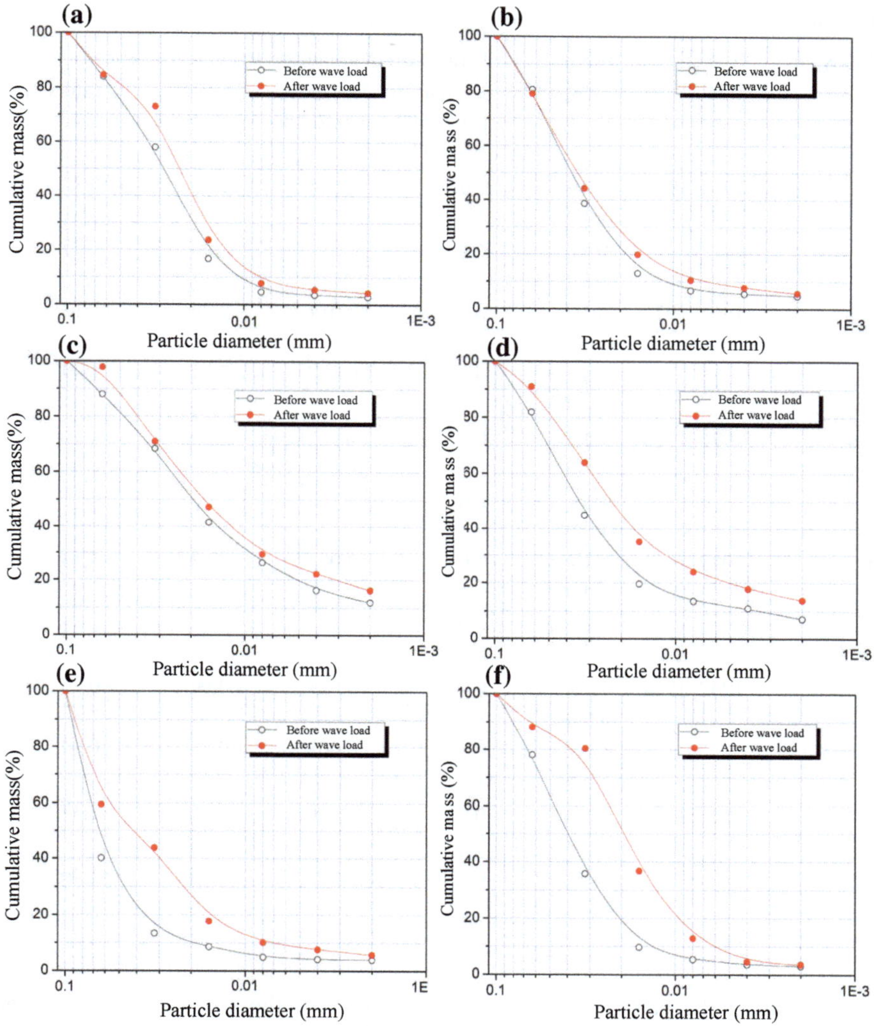

Fig. 6.10 Grain size distribution of surficial sediment in the modern Yellow River delta before and after the dynamic water loading. **a** Study area C2 (Sedimentary age 1904–1929). **b** Study area C3 (Sedimentary age 1964–1976). **c** Study area C4 (Sedimentary age 1976–1996). **d** Study area C5 (Sedimentary age 1929–1934). **e** Study area C6 (Sedimentary age 1947–1964) **f** Study area C8 (Sedimentary age 1964–1976)

The mechanism of the above-mentioned dynamic response of sediments is that pore pressure accumulates under the dynamic load to generate excess pore pressure, thereby generating seepage gradient force, fine particles are transported upward along the seepage channel to the surface layer driven by the seepage gradient force. During this process, excess pore pressure gradually dissipates, and bottom sediments are "roughened" and further compacted, becoming more stable.

Table 6.5 Grain size parameters of surficial sediment due to dynamic loading

Study area	Age of deposition	Test conditions	Parameters of granulometric characteristics			
			Average particle size/mm	Effective size/mm	Uneven coefficient	Curvature coefficient
C2	1904–1929	Before action	0.028	0.014	2.36	0.87
		After action	0.023	0.01	2.60	1.25
C3	1964–1976	Before action	0.039	0.014	3.29	1.13
		After action	0.035	0.007	5.71	1.89
C4	1976–1996	Before action	0.020	–	–	–
		After action	0.017	–	–	–
C5	1929–1934	Before action	0.034	0.003	12.33	5.63
		After action	0.025	–	–	–
C6	1947–1964	Before action	0.088	0.029	3.00	0.73
		After action	0.037	0.008	9.88	0.84
C8	1964–1976	Before action	0.038	0.016	2.56	1.37
		After action	0.019	0.007	3.14	1.46

Comparing the particle size distribution curves of the six study areas, it can be found that seabed with different particle size composition has different ability to transport fine particles upwards under dynamic loading, both the particle size and transport flux. Cumulative curves with slower slope and better sorting have poor transporting ability. The upward transported particles are mainly fine particles (<0.002 mm), upward transport of sediments in C4 (1976–1996) are the most significant among all the other locations. Cumulative curves with steep slope and poor sorting, i.e., the particle size composition is dominated by a certain group, have better transporting ability, also has a relatively larger transported particle size. This was the most significant in the C8 area (1964–1976).

The above results indicate that grain size composition of the seabed has a significant influence on the upward transport process of fine particles driven by the seepage gradient force generated by the waves. The mechanism of the effect is that for fine-grained soil, the well-sorted sediments are closely arranged, and the seepage channels are not easy to produce, under the action of waves, the pore pressure tends to accumulate but is not easy to dissipate. The internal fine particles need higher seepage gradient force for upward transport. However, sediments with poor sorting are relatively loose in particle arrangement, and the seepage channel is easy to form. The seepage gradient force generated by the accumulation of pore pressure under the action of waves is more likely to cause the internal fine particles to transport to the surface layer.

From the above research results, it can be inferred that under the long-term large-wave cyclic loading of the seawater, the fine-grained soil in the modern Yellow River Delta tends to "roughen" during the scouring process of the surface layer. It tends to be stable during the compaction process, but the reduction of fine particles makes the surface layer composed of coarser particles, sorting gets worse (Table 6.5). Therefore, the upward transport capacity of fines is strengthened, which further leads to the "roughening" process again. The formation of the widespread hard crust layer in the modern Yellow River Delta is closely related to the wave-induced dynamic response process of seabed.

The change process of the seabed under the action of ocean dynamics plays an important role in the erosion and resuspension process, not only affecting the change of sediment resuspension but also affecting the particle size composition of resuspended materials. It can be seen that a series of dynamic response processes of sediments under the action of waves play an important role in sediment resuspension.

6.4 Summary

In this chapter, wave-forced dynamic pore pressure response of silty seabed in the subaqueous Yellow River Delta is thoroughly studied. Sediments were found to be sensitive to pore pressure build ups and experienced totally liquefaction under extreme wave conditions. The most commonly used research means including "dynamic triaxial test" and "field experiment" were conducted using sediment sampled from the field or directly in the field. We found that (1) Pore pressure of sediment may increase rapidly in the early stage of dynamic load, and then changes slowly, finally tends to a stable state, showing three stages in the pore pressure curve, or shows four stages under dynamic load, which first rapidly rises, then changes relatively slowly, then rapidly rises, and finally tends to stabilize. (2) Wave-induced pore pressure develop models were constructed, influencing factors on the liquefaction properties were also identified. (3) Clay content has an important influence on the liquefaction properties of the Yellow River sediments. (4) There will be an upward transport of fine sediments under waves in the seabed of the modern Yellow River Delta. (5) Water content will become smaller due to the re-consolidation process after liquefaction.

Seabed response under the action of ocean dynamics plays an important role in the erosion and resuspension process, so this chapter is the transitional section in this book which has important implications for the investigation of sediment erosion and resuspension in the Yellow River Delta.

References

Chang FQ (2009) Study on mechanism of wave-induced submarine landslide at the Yellow River estuary PhD. thesis. Ocean university of China, Qingdao, China

Chang FQ, Jia YG (2012) In-situ test to study silt liquefaction at the subaqueous delta of Yellow River. China Civ Eng J 45(1):121–127

Chu ZX, Sun XG, Zhai SK et al (2006) Changing pattern of accretion/erosion of the modern Yellow River (Huanghe) subaerial delta, China: Based on remote sensing images. Mar Geol 227(1–2):13–30

Jia YG, Liu XL, Shan HX, Zheng JW, Huo SX (2011) The effects of hydrodynamic conditions on geotechnical strength of the sediment in Yellow River Delta. Int J Sedim Res 26(3):318–330

Li AL, Li GX, Lin L et al (2012) Experiment study on pore pressure responses to wave action on silt seabed. Mar Sci Bull 31(1):15–21

Liu XL, Jia YG, Zheng JW et al (2013) Consolidation of sediments discharged from the Yellow River: implications for sediment erodibility. Ocean Dyn 63(4):371–384

Prior DB, Suhayda JN, Lu NZ et al (1989) Storm Wave Reactivation of a Submarine Landslide. Nature 341(6237):47–50

Shan HX, Zhang JM, Jia YG (2006) Study on consolidation process of rapidly deposited seabed soils in yellow river estuary. Chin J Rock Mech Eng 26(8):1676–1612

Xu GH (2006) Study on the landslide of gentle-slope silty seabed under waves-a case of yellow river subaqueous delta abstract. PhD thesis. Ocean university of China, Qingdao, China

Xu GH, Liu HX, Liu JK et al (2012) Role of clay content in silty soil liquefaction. Mar Geol Quat Geol 32(3):31–36

Yang ZS, Ji YJ, Bi NS et al (2011) Sediment transport off the Huanghe (Yellow River) delta and in the adjacent Bohai Sea in winter and seasonal comparision. Estuar Coast Shelf Sci 93:173–181

Zen K, Yamazaki H (1990a) Mechanism of wave-induced liquefaction and densification in seabed. Soils Found 30(4):90–104

Zen K, Yamazaki H (1990b) Oscillatory pore pressure and liquefaction in seabed induced by ocean waves. Soils Found 30(4):147–161

Zeng CN, Liu HL, Chen YM (2008) Test study on influence of fine particle content on dynamic pore water pressure development mode of silt. Rock Soil Mech 29(8):2193–2199

Chapter 7
Physical Mechanisms of Wave-Induced Sediment Resuspension

7.1 Overview

In tidal-dominated environments, sediment resuspension is largely controlled by the bottom shear stress (e.g., Van Rijn 1989), which is related to seabed roughness and mean current velocity. However, considerable observations have found that in shallow environments with frequent waves, sediments are more significantly resuspended by wave action (Paphitis and Collins 2005). According to the research findings of previous chapters in this book, we found the mechanisms of wave-forced resuspension can be summarized in three aspects: enhanced shear erosion due to wave orbital velocity (Lesht et al. 1980), wave pumping of sediments (e.g., Wolanski and Spagnol 2003), and wave-induced seabed liquefaction.

The last two mechanisms are directly related to the wave-induced excess pore water pressure (EPP) response in the seabed. Generally, two types of responses, namely, oscillatory (momentary) pore pressure (OPP) and residual (accumulative) pore pressure (RPP), have been observed in numerous laboratory and field measurements, depending on the manner that the EPP is generated. Essentially, OPP is induced by the elastic volumetric strain of the seabed, while RPP is caused by the plastic volumetric strain of the seabed. Correspondingly, two types of seabed liquefaction, namely, momentary and residual liquefaction (Sumer and Fredsøe 2002) may occur under the extreme conditions, resulting from the oscillatory and unidirectional upward seepage flows, respectively. In the natural environment, residual liquefaction events are probabilistic in the limiting case when the RPP approaches or exceeds the overlying effective stress (e.g., Clukey et al. 1985), whereas the WPS is a more common process, if only the momentary liquefaction occurs in the shallow layers.

In this chapter, we specifically examined the sediment resuspension caused by wave–seabed interactions, i.e., oscillatory and unidirectional upward seepage flows, as well as the attenuation in erodibility under wave actions. To achieve this goal, several test devices were designed and several experiments were conducted with these

Y. Jia et al., *Wave-Forced Sediment Erosion and Resuspension in the Yellow River Delta*,
Springer Oceanography, https://doi.org/10.1007/978-981-13-7032-8_7

specially designed devices. Main research works and achievements are introduced in the following three sections, each section-title corresponds to one mechanism that causes wave-forced sediment resuspension in the Yellow River delta.

7.2 Sediment Resuspension by Wave-Induced Oscillatory Seepage Flows

The influence of wave-induced oscillatory seepage flows was examined in this section. As it is quite difficult for current technology to specially observe wave-induced oscillatory seepage flows, a new experimental device was newly designed and employed to study this issue. We found the wave-induced oscillatory seepage flows have important effects on enhancing sediment resuspension over the silty substrate in wave-dominated hydrodynamic environment.

7.2.1 Methodology

(1) ***Design of the Benthic Chamber***

Sediments can be resuspended by two types of mechanisms, namely, (a) horizontal shearing from oscillating (waves) or unidirectional (currents) seawater flows and (b) vertical pumping from oscillating (OPP) or unidirectional (RPP) pore water flows (seepage). The first mechanism, which is called the shearing effect, has been widely studied around the world; thus, research techniques have been more fully developed, such as field observations with instrumented tripods (e.g., Guillén et al. 2002), moorings (e.g., Schaaff et al. 2006) or in situ flumes (e.g., Maa et al. 1993; Thompson et al. 2011) and considerable experimental works with laboratory flumes (e.g., Widdows et al. 1998). However, studies regarding the second mechanism, the pumping effect, are still mostly conducted in laboratory flumes (e.g., Baldock and Holmes 1999). To the authors' best knowledge, no existing in situ techniques have been reported to specifically study the WPS. However, this topic has received considerable attention in the past but remains of considerable interest. Field observations are of particular interest, which this study seeks to confront. Therefore, we designed a benthic chamber to both enrich new marine geophysical monitoring techniques and progress the scientific fields of cohesive sediment dynamics.

Waves and currents are the primary mechanisms of causing sediment resuspension in estuaries, coastal shelf environments and inner shelf environments (e.g., Wright et al. 1992). To measure the sediment resuspension that is specifically caused by wave-induced oscillatory seepage flows, other causes such as current-induced or wave orbital shear stress should be effectively ruled out, whereas bottom-pressure fluctuations must be free of disturbances because the EPP, which is crucial for inducing seepage flows in substrate, is proportional to the bottom-pressure fluctuations

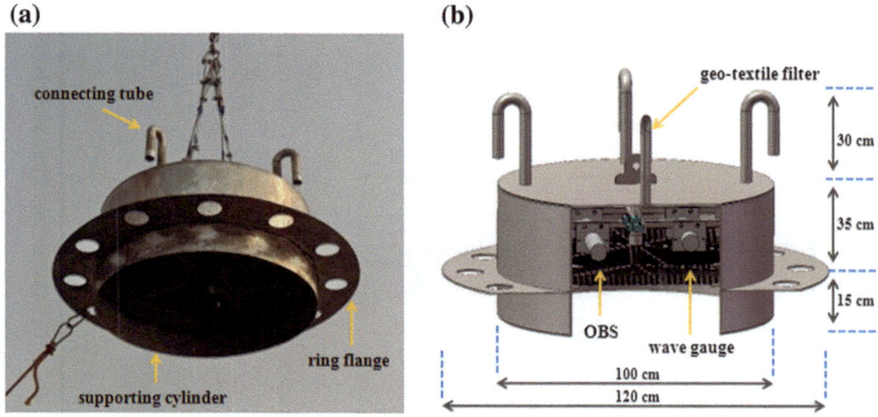

Fig. 7.1 Benthic chamber for the in situ measurement of the WPS at the water-sediment interface: **a** external appearance, **b** internal structure and configuration

acting on the seabed surface. Note that the wave-influenced pore pressure gradient dominates the seepage flows at the water–sediment interface in the YRD, compared to other mechanisms, e.g., topography or groundwater (Santos et al. 2012).

With reference to the design of well-known seepage meters (e.g., Smith et al. 2009), we developed an in situ benthic chamber based on the design requirements and existing theory. A schematic diagram of the chamber is shown in Fig. 7.1. The chamber was expected to achieve the following prerequisites:

(a) Bottom shear stresses from both currents and wave orbital motions could be shielded. As shown in Fig. 7.2, after the chamber was deployed on the seafloor, the supporting cylinder with a sharp edge easily penetrated 15 cm into the seabed because of gravity until the ring flange touched the seabed surface. Consequently, a semi-sealed space was created in the bottom boundary layer to ensure that neither current nor wave shearing affected the seabed below the chamber. The supporting cylinder and ring flange were designed to prevent the horizontal and vertical displacement of the bottom-supported chamber, respectively. The circular holes in the ring flange were designed to lower the difficulty of recovery by reducing the adsorption area between the flange and seabed.
(b) Water pressure fluctuations could be transmitted into the chamber. The water pressure in the chamber was balanced to the external pressure through four bent 4.5-cm diameter tubes. The tubes were welded onto the top of the chamber to provide pathways for the free exchange of seawater, which is why the space in the chamber was described above as “semi-sealed”.
(c) External suspended sediments could be prevented from diffusing into the chamber when water is exchanged through the tubes. The external suspended sediments were always vertically mixed by the convection effect of waves. Hence, the length of the tubes was designed to be 30 cm to effectively attenuate the

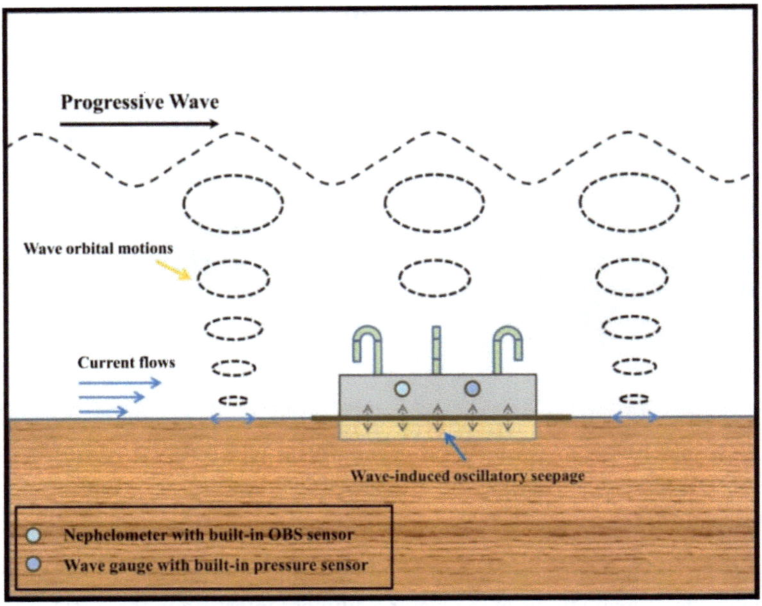

Fig. 7.2 Schematic diagram of the operating principles of the benthic chamber: the seabed below the chamber was shielded from current and wave orbital shearing, while external water pressure fluctuations were successfully transmitted into the chamber through bent tubes, which were crucial to drive the WPS in the chamber

convective kinetic energy. In addition, geotextile filters were added to the outer end of the tubes to further prevent external solids from entering the chamber.

Theoretically, if all the experimental conditions are satisfied, sediment resuspension in the chamber could only be attributed to wave-induced seepage flows. Moreover, if there were no evidence of capturing residual seabed liquefaction events, the mechanism of internal sediment resuspension could be further attributed to the WPS. In this study, sediment resuspension events were inferred from variations in the suspended sediment concentration (SSC) which was measured by the optical backscatter sensors (OBS). A nephelometer with a built-in OBS sensor was mounted in the chamber to record the internal SSC evolutions. In addition, a wave gauge with a built-in pressure sensor was mounted in the chamber to record the water pressures that were transmitted into the chamber. When the chamber was employed in the field, we deployed another platform (e.g., a tripod or quadripod) nearby that was equipped with at least one other OBS and pressure sensor. In this case, the efficiency of the pressure transmission could be evaluated through a comparative analysis of the internal and external pressure records from the wave gauges, while the contribution of the WPS to the total resuspension could also be estimated through a comparative analysis of the internal and external SSC records from the nephelometers.

(2) ***Sea Trial of the Benthic Chamber***

The first sea trial of the benthic chamber was conducted in an abandoned lobe (deposited during 1964–1976) of the subaqueous YRD, China (Fig. 7.3). The surface sediments mainly consist of silt (62.3%) and clay (37.4%). The mean particle size is 0.043 mm. The wave height here is normally smaller than 1.5 m, while prevailing northwesterly winds during the spring or winter could occasionally cause storm waves of 4–7 m in height. The water depth ranged from 7.96 to 10.26 m during the observation period. As the YRD is fairly flat, with an overall slope of 0.3–0.4° (Prior et al. 1989), much of the delta front is subjected to wave action. The study area is currently far from the present mouth of the Yellow River, i.e., has little river sediment supply. Therefore, the variations of SSC in this area are dominated by local resuspension or diffusion afterward. Historically, the study area has suffered from geo-hazards including serious seabed scour and occasional seabed liquefaction. Considering the safety of nearby oilfield platforms, the observation site was selected in this area to study the sediment resuspension due to the wave-induced pore pressure response in the seabed. Additionally, we expected to capture the events of the WPS or residual seabed liquefaction caused by large waves through this fieldwork, so the observations were conducted during the winter.

During the winter from 2016 to 2017, our group at the Ocean University of China successfully deployed and retrieved the benthic chamber and an instrumented quadripod in the northern part of the modern YRD (Fig. 7.4). The OBS (built-in nephelometers, RBR, Canada) and pressure (built-in wave gauges, RBR, Canada) sensors were mounted on the quadripod at 1 and 1.5 m above the bottom (MAB) and in the chamber at 0.4 and 0.3 MAB, respectively. The depth differential between the two pressure sensors was considered during the comparative analysis of the pressure data. Measurements of the seawater turbidity and wave parameters were synchronously recorded inside and outside the chamber at a rate of once per 3 s and a burst sampling rate of 5 min/h (6 Hz), respectively. Turbidity records from the OBS sensors were transformed into the SSC by calibration equations ($C_1 = 0.00635T_1 + 0.24258$ and $C_2 = 0.00472T_2 + 0.44569$, where C is the concentration in g/l and T is the water turbidity in NTU), which were derived from laboratory calibration tests using local silts. In addition, an instrument to profile the suspension turbidity in the benthic boundary layer (Argus Surface Meter IV, Argus, Germany), with an array of built-in OBS sensors at a separation distance of 1 cm, was mounted on the quadripod to record the turbidity of the water column from the bottom to 1.8 MAB at a burst sampling rate of once per 30 min. As the single-point nephelometers inside and outside the chamber were not located at the same elevation along the water column, data from the OBS sensor in the ASM IV at the same elevation with the inner OBS were employed for comparative analysis. The OBS sensors in the ASM IV were also calibrated using local silts. Moreover, a scientific diver was employed to make daily checks for the condition of the two platforms during the observation period, if only the weather and sea conditions allowed.

Fig. 7.3 Study area and observation site: the modern Yellow River Delta (YRD), Bohai Bay, China and the observation site in the northern YRD (38°10′15.40″N, 118°54′55.37″E)

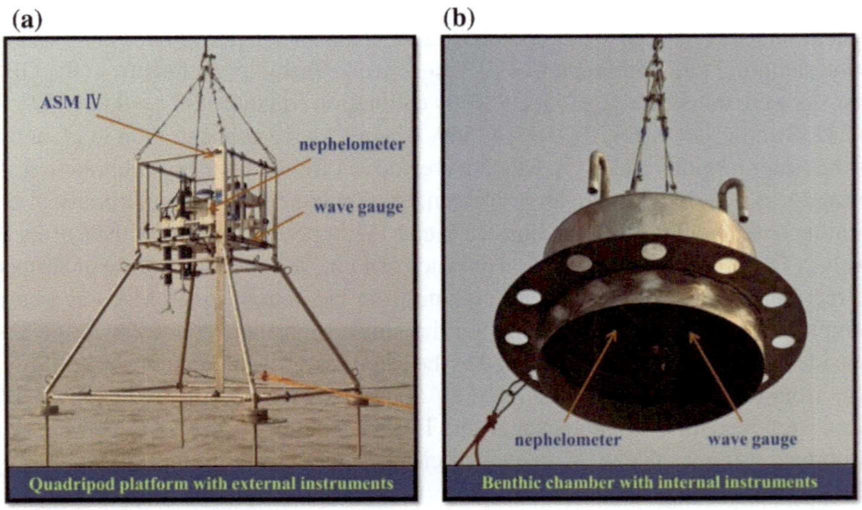

Fig. 7.4 Instrumentation and deployment of the observational system: **a** the instrumented quadripod platform, **b** the benthic chamber

7.2.2 Results

From the sea trial, we successfully obtained a complete dataset of the time series of the significant wave height (H_s), depth (linearly to pressure) and SSC both inside and outside the chamber. Before further analysis, the experimental prerequisites were examined using the dataset.

(1) ***Efficiency of Transmitting Wave-Pressure Fluctuations***

As it was shown in Fig. 7.5, a strong linear relationship ($R^2 = 0.98568$) was identified between the internal and external H_s through regression analysis. In addition, the temporal evolutions of the H_s inside and outside the chamber also showed quite good consistency (Fig. 7.6a), which proved that the chamber effectively transmitted the wave parameters. The consistency of the water pressure records inside and outside the chamber further demonstrated the high efficiency of water pressure transmission (Fig. 7.6b).

The regularity of the wave events was also found in this observation, wave events usually occurred every 1–4 days and each lasted approximately 1–2 days. Variations of the external SSC strongly corresponded to the occurrence of wave events, which indicated that the sediment resuspension was truly caused by waves at this location.

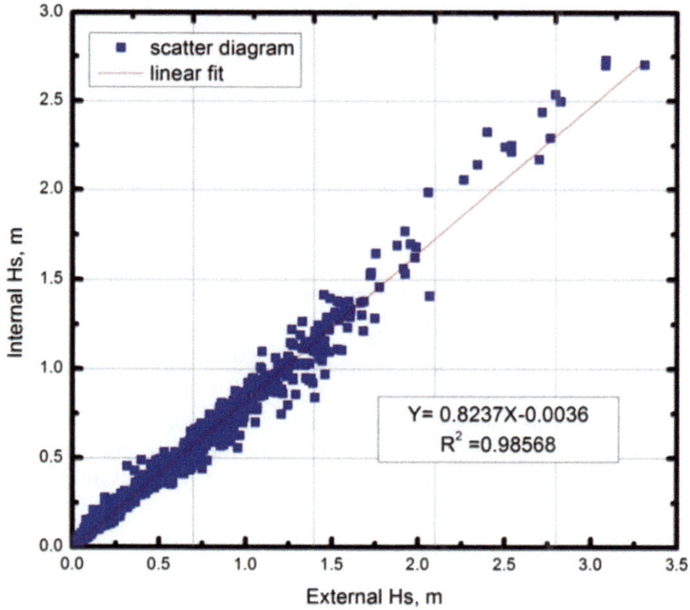

Fig. 7.5 Linear fit of the measured significant wave heights (H_s) inside and outside the benthic chamber

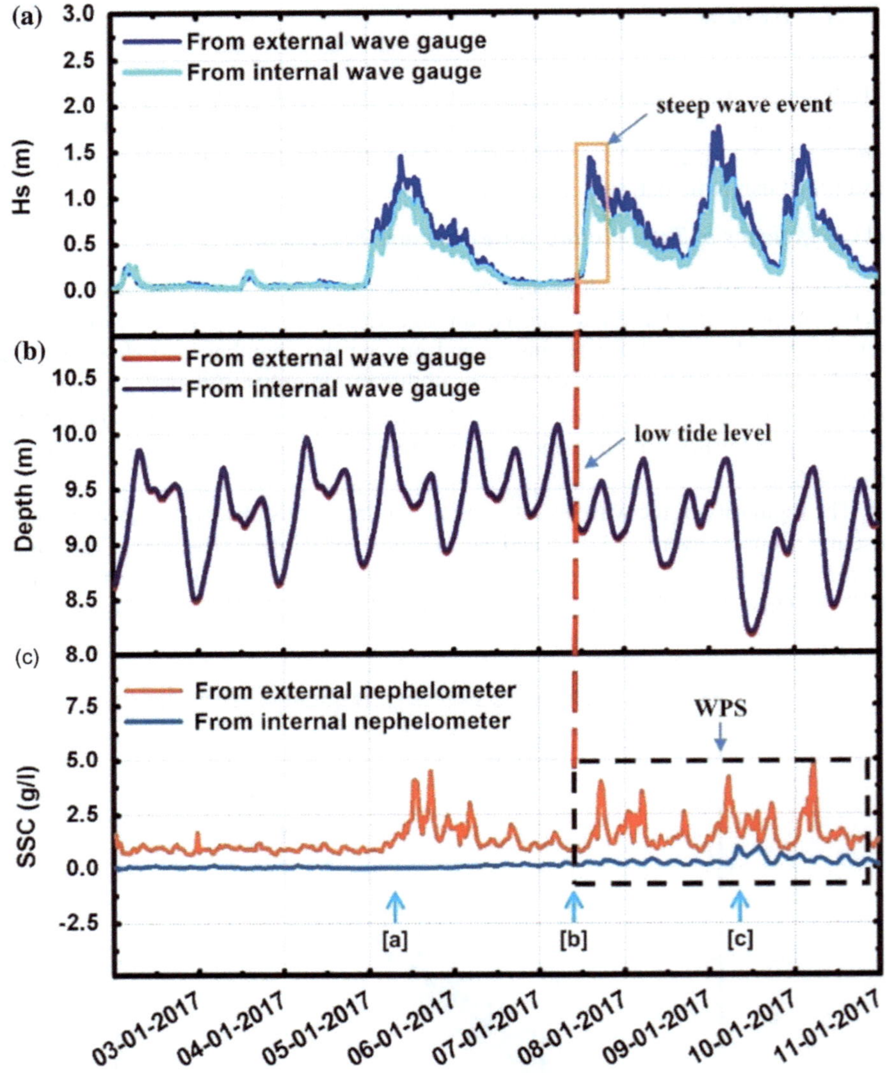

Fig. 7.6 Time series of **a** the significant wave height (H_S), **b** the water depth (linearly to pressure), and **c** the suspended sediment concentration (SSC) inside (i.e., from internal instruments) and outside (i.e., from external instruments) the chamber throughout the observation period

Table 7.1 Diver's daily reports on the condition of the chamber

Date	Time	Altitude	Reliability
Jan. 02	13:09–13:36	No change	√
Jan. 03	*	*	No diving
Jan. 04	12:40–13:12	No change	√
Jan. 05	11:12–11:30	Sink for 10 cm	Misjudgment
Jan. 06	12:53–13:40	No change	√
Jan. 07	11:48–12:23	No change	√
Jan. 08	13:36–14:07	No change	√

Note The apparent sinking on January 5 was determined to be a misjudgment that resulted from siltation of sediments. Because no signal anomaly was reflected in the depth differentials in Fig. 7.6b. The synchronous sinking of the two platforms without any reflection in the depth differentials was improbable

(2) ***Efficiency of Shielding the External SSC***

According to Fig. 7.6c, we found a consistent offset exist between the internal and external SSC curves, the SSC level inside the chamber was low to zero continuously in the first 5 days. This well demonstrated a successful effect on shielding the external suspended solids. In addition, checking the instrument before and after the deployment indicated that the geotextile filters were undamaged throughout the observation period. The daily descriptions from the diver also indicated no significant seabed scour ever occurred around the device in this observation (Table 7.1), which would have resulted in sinking or tilting of the bottom-mounted benthic chamber.

In this case, the newly developed device was found to transmit pressure successfully, besides, no intrusions of external SSC into the chamber were observed. Therefore, all the experimental prerequisites were truly satisfied in this observation. According to the observational data and existing literature, in this part, we propose the physical mechanism of the WPS, which could explain all the results in this study. Then, we evaluate the quantitative contribution of the WPS to the total resuspension of the Yellow River silts under various wave conditions. Finally, an improved conceptual model for the resuspension of silty sediments is proposed, in which we fully consider the role of wave-induced seepage flows in affecting sediment dynamics.

(3) ***Deficiencies and Future Improvements***

As a newly developed device, the deficiencies of and possible improvements for the in situ benthic chamber is discussed, in anticipation of future studies. At the moment, the chamber shows satisfying work efficiency for a short week-long observation period without extreme large waves. As the WPS is assumed to be more intense under extreme storm waves, longer term observations are planned for the near future. When the geometric shape of the chamber is modified from its present cylindrical shape to a domed shape, because streamlined designs are helpful to avoid erosion near the device under energetic storm waves also over more cycles of tidal currents.

In addition, changing the bend tube design into a chamber top that consists of a rubber membrane is also under consideration, to improve the pressure transmission while also avoid the influence of injected currents. We suggest that this chamber can be used in other wave-dominated regions with sandy or silty substrates, because the clayey seabed has a low possibility of liquefaction. Additionally, the benthic chamber is not necessary in tidally dominated environments because of the weak effect of wave pumping, where the popular in situ annular flow flumes (e.g., Maa et al. 1993) are sufficient for most erosion issues.

7.2.3 Physical Mechanism for Sediment Resuspension by Transient Seepage Flows

Oscillatory seepage has been described by many terms: subtidal pumping, seepage across the sediment/fluid interface (Baldock and Holmes 1999), advective pore water exchange (Precht and Huettel 2004), infiltration (Willets and Drossos 1975), exchange process at the sediment–water interface (Friedrichs et al. 2006), filtration flow or generally referred to as seepage flows (Myrhaug et al. 2014). In fact, these descriptions all have a close relationship with wave pumping, which is driven by pressure gradients that are generated by the different hydrostatic pressure underneath wave crests and troughs. Pore water and nutrient elements in it can be frequently released or exchanged by this process. When sediments at the interface or in shallow layers are mobilized and therefore resuspended by the oscillatory seepage flows, WPS occurs (Wolanski and Spagnol 2003).

In this section, we attempt to detailed explain the physical mechanism of the WPS using the current knowledge of wave-induced momentary seabed liquefaction from the fields of marine geotechnical engineering. As shown in Fig. 7.7, the soil skeleton is compressed and rebounds with the cyclic passing of the wave crests and troughs when bottom-pressure fluctuations are applied on a porous but low-permeability silty seabed. This phenomenon occurs because an excess in pressure with respect to the mean water level (MWL) is generated in the sediments from the passage of a wave crest, which creates downward seepage forces, and a reduction in pressure with respect to the mean water level is induced by the passage of a wave trough, which then causes upward seepage forces. In natural submarine environments, sediments in shallow layers are frequently exposed to these seepage forces, which have a cyclically alternating magnitude and direction with the same frequency as surface-wave oscillations.

For each wave oscillation cycle, the hydrodynamic forces result in two vertical components, namely, vertical forces that act upwards to pull the sediment out of the seabed (mobilizing the sediments) and vertical forces that act downwards to press the sediment down (stabilizing the sediments). Before this seepage force exceeds a certain threshold, only water is exchanged at the interface (i.e., wave pumping), and the sediments only have a tendency for motion. However, momentary liquefaction

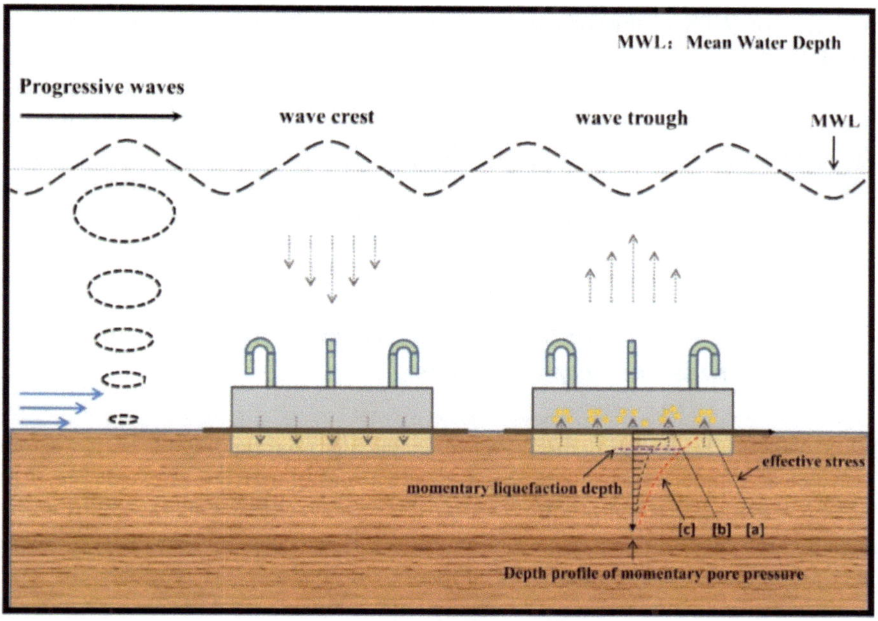

Fig. 7.7 Schematic diagram of the physical mechanism of the WPS. The dashed black line is the depth distribution of the wave-induced momentary pore pressure (MPP), arrows in the chamber refers to the resultant oscillatory seepage flows and the yellow points in the chamber refers to the resuspended sediments by WPS. Note that the red line refers to a significantly increased MPP profile that was caused by the arrival of large steep waves, hence the purple line, pointing to the intersection of effective stress and increased MPP, refers to the depth of momentary sediment liquefaction

occurs as this force abruptly increases or is even transiently sufficient to offset the effective overburden (including the gravitational force and ambient restraint forces), and some of the sediments (especially fine sediments) are transported out of the shallow layer and become resuspended material (i.e., the WPS). We argue that the initiation of WPS is a direct consequence of wave-induced momentary seabed liquefaction in the shallow seabed. Afterward, the permeability of seabed increases significantly and sediments become more susceptible to be pressed into or pulled out of the interface by the wave-induced oscillatory seepage flows.

The observational results could be adequately interpreted using the above physical mechanism. Three sub-periods (a, b, and c in Fig. 7.6) corresponded to the three seabed states in Fig. 7.7. No WPS occurred during sub-period [a] because the MPP did not exceed the effective stress of the overlying seabed. However, the effective stress of a seabed is not constant and dynamically changes because of wave-induced excess pore pressures (the effective stress equals the total stress minus the excess pore pressure). Therefore, continuous wave action gradually increased the excess pore pressure and thus attenuated the effective stress until a threshold stage [b] was reached, after which the momentary pore pressure exceeded the overlying effective stress and the WPS was detected during sub-period [c]. It is worth noting that the initiation of WPS in this observation was triggered by a large steep wave event corresponding to the low tide level (Fig. 7.6c). Large waves generated or arrived abruptly will increase pore pressure sharply and this influence would be especially important when the water depth is shallow, because the excess pore pressure over the hydrostatic pressure is enlarged. When the significantly increased MPP profile due to the arrival of large steep waves intersects with the attenuated effective stress of seabed, momentary sediment liquefaction occurs and thus WPS is triggered.

7.2.4 Quantitative Contribution of Sediment Resuspension by Transient Seepage Flows

After successfully detecting the WPS events, the contribution of the WPS to the total resuspension under various wave conditions was preliminarily estimated (Fig. 7.8). The contribution was simply provided as $C = S_w/S_t$, where C is the contribution, S_w is the internal SSC that is caused by the WPS, and S_t= 5 g/l and 1.9 g/l are the maximum and mean external SSC in this observation, respectively. The most conservative estimate was calculated when the maximum external SSC was employed, and the most exaggerated estimate was calculated when the mean external SSC was used.

According to Fig. 7.8, it is clear that no WPS events were detected during the early stage of the first wave event from January 5 to 6, indicating the corresponding waves never caused the underlying seabed to reach the threshold state [b]. Nevertheless, this event contributed to the later occurrence of the WPS, because we found that small WPS events had begun in the late stage of this wave event, although remained insignificant. However, when a wave event with similar magnitude arrived one day later on January 7, more significant WPS events were triggered. We believe that one important reason for the triggering effect is that the second wave event grew much more rapidly than the previous one, which is consistent with the mechanism of the wave-induced transient seabed liquefaction (i.e., excess pore pressure increases abruptly and even transiently sufficient to offset the effective stress of overburden). The second reason is the partially mobilized seabed by the first wave event had not recovered to its original state by the arrival of the latter wave event. Therefore, WPS events with larger magnitude were detected during January 7–11. In this case, the observational results suggested that wave height was not the only controlling factor of triggering liquefaction, large numbers of normal wave cycles also could induce the liquefaction of seabed (Green and Coco 2014).

In this observation, the WPS comprised less than 10% of the total resuspension under a single normal wave event (H_s ~ 1–1.5 m). However, 20–60% of the suspended sediments originated from the WPS in the presence of continuous normal waves. Such a considerable contribution demonstrates the significance of the WPS in contributing to sediment resuspension and thus invalidates a commonly accepted opinion that silty sediments are purely eroded by the shearing effect from either wave orbital or the

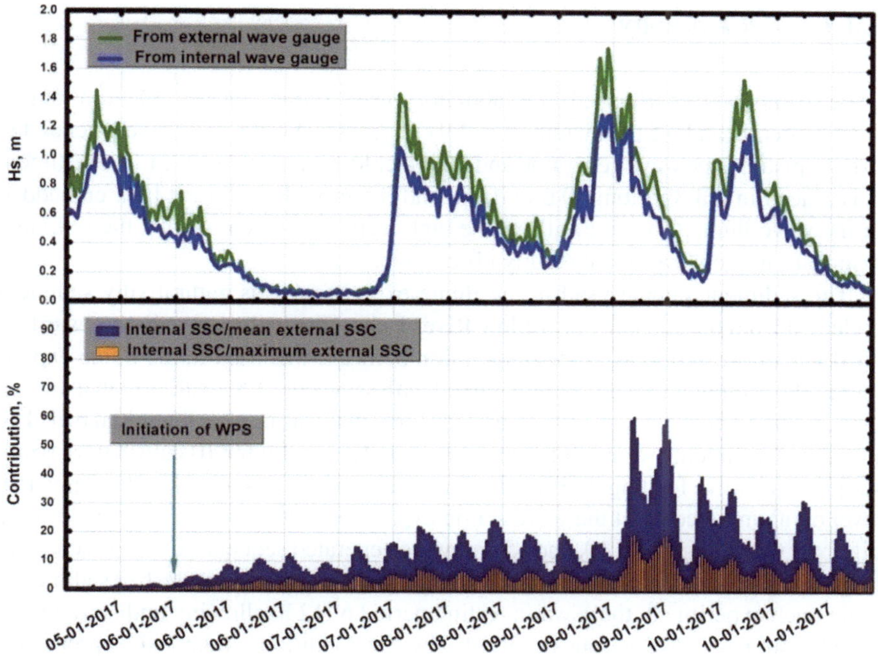

Fig. 7.8 Quantitative contribution of the WPS to the total sediment resuspension under various wave conditions in the subaqueous Yellow River Delta

current velocity, only slightly affected by wave pumping effects because of their low permeability. In fact, the wave pumping of silts is quite important when the seabed is momentarily liquefied, after which considerable sediments can be resuspended by the wave pumping effects under certain conditions in the natural environment.

7.3 Sediment Resuspension by Wave-Induced Residual Seepage Flows

The influence of wave-induced residual seepage flows was specially examined in the following two sections. Wave flume experiments were employed to study this issue. We found that wave-induced residual seepage flows also have important effects on influencing sediment resuspension over the silty substrate in wave-dominated hydrodynamic environment. More complicated, on the one hand, it will cause vertical migration of fine-grained sediments from internal seabed driven by residual seepage flows which is specially discussed in the present section; one the other hand, it will lead to attenuation of surface erodibility which will be specially discussed in the next section.

7.3.1 Methodology

Experiments were performed in a laboratory wave flume that was 14 m in length, 0.7 m in depth, and 0.5 m in width (Fig. 7.9). Progressive water waves were generated using a piston-type wave generator, and the water level was set to 40 cm. The sediment was placed in a 3.8 m long, 0.6 m deep, and 0.5 m wide sediment tank embedded in the wave flume. The sidewalls of the tank were transparent so that the sediment responses to the waves were readily observable.

The sediment sample used in the flume experiment was natural silty sediment collected from the subaqueous Yellow River Delta. The collected sediment was first air-dried and stirred to avoid agglomeration before it was used in the flume experiment. The sediment was thoroughly mixed with seawater (3.5% salinity) in a mixing tank and homogenized to form a saturated slurry that was used as the initial bed. The initial bed sediments were left to consolidate under gravity for 10 days before being subjected to wave loading. Lab-made standard seawater was added to the flume to a level of 40 cm to conduct the wave action experiment. Progressive waves with three different wave heights (5, 10, and 15 cm) were generated sequentially, and each wave height series consisted of two runs of wave generation and tests of the bed sediments. After each experiment, there was a resting period of 12 h, allowing undisturbed settling before new wave actions occurred. The test conditions and physical properties of the bed sediment are summarized in Tables 7.2 and 7.3, respectively.

Pore water pressure measurements were conducted throughout the experiment in the center of the sediment tank at depths of 10, 20, 30, and 40 cm below the bed surface (Fig. 7.9) using four pore pressure transducers (Φ 20 × 60 mm, piezoresistive type, model 86 of Joint Sensor Instruments, Shenzhen, China). At the same section as the pore water pressure measurements, the water-surface elevation was measured using a wave gauge to determine the corresponding water pressure at each depth. The pressure and wave signals were acquired simultaneously using DASYLab data acquisition software.

Four numbered PVC tubes (7.5 cm in diameter and 1.0 m in length) were used for undisturbed sediment sampling to evaluate the particle size composition, microstruc-

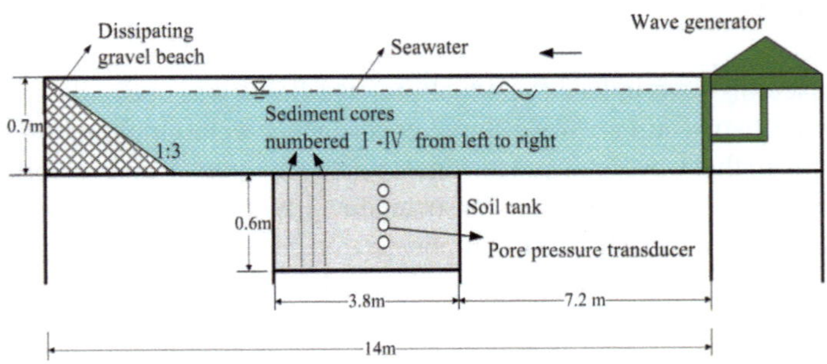

Fig. 7.9 Setup for the laboratory wave flume

Table 7.2 Summary of the test conditions used in the flume experiments

Run no.	Wave height H (cm)	Wave period T (s)	Water depth h (cm)	Duration of waves (h)	Post-experiment rest time before the next run (h)	Bed response to waves	Sediment layering	Cores
0	–	–	0	–	240	–	None	I
1	5	3.2	40	10	12	Unliquefied	SFL	–
2	5	3.2	40	9	12	Unliquefied	SFL, SAB	II
3	10	1.7	40	8	12	Liquefied	SFL, SAB, ILL	–
4	10	1.0	40	8	12	Liquefied	SFL, SAB, ILL	III
5	15	1.0	40	6	12	Liquefied	SFL, SAB, ILL	–
6	15	1.0	40	6	–	Liquefied	SFL, SAB, ILL	IV

SFL superficial fluid mud layer; *SAB* subsurface alternating bands; *ILL* internal liquefied layer. – data does not exist

Table 7.3 Physical properties of the sediment sampled before and after wave loading. All values are taken from the average of data obtained in the vertical (i.e., parallel to the bed depth) direction

		Value			
			After wave loading		
Parameter	Symbol	Before wave loading (*Run* 0)	Wave height, $H = 5$ cm (*Run* 2)	Wave height, $H = 10$ cm (*Run* 4)	Wave height, $H = 15$ cm (*Run* 6)
Mean grain size	d_{50}	0.036 mm	0.036 mm	0.037 mm	0.040 mm
Unit weight	γ	18.2 kN/m^2	19.8 kN/m^2	20.2 kN/m^2	20.5 kN/m^2
Water content	w	32.6%	23.4%	20.7%	18.9%
Wet density of sediment	ρ_{wb}	1.86 g/cm^3	2.02 g/cm^3	2.06 g/cm^3	2.09 g/cm^3
Dry density of sediment	ρ_{db}	1.41 g/cm^3	1.64 g/cm^3	1.71 g/cm^3	1.76 g/cm^3
Specific gravity	G_s	2.70	2.70	2.70	2.70
Void ratio	e	0.92	0.65	0.58	0.53
Porosity	n	0.48	0.39	0.37	0.35
Critical hydraulic gradient	i_c	0.89	1.03	1.08	1.11

ture, and physical and mechanical properties (Fig. 7.9). At the end of each wave height series, a PVC tube was gently pushed into the sediment until it reached the bottom of the tank, and the uppermost part of the tube, which was approximately 5 cm above the bed surface, was sealed with a tube cap to ensure that the sediment core was protected from high-frequency pressure fluctuations due to the waves. To decrease the disturbance caused by the sampling, the PVC tubes situated at different stages were kept inside the bed and were drawn out at the end of the flume experiment. The undisturbed sediment cores were first cut in half longitudinally. Half of the sediment cores were sliced longitudinally at 2-cm intervals for bulk density, water content, compactness, and particle size composition tests, and two sets of parallel tests were made for each sediment sample. The other half of the sediment cores were sealed with adhesive tape and plastic wrap to ensure that there were no leaks or deformations inside the sample before the microstructure tests were performed.

Following the standard sediment-test method (GBT 50123-1999), sediment wet density, ρ_{wb}, was measured using a steel ring sampler (Φ 6 cm, height 2 cm), which was pushed into the sediment samples. The sediment core was weighed to determine ρ_{wb}. It was then dried in an oven for at least 8 h at 105 °C and reweighed to determine the water content, w. Sediment dry density, ρ_{db} (the ratio of total dry mass to total volume), void ratio, e (the ratio of the pore volume to the volume of the solids), and porosity, n (the ratio of the pore volume to total volume) were determined as follows:

$$\rho_{db} = \frac{\rho_{wb}}{1 + w} \tag{7.1}$$

$$e = \frac{\rho_w}{\rho_{db}} \cdot G_s - 1 \tag{7.2}$$

$$n = \frac{e}{1 + e} \tag{7.3}$$

where ρ_w is the density of water (approximately 1.0×10^{-3} kg/m^3) and G_s is the specific gravity of the particles. The particle size composition was determined using conventional sieve analysis for the fraction of the sediment with a grain size greater than 0.063 mm, and the hydro-suspension method was used for the fraction of the sediment with a grain size less than 0.063 mm. Using a pocket penetrometer designed by Shenyang Jianke Instrument Research Institute, sediment compaction was measured in the vertical (i.e., parallel to the bed depth) and horizontal (i.e., parallel to the bed surface) directions by slowly pushing the probe into the sediment to the required depth, 10 mm from the end of the probe. Micro- and meso-structure measurements were obtained using a DR-type X-ray scanner (Koninklijke Philips N.V., Eindhoven, Netherlands) and a stereoscopic microscope (SMZ 1500, Nikon, Japan). The X-ray scanner uses X-rays to measure the mean absorption coefficient of the materials (Yamada et al. 2011). The cross-sectional images are displayed in greyscale with darker and lighter zones corresponding to X-ray attenuation. In addition to the above-mentioned measurements, digital photos were obtained, and graduated scales were created to accurately record the development of the studied features.

7.3.2 Results

Over the course of the experiments, a sequential pattern of stratification development for an originally uniform bed was noted in the unliquefied, liquefied, and post-liquefied phases of the wave-induced bed sediment's response, as shown in Fig. 7.10a–c. Each of these phases is briefly described and illustrated below and are then discussed in subsequent sections. The order of the experiments was established such that each experiment had wave conditions that produced a stronger sediment response than the previous experiment. This allowed the sorted deposit from a previous run to be easily eroded within the first several minutes of the next experiment; therefore, the wave–bed interactions were theoretically unaffected by the initial bed conditions.

As noted through the sidewall of the flume, under a 5-cm-high wave action, sediment ripples formed on the bed surface (Fig. 7.10d). In localized regions approximately 10 cm below the surface, a series of tiny vertical cracks emerged within the internal bed, serving as seepage channels through which a small amount of fine sediment particles moved in an upward direction (Fig. 7.10e). However, no notable instability event occurred until the generated wave height was increased to 10 and 15 cm, when a liquefied response occurred on the bed surface that was characterized by an obvious arc-shaped boundary surface (Fig. 7.10b). The post-liquefied bed presented a complex internal structure (Fig. 7.10f) in contrast to the originally uniform silty bed. The bed sediment's responses to waves evolved throughout the experiments, producing distinct bed sequences, including superficial fluid mud layers, subsurface alternating bands, and internal liquefied layers. Each of these features is described separately, with corresponding analyses of the grain-size composition and texture of the bed sediment.

(1) ***Superficial Fluid Mud Layer***

Not all of the fine-grained matter is suspended in the upper water by regular water waves; instead, most are preserved at the bed surface, therefore forming a superficial fluid mud layer. Figure 7.11 illustrates this result along with the color difference, facilitating the observation between the fluid mud layer (yellow), the upper water with a high suspended sediment concentration and the lower bed (brown).

The superficial mud layer formed by wave action during the experimental processes presents two sedimentary features (i.e., wave ripples and horizontal bedding) according to the status of the lower bed. For a freshly deposited bed, the turbulent motion of water particles due to regular waves (*Run 1*) formed large-scaled sediment-laden vortices, which built up the crests by scooping coarse sand toward them. As the bed sediments consolidated under the continuous action of wave loading, the waveform of the bed surface maintained equilibrium, and the bed ripples were fully developed (Fig. 7.10d). A mud layer was then draped over the wave ripples after the overloading wave had subsided (Fig. 7.11a). When the bed was exposed to progressively higher waves, a liquefied response occurred on the bed surface, and the sediment particles were resuspended from the previously formed ripples into the

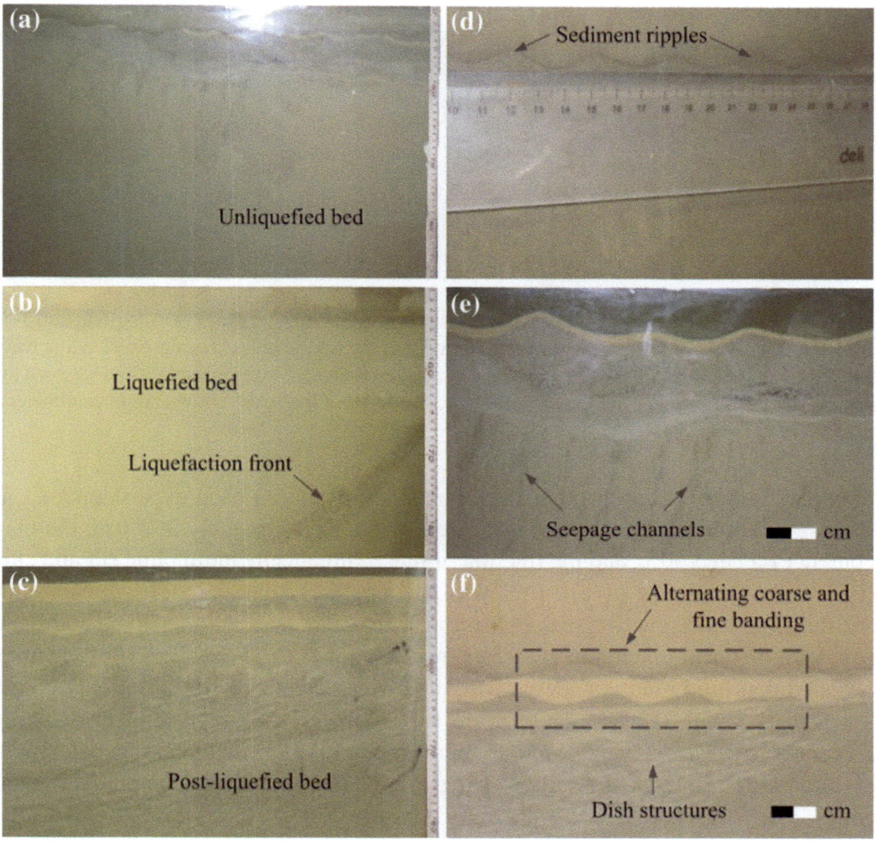

Fig. 7.10 Photos of the characteristics of bed responses to waves during the flume experiment and the resulting features, which are labeled and marked by the arrows. **a** *Run* 2 at 1.5 h. Size sorting of the sediment occurs on the bed surface, but the bed is not liquefied by the waves. **b** *Run* 5 at 1 h. Bed liquefaction is caused by the surface wave loading, with the liquefaction front changing over time. **c** *Run* 4 post-experiment, 12 h after the waves are switched off, nonuniform bed visible. **d** Surface sediment ripples from *Run* 1 at 2 h. **e** Vertical seepage channels from *Run* 2 post-experiment, 10 h after the waves are switched off, each separated by unliquefied bed sediment. **f** Internal structure of the post-liquefied bed from *Run* 4 at 1 h. All of the scales shown are in centimeters

water, resulting in a flat topography on the surface. This allows time and space for the fine-grained sediment inside the bed to accumulate before being sealed by the settling of material held in suspension, therefore leaving the fine-grained material trapped in a horizontal fluid mud layer (Fig. 7.11b).

Consistent with the results of Hooshmand et al. (2015), although our initial sediment bed contained primarily silt-size particles, sand differentiated as a result of the suspension of clays and fine silts that coarsened the sediment bed, resulting in a significant active bed load layer, and supporting the formation of ripples (height 0.6–1.2 cm, wavelength 3.4–6.5 cm, and steepness 0.18–0.24, estimated from mul-

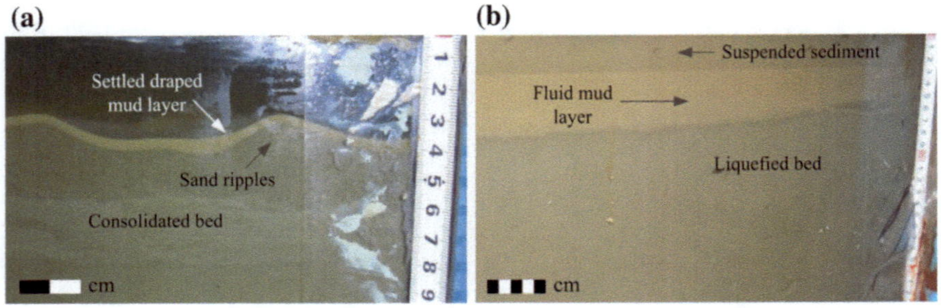

Fig. 7.11 The mud layers formation at the bed surface from 2 runs. **a** *Run* 1 post-experiment, 10 h after waves are switched off. A mud layer is draped over the wave ripples after the overloading wave has subsided. **b** *Run* 6 at 1 h. A clearly defined horizontal fluid mud layer is sandwiched between the upper suspended sediment and the lower liquefied bed. All of the scales shown are in centimeters

tiple photographs taken through the flume sidewall during each experiment). Concerning the fluid mud in the experiment, d_{50} was approximately 3–4 μm, the clay content exceeded 70%, and the fraction of fine sediments including fine silt and clay (i.e., with a grain size of <10 μm) was nearly 90% (Table 7.4). The thickness of the fluid mud, H_{fm}, is proportional to the wave height, H, on the flume sediment bed, with the local maximum H_{fm} reaching 5 cm for $H = 15$ cm. Field observations revealed that fluid mud is widely distributed over seabeds of silty or muddy coasts and estuaries with thicknesses of up to approximately 1–5 m. Sediment dynamic responses to waves may substantially contribute to the formation and development of surficial fluid mud, especially in wave-dominated, fine particle-supplied coastal zones.

(2) ***Subsurface Alternating Bands***

Experimental observations in *Runs* 2 to 6 also revealed that a composite morphology of beds containing one or more groups of alternating coarse and fine sediment banding was distributed below the superficial fluid mud layer. These alternating bands have also been documented in the field in the modern Yellow River Delta and are associated with hydraulic instability and the gravitational gliding of the seafloor.

Figure 7.12 records the complete process of the subsurface alternating bands. Fine particles are separated from coarse particles by wave sorting. Next, coarse particles form bed ripples on the surface, and the suspended fine particles settle under weak wave conditions and cover the bed ripples (Fig. 7.12a). When the action of another set of waves (*Run* 2) is exerted on the sediment bed, erosion occurs on the crest of the ripples due to the waves' intense turbulence, and another sorting event occurs in the reconstructed bed, causing a new layer of coarse particles to form above the fine particle layer (Fig. 7.12b). When the wave action stops, suspended fine particles settle on the coarse particles and form a thick fluid mud layer (Fig. 7.12c). Coarse grains are separated out by larger waves (*Runs* 3 and 4), and these repeated wave actions result in multiple alternating coarse and fine sediment banding (Fig. 7.12d). Fine particle migration pipes are observed when thicker, fine-grained sediment band

Table 7.4 Detailed results of the particle size distributions of typical sediment layers, including the initial bed (IB, sampled after *Run* 0), the superficial fluid mud layer (SFL, sampled after *Run* 4), and the internal liquefied layer (ILL, sampled after *Run* 4)

Samples	% Fine sand (0.25–0.075 mm)	% Coarse silt (0.075–0.05 mm)	% medium silt (0.05–0.01 mm)	% Fine silt (0.01–0.005 mm)	% Clay (<0.005 mm)	d_{30} (mm)	d_{50} (mm)	d_{60} (mm)
IB 1#	21.8	19.5	37.3	3.8	17.6	0.027	0.036	0.051
IB 2#	21.3	18.9	39.4	2.6	17.8	0.024	0.036	0.05
SFL 1#	–	1.8	7.6	15.4	75.2	0.002	0.003	0.004
SFL 2#	–	1.4	15.3	16.7	66.6	0.002	0.004	0.004
ILL 1#	25.7	20.8	37.2	1.5	14.8	0.027	0.046	0.058
ILL 2#	21.5	24.3	38.8	0.7	14.7	0.029	0.046	0.056

Grain sizes were measured both by sieving and a sedimentation process using a hydrometer. Two sets of parallel tests were made for each sediment sample exists in the subsurface bed, from which pressurized pore water is vented, fluidizing sediment along its path (Fig. 7.12e). The fine-grained sediment is transported upward through the coarse-grained layer into the overlying fluid mud layer, deforming the cross-bedding that previously formed during deposition on the bed ripples (Fig. 7.12f). During a long sedimentary history, this type of deformed cross-bedding could be preserved in the formation as a water-escape structure, indicating a sedimentary environment of rapid deposition of high-porosity sediment followed by settling and compaction-driven expulsion of the water. However, there is a clear correlation between its formation and the wave-induced sediment response.

(3) ***Internal Liquefied Layer***

During the experiments using larger waves ($H = 10$ and 15 cm), a liquefied response occurred on the bed surface that was characterized by a prominent arc-shaped boundary surface. This instability event is due to the cyclical movement of the liquefied sediment under wave-induced shear forces. Jeng and Zhao (2015) also conducted a series of numerical simulations to investigate the development of the liquefaction zone with different types of waves. In the case of progressive waves, a small area of sediment will be liquefied initially in the region where the initial wave shear stress is largest. This liquefaction zone then extends laterally and vertically and reaches a maximum value after a number of wave cycles. The boundary surface of the liquefaction zone tends to be an arc shape, with its maximum value being restricted to the experimental setup geometry of the sediment tank. Then, resolidification of the liquefied bed occurs with a gradually diminishing liquefaction zone. The internal liquefied layer here refers to the sediment layer that has gone through the downward extension of liquefaction and upward progression of resolidification in each run. Sediment grading analysis demonstrated that the fine particle content decreased substantially in the internal liquefied layer compared to the initial bed (Table 7.4), indicating that the sediment was coarsened in the internal liquefied layer.

Another interesting phenomenon is that the interface of the liquefied bed provides new channels for fine particle transport. We observed the separation of fine particles from the original grain skeleton that then entered into the pore fluid. Coarsening occurred around the boundary of the liquefied layer where a deep-colored arc boundary appeared (Fig. 7.13a). With the sustained action of wave loadings, pore water was discharged along the failure boundary and the vertical microchannels, resulting in decreased pore water pressure. When the dissipation rate exceeded the accumulation rate of the pore water pressure, the depth of the liquefaction zone was reduced, and the failure boundary gradually migrated upwards (Fig. 7.13b). During this process, some of the particles moving along the boundary remained nearby, showing a discontinuous light-colored banding distribution (Fig. 7.13c). With the gradual upper shift of the interface, the internal liquefied layer was characterized by dish structures on the glass wall of the wave flume (Figs. 7.12f and 7.13d).

(4) ***Particle size Composition***

The particle size characteristics of seabed sediment are of fundamental importance not only to understand the sediment's dynamic behaviors in various marine envi-

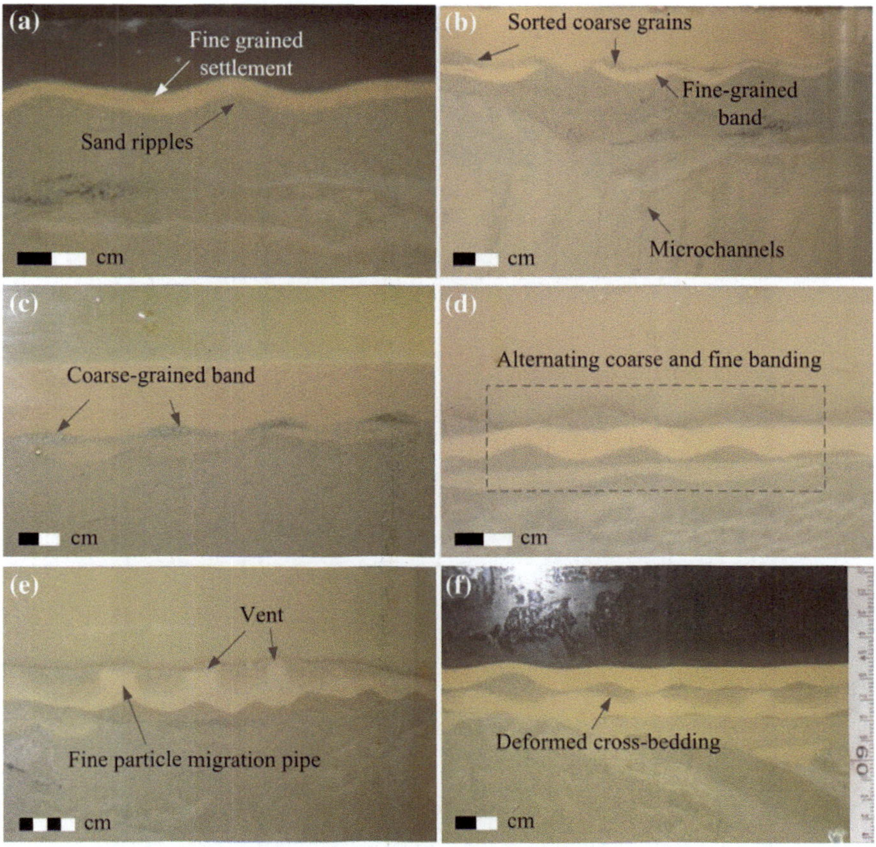

Fig. 7.12 The formation process of alternating coarse and fine bands at the bed subsurface. **a** *Run* 1 post-experiment, 10 h after the waves are switched off. Fine particles are separated from coarse particles by wave sorting. **b** *Run* 2 at 1.5 h. Fine-grained band forms due to the re-imposed wave action. **c** *Run* 2 post-experiment, 2 h after the waves are switched off. Suspended fine particles settle over residual coarse-grained band and form a thick fluid mud. **d** *Run* 4 at 1 h. Repeated wave-induced sediment sorting on the bed surface produces multiple alternating coarse and fine bands. **e** *Run* 4 at 1.5 h. Pressurized pore water pipes can be observed in the subsurface bed, fluidizing sediment along its path. **f** *Run* 4 post-experiment, 12 h after the waves are switched off. Multiple pipes result in deformed cross-bedding

ronments but also to determine the local sedimentary structures on the seabed (Grabowski et al. 2011). The sediment structure of the grain-size composition as a function of time may have direct implications for the significant role played by wave action in controlling the vertical sediment differentiation.

Figure 7.14 displays the vertical profiles of the mean grain size, d_{50}, of the sediment samples collected before (*Run* 0) and after being subjected to wave loadings (*Runs* 2, 4, and 6). The results illustrate that 5-cm-high waves (*Run* 2) primarily affect the grain-size distribution of the surface sediments within a depth of 10 cm, where the

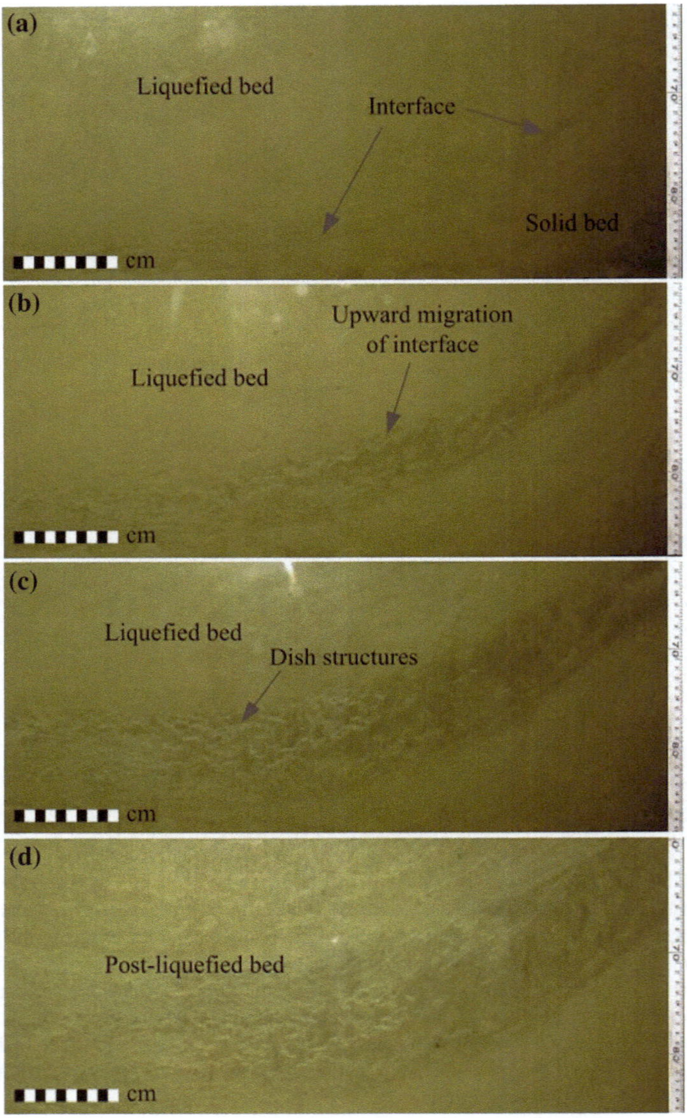

Fig. 7.13 Series of photographs from *Run* 5 demonstrating the development of an internal liquefied layer. The right margin of the photographs has been aligned to the side edge of the tank containing the sediment for clarity. **a** At 0.5 h, the liquefaction zone reaches its maximum depth of 28.5 cm for this run, producing a deep-colored arc interface (the liquefaction front) between the liquefied and solid bed. **b** At 1 h, following the discharge of pore water and resolidification of the liquefied bed, the interface migrates gradually in the upward direction. **c** At 2.5 h, dish structures emerge due to fine sediment amalgamation along the migrating interface. **d** At 6 h, the interface reaches the bed surface and the layer remains stable for the duration of this run. All of the scales shown are in centimeters

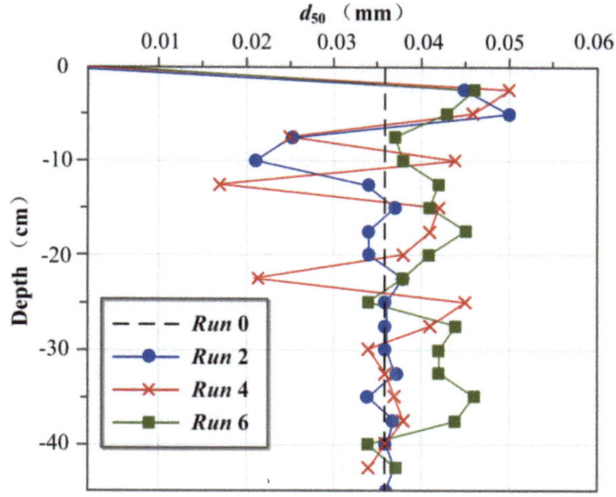

Fig. 7.14 Vertical profiles of the mean grain size d_{50} in sediment cores sampled after *Runs* 0, 2, 4, and 6

minimum and maximum values of d_{50} are 0.021 mm and 0.050 mm, respectively. In the lower part of the flume bed, d_{50} stabilizes at a value of 0.036 mm, which does not represent a substantial change from its previous value (*Run* 0). The depth affected by 10-cm-high waves (*Run* 4) is significantly larger than that of 5-cm-high waves. The d_{50} of the sediment layer above a depth of 30 cm clearly increases toward the surface, only decreasing at individual depths, such as 12.5 and 22.5 cm. This indicates that under sustained wave action, the bed is coarsened, and fine-grained interlayers develop within it. Following 15-cm-high waves (*Run* 6), the depth affected by the coarsening further increases to 40 cm, and the bed appears more uniform, with a d_{50} ranging from 0.040 to 0.045 mm, with very few exceptions.

Similarly, Fig. 7.15 displays the vertical profiles of the clay content (i.e., grain size <5 μm) in the sediment samples during the flume experiments. Under the action of 5-cm-high waves (*Run* 2), the sediment within 10 cm of the bed surface (except for the uppermost part) has a significantly lower clay content than the original bed (*Run* 0), illustrating that surface fine grains are separated from the bed by the waves, as reported in Jia et al. (2014). The sediment within 30 cm of the bed surface shows a relatively nonuniform clay content under the action of 10-cm-high waves (*Run* 4); this content is higher at some depths and lower at others. These observations indicate that vertical migration of clay components occur in a similar manner as pore water in permeable sediments, as described in Santos et al. (2012). The greatest changes in the clay fraction of the bed sediment occur after the action of 15-cm-high waves (*Run* 6). Divided by the 20-cm-depth boundary, the upper sediment has a larger clay fraction, with a maximum value of 28% at approximately a 10 cm depth. In contrast, the clay fraction of the lower sediment is obviously reduced compared to the initial value. The ratio of the upper bed's clay percentage to that of the lower bed approaches 2:1, further demonstrating that fine-grain particles migrate from the bottom to the top under cyclic wave loading.

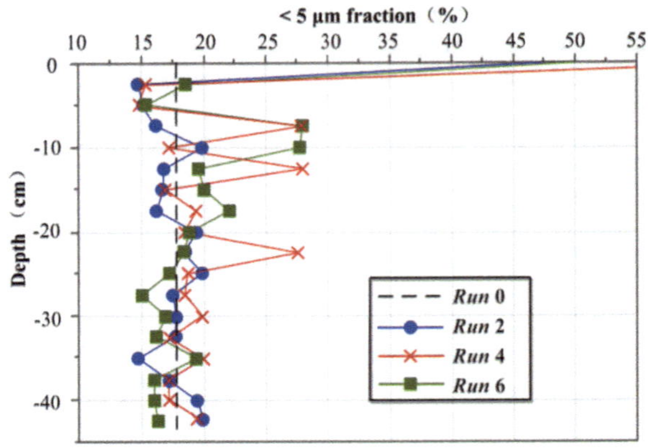

Fig. 7.15 Vertical profiles of the clay (<5 μm) fraction of sediment cores sampled after *Runs* 0, 2, 4 and 6

(5) ***Micro- and Meso-structure***

Photographs, X-radiographs, and physical properties, including the horizontal and vertical compactness (R_{hor} and R_{ver}) and the water content (w), obtained from bed sediments at different stages of the laboratory flume experiment are shown in Fig. 7.16. Many water-filled voids are seen in the photograph of the sediment sample under the action of 5-cm-high waves (Fig. 7.16a); however, these voids almost disappear under the action of larger waves, suggesting a consolidation process during the continued wave loading that results in the expulsion of water from pores between the sediment particles and the gradual increase in sediment density. This pattern is clearly illustrated in sediment compactness and water content profiles that compare the test results under different wave heights (Fig. 7.16). The final bed sediment was significantly compacted with increased compactness (mean values of 5.8 N in the horizontal and 6.3 N in the vertical directions) and reduced water content (mean value of 18.9%) after six wave action runs of repeated liquefaction and resolidification.

X-ray photographs in which the lighter portion indicates a higher density can be used to acquire the high-resolution distribution of microstructures and beddings in marine sediments. Fine textures characterized by alternating dark- and light-colored streaks were identified in the X-ray photographs (Fig. 7.16b, c). This pattern is likely caused by the migration of fine-grain particles under the driving force of wave-induced pore water seepage, which is inferred from the above-mentioned temporal and spatial variations in the sediment particle size composition.

To better understand the wave-induced sediment response in the microstructure characteristics during the experimental process, stereoscopic microscope photos were obtained at 5-cm and 35-cm depths in each sediment sample. The micrograph taken at a 5-cm depth under the action of 5-cm-high waves reveals that surface sediment particles are relatively large, with a very small amount of fine particles filling in the pores (Fig. 7.17a), whereas bottom-sediment samples are less sorted and have smaller pores between the sediment particles (Fig. 7.17b). This finding may be related to the erosion of surface fine particles from the sediment skeleton into the seepage due to the amplification of excess pore pressure induced by sustained wave forces,

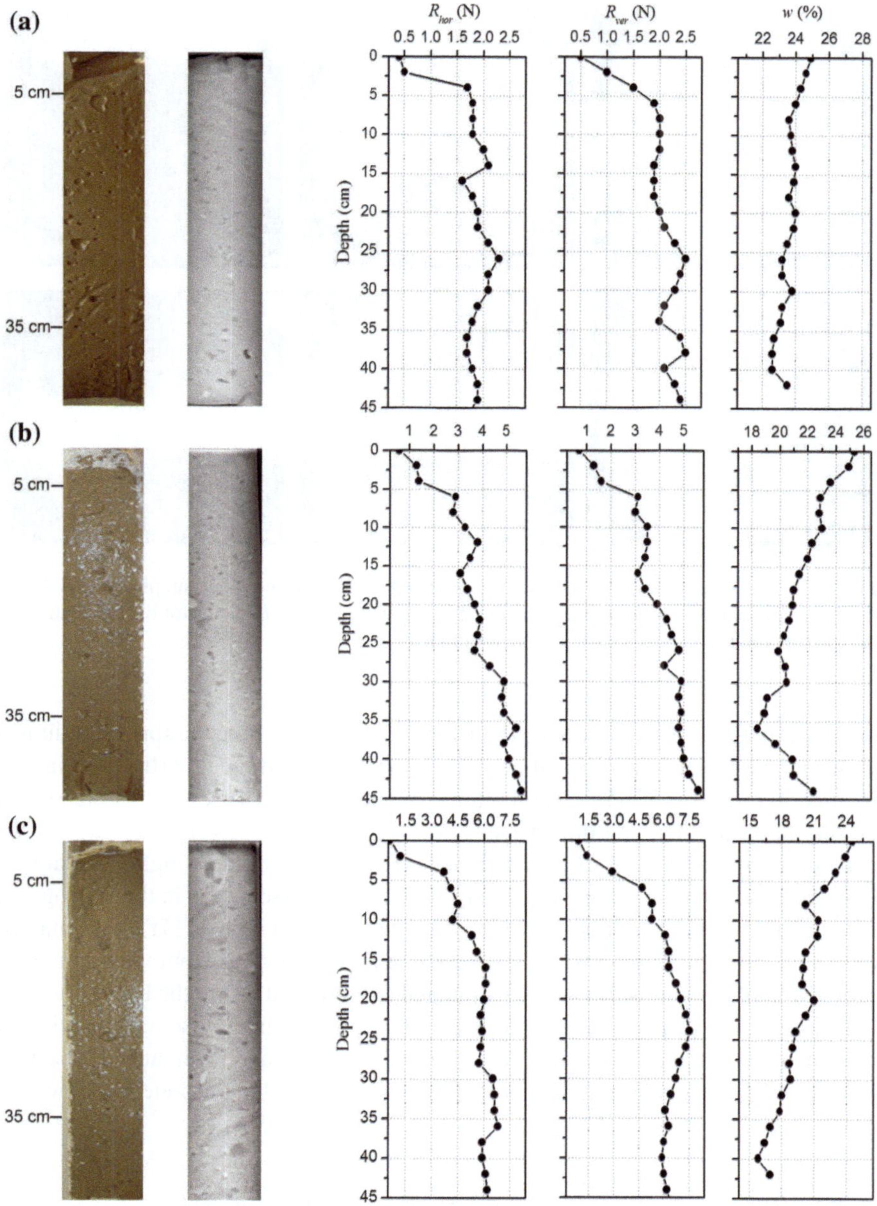

Fig. 7.16 Photographs (left), X-ray radiographs (center), and physical property data (right), including the horizontal and vertical compactness (R_{hor} and R_{ver}) and the water content (w) of the sediment cores sampled after **a** *Run* 2 ($H = 5$ cm), **b** *Run* 4 ($H = 10$ cm), and **c** *Run* 6 ($H = 15$ cm)

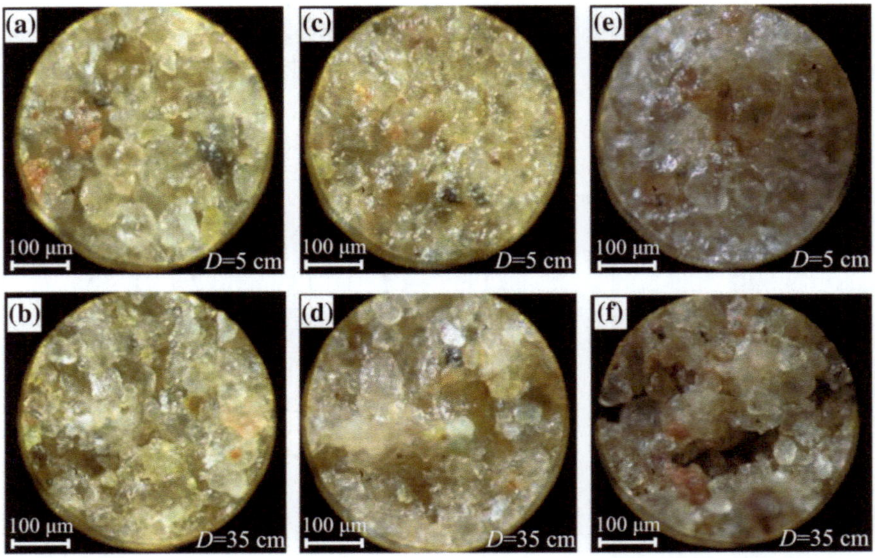

Fig. 7.17 Stereoscopic microscope photos at selected sections of sediment samples after **a–b** *Run* 2 ($H = 5$ cm), **c–d** *Run* 4 ($H = 10$ cm), and **e–f** *Run* 6 ($H = 15$ cm). *D* stands for the depth at which the micrograph was taken in each sediment sample

as shown in Fig. 7.10e. Following the 10-cm-high wave action, the surface sediment is further compacted and it is difficult to observe clear sediment particle outlines in the micrograph (Fig. 7.17c). The bottom-sediment particles are also arranged very closely to each other, and the coarser grains account for a larger percentage of the content compared to the surface sediment (Fig. 7.17d). Interestingly, a relatively large pore is formed between the widely distributed coarse grains in the micrograph taken at a 35-cm depth after the 15-cm-high wave action (Fig. 7.17f). This may be due to the removal of pore water and very fine grains upwards from within the bed, provided that the pore water has accumulated (to some extent) in the bed sediments; therefore, fluid escape structures can form. Pore formation has also been previously documented in studies of seepage (e.g., Mörz et al. 2007) and form during processes of increasing and decreasing hydraulic gradients without the introduction of waves.

7.3.3 Physical Mechanism for Sediment Resuspension by Residual Seepage Flows

According to the wave flume experimental results presented in this section, along with the results presented by Mörz et al. (2007), who provided detailed descriptions of conduit and pipe formation under different hydraulic gradients in saturated granular

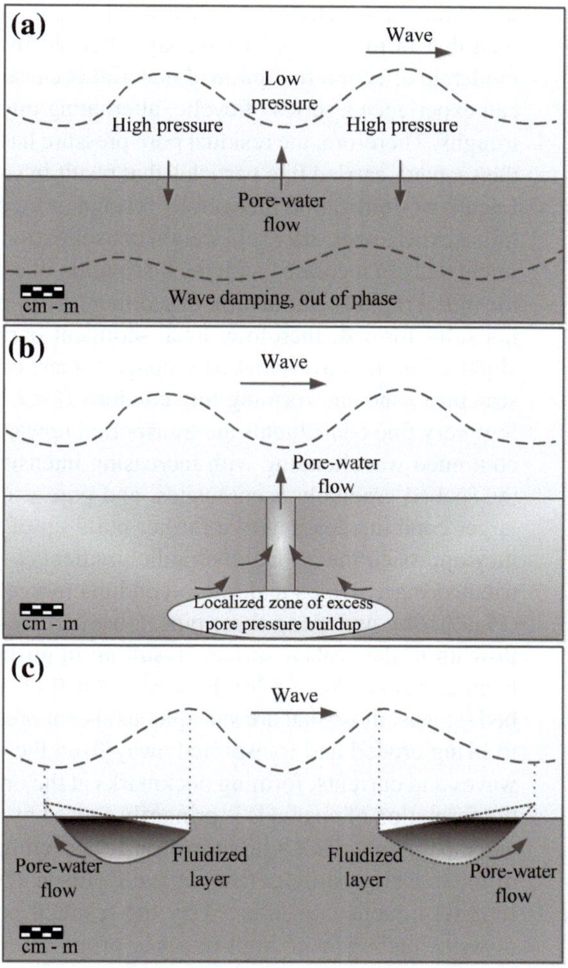

Fig. 7.18 Schematic representation of wave-induced pore water flow regimes, including seepage induced by the transient pore pressure response (**a**), localized fluidization induced by residual pore pressure response (**b**) and internal liquefaction induced by residual pore pressure response (**c**)

media, the driving mechanism of wave-induced pore water seepage within the seabed can be classified into three regimes that pertain to the natural seabed (Fig. 7.18).

(1) Seepage induced by the transient pore pressure response. Waves propagating across a permeable seabed lose energy to percolation through the bed, where damping reduces their magnitude with increasing sediment depth. Pore water may be driven by pressure gradients generated by the various hydrostatic pressures underneath wave crests and troughs (Fig. 7.18a, from Santos et al. 2012). The magnitude of wave-induced pore water seepage is directly proportional to the amplitude of the surface wave, the permeability of the sediments, and the depth. Simulation results from laboratory experiments indicate that the pore water seepage rate caused by transient pore pressure is $0.11\ \mathrm{m} \cdot \mathrm{d}^{-1}$ under wave conditions with $T = 1$ s, $H = 5$ cm, and $h = 50$ cm. However, the grain size of

the sediment selected for calculation (1-mm diameter) was considerably larger than that in this study. The seepage induced by transient pore pressure has a moderate effect on fine-grained material because each unit of surface sediment can experience a series of cyclic, alternating interactions with wave crests and troughs. Therefore, the residual pore pressure has the primary responsibility for the seepage-carried fine particles that result in seabed stratification.

(2) Localized fluidization induced by residual pore pressure response. In the wave flume experiment, after self-weight consolidation for ten days, the seabed's sediment density increases with the dissipation of excess pore pressure induced by the initial rapid accumulation of sediment. However, the sediment structure has not fully formed; therefore, local sediment particles within the seabed move slightly due to wave-induced seepage forces, connecting the voids in a weak structure zone and forming tiny conduits ($i < i_c$). In that seepage field, only a few, very fine constituents are transported upward via the seepage force. Under continued wave loading with increasing intensity, excess pore pressure inside the seabed gradually accumulates, and pore water can flow intensively to the larger conduits, leading to a further build-up of localized pore pressure. Once they approach the critical hydraulic gradient ($i - i_c$), the discontinuously distributed macro-voids and micro-conduits merge to form larger diameter pipes (Fig. 7.18b), and along these pipes, fine particles are carried by the pore water flow up to the seabed surface, resulting in gushing water and extruding sediment as vents (Fig. 7.12e). However, much of the material extruded from the bed (at least those that are subaqueous) is not preserved at the vents but is prone to being eroded and transported away from the vent site by the joint action of waves and currents, forming pockmarks at the original vent site. In this regime, the formation of channels is primarily caused by the movement of fine particles and the fracture of clay structures, and connecting forces remain that affect most of the fine clay particles (Simon and Collison 2001).

(3) Internal liquefaction induced by the residual pore pressure response. Cycles of wave loadings with high-intensity propagation through sediments can cause pore water inside the seabed to continuously gather into conduits and pipes. When excess pore pressure approaches the initial mean effective normal stress $\left[p \sim p' = \gamma' z(1 + 2K_0)/3\right]$ (e.g., Sumer et al. 2006), the connecting forces between particles and the sediment strength almost vanish, transforming the body from a solid state to a mobile, liquefied state. Then, mass instability can easily occur under wave-induced shear stresses. It was observed in the experiment that the venting sediment in the fluidization zone gradually expanded to a maximum depth of 28.5 cm, linking to a liquefied bed with an arc-shaped boundary surface. Most of the fine particles were scoured by the pore water flow along the arc-shaped sliding surface because the seepage velocity of the pore water is relatively high near the sliding surface (Fig. 7.18c). Upon regaining sediment strength due to the resolidification of the liquefied bed and the upward migration of the sliding surface (Fig. 7.13), a portion of the fine particles remained at the sliding surface and formed dish structures. In this process, fine constituents were

extruded by the upward pore water discharge and accumulated on the seabed surface to form a movable or stable superficial fluid mud layer.

In the modern subaqueous Yellow River Delta, which is formed by the rapid deposition of sediments discharging from the Yellow River and experiences strong wind-induced waves that act on the seabed, considerable residual pore pressure could be widely generated within the natural deposits. Therefore, the internal liquefaction and resolidification of the seabed subjected to continued wave loading is likely the main driving mechanism for the removal of fine particles upwards with the pore water flow. Unfortunately, there is insufficient quantitative data on the actual behavior of pore water flows within the seabed to compare the magnitude of the three different processes for grain-size differentiation under waves. This inference needs to be validated by field observations, although this is difficult for technical reasons.

With the passage of an ocean wave, the seabed is subjected to a bottom-pressure wave. The mechanisms of the wave-induced seabed response can be classified into two categories, depending on how the excess pore pressure is generated. The first category is generated by transient or oscillatory pore pressures, which are usually accompanied by damping amplitude and phase lag in the pore pressure. The second category, similar to that induced by earthquakes, is caused by the residual or progressive nature of excess pore pressure (Seed and Rahman 1978). When wave loading exposes the seabed, transient pore pressure and residual pore pressure appear simultaneously, and only the latter contributes to the decrease of vertical effective stress and the build-up of excess pore pressure. With excess pore pressure and diminishing vertical effective stress, a portion of the seabed may become unstable or even liquefied. Once liquefaction has occurred, wave action dictates that the sediment particles will likely be carried away as a fluid by any prevailing bottom current or mass transport (Sassa and Sekiguchi 2001). An internal liquefied layer can be observed in the flume experiment (Fig. 7.13). However, it appears that pore water seepage features within the seabed observed in the experiment are not the product of mass liquefaction; instead, they are characterized by a broad spatial evolution of locally fluidized sediment. This phenomenon is closely related to the dynamic response of pore water pressure and controls the variation of the composition and texture of the bed sediments.

During wave–seabed interactions, wave-induced shear stress is the driving force for the horizontal introduction and migration of seabed sediments, while the pore water pressure gradient, i, and the seepage force, j, are the forces driving upward sediment particle transport according to the following equation:

$$i = (h_1 - h_2)/L \tag{7.4}$$

where h_1 and h_2 are the pore water pressure heads of the seabed sediment at different heights and L is the vertical distance between the two heights (both in m). The seepage force acting on the transported particles within the sediment skeleton due to the viscous friction of the pore water flow can be expressed as follows:

$$j = i\gamma_w \tag{7.5}$$

where j is the seepage force of pore water (in kN/m^3) and γ_w is the unit weight of the pore water, which is 10.3 kN/m^3 in this experiment because the pore water is standard seawater. The body force, a force that results from the joint action of the seepage force and gravity on the particles, controls the stability of the sediment skeleton. When the upward seepage force exceeds a critical value, the sediment skeleton experiences a hydraulic failure and particles begin to be transported upward, which is characterized by macroscopic fluidization structures within the seabed (Nichols et al. 1994; Mörz et al. 2007. The hydraulic gradient (dimensionless) at the point of hydraulic failure (occurrence of fluidization structures) is called the critical hydraulic gradient, i_c, and is expressed as follows:

$$i_c = \frac{G_s - 1}{1 + e} \tag{7.6}$$

where G_s is the specific gravity of the particles. The critical hydraulic gradient in this experiment can be estimated based on the physical parameters of the bed before the wave loading, as summarized in Table 7.3.

The distribution of the excess pore pressure in the wave flume experiments shows that the pore pressure gradient pointed upward at a depth of 30 cm and that the pore pressure gradient varied when the bed sediments were exposed to different wave loadings (Fig. 7.19). During the series of tests with wave heights $H = 5$ cm, 10 cm, and 15 cm, the greatest pore pressure gradients appeared at depths ranging from 20 cm to 30 cm, reaching 0.65 kPa, 1.19 kPa, and 1.47 kPa, respectively, which are equivalent to hydraulic gradients of 0.63, 1.16, and 1.43, respectively. The i_c for these three test conditions was 0.89, 1.03, and 1.08, respectively (Table 7.3). This explains why only microchannels—but no piping seepage—occurred under the lowest wave loading condition ($H = 5$ cm) because the measured hydraulic gradient did not reach the calculated critical hydraulic gradient. In the case of the two test series with $H =$ 10 and 15 cm, the critical hydraulic gradients were exceeded, causing the initiation of pipes and sediment venting to the surface (i.e., localized fluidization). As the excess pore pressure (p) caused by cyclic loading continued to increase the mean effective normal stress (p'), the localized sediment failure expanded to bed liquefaction on a large scale, as described in the previous section.

7.3.4 *Quantitative Contribution of Sediment Resuspension by Residual Seepage Flows*

Because of wave-induced localized fluidization and liquefaction, fine-grained materials inside the seabed will continuously move upward via the pore water pressure gradient force to thicken the fluid mud layer; this theory is supported by an analysis of sediment particle compositions and microstructure characteristics. It has been

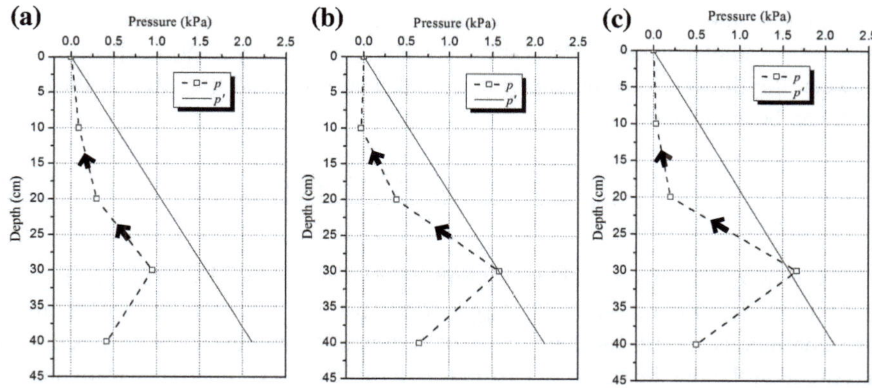

Fig. 7.19 Measured maximum pore pressure gradient for the bed under the action of 5-cm (**a**), 10-cm (**b**), and 15-cm (**c**) high waves, respectively. The excess pore pressure, p, is plotted against the calculated initial mean normal effective stress, p', which is indicative of mass liquefaction onset (e.g., Sumer et al. 2006). The arrows represent the relative direction of pore water seepage between the layers where the excess pore water pressure is measured

hypothesized that significant sediment transport on continental shelves is caused by wave-loaded fluid mud that flows downslope under the influence of gravity. The question is as follows: What is the contribution of this upward transport of fine sediment to the previously mentioned superficial fluid mud layer? That contribution can be calculated from the measured thickness and sediment concentration of the fluid mud layer and the liquefied layer, following Eqs. (7.7)–(7.9):

$$M_{fm} = S \times \int_0^h dz \times C \tag{7.7}$$

where S is the area of the bed surface; h is the calculated height of the fluid mud layer and C is the mean sediment concentration of the fluid mud layer. The calculated quantity of the superficial fluid mud layer, M_{fm}, is 9.5 kg, based on Eq. (7.7). The total quantity of fine particles inside the flume bed affected by waves can be estimated as follows:

$$M_{fp} = S \times \int_0^h dz \times \frac{\rho}{1+w} \times \psi_{0.01} \tag{7.8}$$

where $\Psi_{0.01}$ is the mass fraction of the fine particles with diameters less than 0.01 mm (20.9%, according to Table 7.4) and M_{fp} is the quantity of fine particles within the internal liquefied layer (168 kg, as estimated by Eq. (7.8)). Comparing M_{fm} and M_{fp} in the wave flume experiment, it can be observed that only 6% of fine (silt and clay) particles were transported to the bed surface during the wave-induced liquefaction process. We assume that the decreased fine particles within the liquefied layer are sorted due to sufficient liquefaction and fluidization and contribute entirely to the formation of the surficial fluid mud layer (i.e., $\Delta M_{fp} = \Delta M_{fm}$). Substituting the relevant parameter values from Tables 7.3 and 7.4 into Eq. (7.9), a quarter ratio between the thickness of the superficial fluid mud layer and the internal liquefied layer can be estimated:

$$\frac{H_{fm}}{H_l} = \frac{\frac{\rho'}{1+w'} \times \psi'_{0.01} - \frac{\rho}{1+w} \times \psi_{0.01}}{C} \approx \frac{1}{4} \tag{7.9}$$

There are relatively few field measurements of wave-induced seabed liquefaction and fine particle migration during large storms because it is extremely difficult to take continuous in situ measurements of sediment from the upper near-bed to a certain depth below the surface. This makes it difficult to compare our estimates to field measurements. Theoretical calculations in the subaqueous Yellow River Delta indicate that maximum liquefaction occurs at a water depth ranging from 7 to 8 m, and the maximum depth of the liquefied layer induced by accumulated excess pore pressure reaches 4.2 m. We estimate that the maximum thickness of the in situ fluid mud layer can reach approximately 1 m. This magnitude of upwelling fine-grained sediments under waves is equivalent to the sedimentary layer formed during floods with extremely high sediment concentrations in the Yellow River estuary delta. Thus, wave-induced seabed responses play a significant role in pore water flow with upwelling fine-grained sediments and the corresponding seabed inhomogeneity and stratification processes.

7.4 Sediment Erodibility Attenuation Due to Wave-Induced Seabed Liquefaction

As it has been referred at the beginning of the previous section, the other aspect of the influence of residual seabed liquefaction on sediment erosion is discussed in this section. When sediments were liquefied by waves, seabed will lose its original strength and be transferred from the solid state into the liquid state, so is it normal to suspected that sediment erodibility will experience and attenuation due to the wave-induced residual liquefaction. But how much is the attenuation degree under different wave parameters also with the period of wave loadings, is still unclear and merits further investigation. Therefore, a series of tests and experiments were conducted using sediment cores from different sites of the Yellow River Delta.

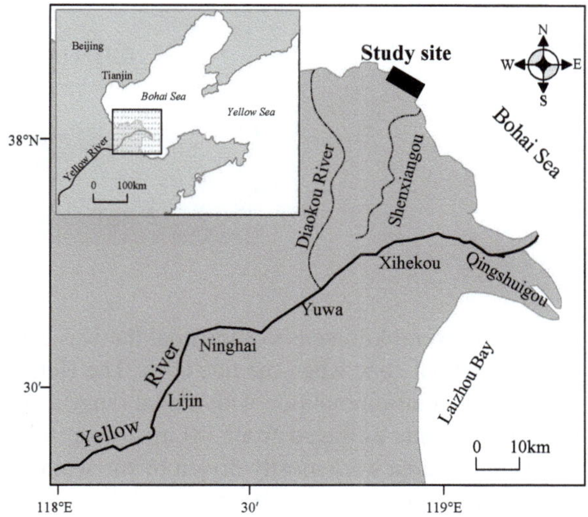

Fig. 7.20 Location of the study area

7.4.1 *Methodology*

(1) ***Study Area***

The Haigang mudflat, which located in the northeast part of the modern Yellow River Delta (Fig. 7.20), was chosen as the typical site to conduct a series of field experiments and collect samples for laboratory experiments. For the purpose of investigating the erodibility variation of silty sediment under waves. In the coastal section of Haigang, relatively strong scouring occurs and the coastal line retreats to Haigang, where a wave-induced longshore current is the major driving force transporting sediment from eroded areas (Chu et al. 2006), and an isobath of 3–5 m annually retreats 150–400 m shoreward.

Sediment in the study area was mainly composed of silt particles (~69.5%), clay particles (~19.5%), and a small fraction of sandy particles (~11.0%). The averaged grain diameter was 0.026 mm, the liquefy limit ranged from 25.8 to 27.4%, and the plasticity index was between 8.5 and 8.6. Therefore, the sediment was classified as clayed silty sediment. The water content was moderate (20–30%), but the degree of saturation was so high that it is easily liquefied when an external dynamic load exerted pressure on it.

The detrital minerals of the surface material were mainly built up by quartz and clay and showed an abundance of illite, reflecting the source of the sediment to some degree. The benthos distributed in the surficial sediment were primarily crabs; the number of burrows was less than 40/m^2 with a diameter of ~5 cm. Moreover, the study area experienced a large post vortex ripple morphology, which developed on the bed of a previous plane, and around high tide, the ripples are fully developed, with a height of ~0.01 m and a width of approximately 0.03–0.04 m. The Haigang tidal flat is a micro-tidal, semi-enclosed area with a mean depth of ~0.5 m at high

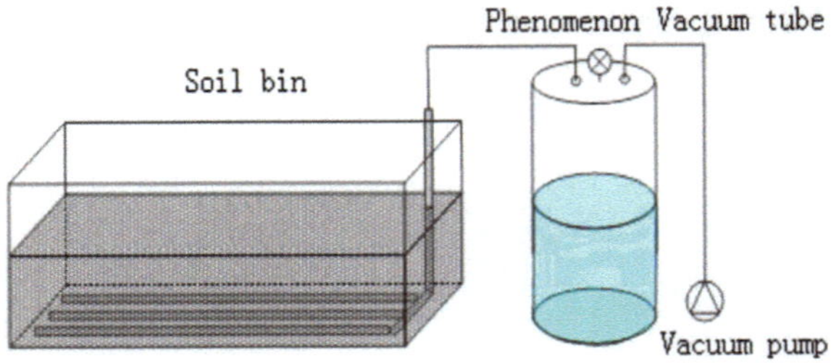

Fig. 7.21 Illustration of the vacuum preloading system

tide when several kilometers of the tidal flat was exposed due to the gentle gradient of the intertidal zone when the tide ebbs. The Haigang tidal flat was dominated by an irregular semi-diurnal tide with a tidal range averaging 0.6–0.8 m. The speed of the tidal currents averaged from 1.0 m/s (neap tide) to 2.0 m/s (spring tide). The waves in this area were mostly driven by the winds in the Bohai Sea and thus wave characteristics had distinct seasonal variability associated closely with the activities of the monsoon (Wang et al. 2007). In normal sea conditions, the wave height remains at about 20 cm.

(2) ***Experimental Procedure***

Firstly, sediment core samples were collected from the Haigang mudflat for laboratory tests. Furthermore, field wave simulation and recirculating flume experiments, were conducted, as well as sediment strength tests, and the other geotechnical parameters tests.

Laboratory Experiments

Samples were prepared using a vacuum preloading system (Fig. 7.21), which allowed the experimental samples to be made quickly and fairly homogeneously.

The samples were prepared as follows: (1) The air-dried silt was ground up and uniformly mixed with water using a stirring mill; (2) The prepared mud (with a water content of approximately 45%) was spread in a soil bin and vacuum preloading was conducted; this was proceeded by a grading analysis and a test of the liquid limit and plastic limit in the consolidation process; (3) Twenty-two samples were taken simultaneously using thin wall samplers when the prepared samples in the soil bin in plastic state (moisture content was ~21% in this experiment). Eighteen of the 22 samples were subjected to triaxial tests, direct flume experiments, and strength tests in succession; two samples were tested for bulk density and moisture content; and the other two samples were saved. The properties of the sediment samples after vacuum preloading are presented in Table 7.5.

Prior to the triaxial tests for further consolidation and wave simulations, the samples were vacuum saturated, which only required that the degree of saturation be more than 0.85. All the triaxial tests were conducted in equal pressure drainage consolidation with a consolidation stress ratio of 1.0 and a consolidation time of 2 h. Parallel determinations in the same wave loading condition were carried out for the strength tests in the dynamic triaxial device and the scouring experiment

Table 7.5 Properties of the sediment samples after vacuum preloading in the laboratory experiments, and properties of the sediment samples after drainage consolidation in a dynamic triaxial device

Clayed silt sediment					
Properties of the sediment samples after vacuum preloading in the laboratory experiments					
Physical parameters	Specific weight	Bulk density (g cm^{-3})	Water content (%)	Degree of saturation	Pore ratio
	2.70	1.95	19.20	0.75	0.65
Consistency and related parameters	Consistency	Liquefy limit/%	Plastic limit/%	Liquefy index	Plastic index
	Flexible state	26.30	16.50	0.28	9.80
Graduation and related parameters	Graduation	Average grain diameter	Effective grain diameter	Non-uniformity coefficient	Curvature coefficient
	Homogeneity	0.048	0.006	10.03	1.57
Properties of the sediment samples after drainage consolidation in dynamic triaxial device					
Bulk density (g/cm^{-3})	Water content (%)	Degree of saturation	Pore ratio	Liquefy limit	Consistency
1.97	24.9	0.92	0.89	0.86	Soft flexible state

followed the penetration tests. In detail, strength tests in a dynamic triaxial device both included the shear strength tests and residual strength tests through static tests, and the penetration tests were conducted using a pen-penetrometer with a measuring range of 20 N, on the top of the sediment samples in one parallel group and three measured penetration value for each sample can be gained and the average value of them are adapted. Moreover, a drainage method was adapted to measure the sample volume; the sample mass was obtained using a balance; and then the bulk density was calculated by dividing the sample mass by the sample volume. Moisture content was measured using drying ovens. Table 7.5 shows the properties of the sediment samples after drainage consolidation in the dynamic triaxial device (i.e., before being subjected to the wave simulations). The design parameters in the wave simulations for the samples are shown in Table 7.6.

A large flume with graduation, a direct flume, and a water pump constituted the scouring system (Fig. 7.22), which supplied circulating water for the erodibility measurements. The current velocity, v, for the scouring experiment was obtained from the distance between the bottom of the direct flume and the water surface of the large flume h, and the level distance that the scouring water runs s based on the formula $\upsilon = s\sqrt{g/2h}$ (g is the gravity acceleration). The vertical distance h was considered a constant in this experiment because the scouring current was circular and the water level had few fluctuations.

In the scouring experiment, the threshold condition was defined as the scouring current velocity (cm/s) at which the current could cause the samples local failure

Table 7.6 Design parameters of the wave loadings in the laboratory experiment

Parameters	Serial number of sediment sample								
	0–1	1–1	2–1	3–1	4–1	5–1	6–1	7–1	8–1
	0–2	1–2	2–2	3–2	4–2	5–2	6–2	7–2	8–2
T/s	8	8	8	8	8	8	8	8	8
f	0.125	0.125	0.125	0.125	0.125	0.125	0.125	0.125	0.125
N	0	30	30	30	30	30	60	90	110
σ_d/kPa	0	15	20	25	30	16	16	16	16

Note T, f, and N stand for the period, frequency, and cycle number of the simulated wave loadings, respectively

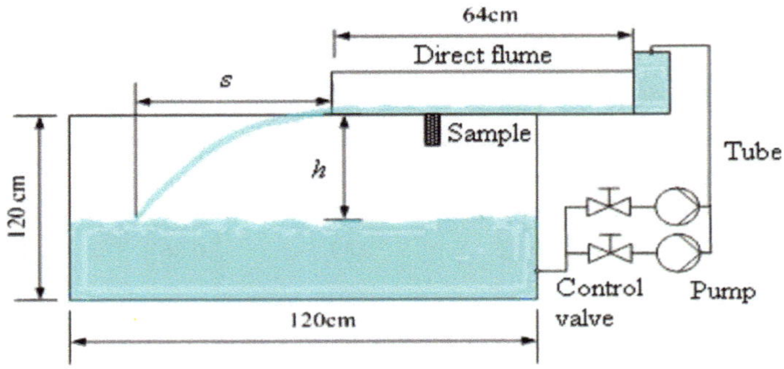

Fig. 7.22 Illustration of the scouring system

and visually suspended particles, and *erosion rate* [g/(cm^2 s)] was defined as the mass of silt removed from the samples by the current per unit area and per unit time. The water pump was slowly turned up until local failure appeared on the samples and visual particles were suspended; the current velocity at that point was defined as the threshold velocity. Then, the current velocity was increased until it reached 1.37 m/s, at which point the scouring time per 10 mm of the sample was recorded. Finally, 6–7 sets of data were obtained in the scouring experiment; three of these sets of assembled data were used to calculate the average erosion rate.

The Prandtl empirical formula was applied to determine the critical shear stress of the samples exerted by wave loadings under different conditions. The formula was described as follows:

$$\frac{1}{\sqrt{\lambda_1}} = 2.0 \log R_e \sqrt{\lambda_1} - 0.8 \tag{7.10}$$

where λ_1 is the drag coefficient that was determined by the formula $\lambda_1 = 8\tau/\rho U^2$, ρ is the medium density that corrects to 1.10 g/cm^3; U is the scouring current, cm/s; τ is the shear stress caused by the scouring current of U, Pa; R_e is the Reynolds number that was determined by the formula $R_e = Ud/\upsilon$; d was the width of the

Fig. 7.23 Photo of the wave loading simulations

scouring flume, which was 8 cm in this experiment; υ is the kinetic viscosity, which was 0.01 cm^2/s in this experiment. The Prandtl Empirical Formula was applied in the condition where the Reynolds number was in the turbulance range of 4×10^3–2×10^6, which was satisfied by the scouring experiment with a Reynolds number in the range of 2×10^4–4×10^5.

Field Measurements

In the field, wave loadings were simulated using a wave-producer device (Fig. 7.23) that was applied to the undisturbed surface sediment where little bioturbation was distributed, and centered on a small region at 38°04.079 N, 118°56.409 E. The variability of the characteristics of the seven test points were not significantly different based on the geotechnical parameters that were tested (Table 7.7) before wave simulations with cycle numbers of 0, 10, 20, 30, 40, 50, and 60, respectively. When proceeding experiments, driving the handle downward artificially at a constant speed until it reached the baffle in the middle of seawater height in the inner housing of the wave-producer device, a wave-pressure with nearly uniform amplitude could be applied on the sediment surface. The magnitude of simulated wave loadings exerted on the superficial sediment, depending largely on the height of seawater in the housing of wave producer, was indirectly measured by a piezoresistive pressure sensor

Table 7.7 Results for the field experiment, including physical and mechanical parameters, sediment components, and threshold conditions for each experiment. The threshold velocity was determined using a turbulent formula. The other geotechnical parameters were evaluated with the average value of no less than three measured values

No. of cycles	Physical parameters				Sediment components				Mechanical parameters		Threshold conditions	
	Wet bulk density (kN · m^{-3})	Dry bulk density (kN· m^{-3})	Moisture content (%)	Porosity ratio	Average grain diameter (mm)	Clay content (%)	Silt Content (%)	Sand Content (%)	Undrained Shear strength (kPa)	Penetration strength (N)	Threshold velocity (cm · s^{-1})	Critical shearing stress (Pa)
Natural values of the surficial sediment before wave loadings action												
0	19.5	15.4	26.67	0.75	0.026	19.5	75.2	5.3	14.20	4.89	16.22	0.086
10	19.9	15.7	26.5	0.72	0.03	19.5	75.0	5.5	14.62	5.02	–	–
20	19.6	15.6	26.53	0.75	0.036	19.4	75.3	5.3	13.90	4.90	–	–
30	19.2	15.3	26.5	0.78	0.037	19.4	75.1	5.5	14.00	4.90	–	–
40	19.5	15.4	26.81	0.67	0.032	19.5	75.2	5.3	14.00	4.49	–	–
50	19.4	15.3	26.44	0.63	0.033	19.7	75.3	5.2	13.82	4.95	–	–
60	19.2	15.2	26.42	0.68	0.032	19.6	75.4	5.2	14.72	4.70	–	–
Changed values of the surficial sediment exposed to wave loadings												
0	19.5	15.4	26.67	0.75	0.026	19.5	75.2	5.3	14.20	4.89	16.22	0.086
10	19.8	15.7	26.06	0.72	0.030	20.6	74.4	5.0	10.00	2.47	13.99	0.066
20	19.7	15.4	27.63	0.75	0.036	19.5	66.3	14.2	3.38	1.20	8.53	0.028
30	19.3	15.2	26.99	0.78	0.037	20.1	76.9	3.0	3.67	1.08	7.58	0.023
40	20.2	16.2	25.07	0.67	0.032	22.2	72.7	5.1	6.50	2.47	12.93	0.058
50	21.0	16.6	26.43	0.63	0.033	16.7	77.3	6.0	2.84	0.48	7.71	0.024
60	20.1	16.1	25.11	0.68	0.032	18.2	67.8	14.0	4.00	0.90	10.43	0.040

that was connected to a pore water pressure-measuring device. In this experiment, the period of the wave loading was 8 s, and the magnitude was ~4 kPa. The wave loading simulated in the experiment was equivalent to the real wave with the period of 8 s, the wave height of 0.8 m, on the condition that the ratio of water depth to wavelength was very small.

Sediment erodibility was determined in a 1.2 m long aluminum-recirculating flume filled with seawater that was designed and constructed based on a mobile recirculating flume (see more details in Fig. 4.2). The main channel had a width of 0.1 m and a height of 0.15 m. The test section was 0.25 × 0.10 m, into which 2 cm of sediment was inserted, and was located on the surface sediment exerted by the wave loadings. The flow rate, which ranged from 0 to 50 cm/s, was induced by rotation of a propeller located on the other side of the flume and connected with an adjustable electric motor. Flow measurements were conducted with an LGY-II current meter, which auto-recorded the average velocity and was placed in front of the test section. The speed of the current running in the channel made of aluminum was changed at least three times, and the erosion process at each speed lasted about 5 min until the turbidity in the channel was not fluctuating or was decreasing, indicating that the point had been reached where deposition was greater than erosion.

The suspended particulate mass (SPM) was used to determine the sediment erosion flux at each current velocity. Interval turbidity measurements were made with a TURB 335IR-type flow-through turbidimeter. The interval of the collected suspended sediment samples was around 20 s each time and was recorded exactly with a solar stopwatch. Based on the background turbidity, the volume of seawater in the recirculating flume, and the calibration curve between SSC (suspended sediment content) and NTU (turbidity) in this experiment, we got all the background SPM of the sediment, upon which simulated wave loading was applied with cyclic numbers of 0, 20, 30, 40, 50, and 60. In addition, the data from the laboratory tests indicated that SSC had a good linear relationship with NTU; the regression function of (SSC) was 0.006 × (NTU)-0.023. SSC can be determined by the measured corresponding turbidity based on the regression function. Finally, SPM can be obtained by calculating the product of SSC and seawater volume in the recirculating flume. Methods for determining the threshold velocity have been illustrated in some reviews (Aberle et al. 2004). In this study, the linear correlation between the current velocity and SPM (i.e., erosion flux) was determined following Neumeier et al. (2006). Then, the corresponding velocity of the background SPM could represent the threshold velocity of the seabed.

For an open channel, the current can be taken as turbulence on the condition that $R' > R_{ec}$, and R_{ec} can be calculated as shown in Eq. (7.11):

$$R_{ec} = V_c R/\upsilon = 580 \tag{7.11}$$

where R' is the Reynolds number; R_{ec} is the critical Reynolds number; V_c is the critical velocity; υ is the kinematic coefficient of viscosity, $\upsilon = 10^{-6}$ m²/s; R is the hydraulic radius, $R = A/\chi$, χ is the wetted perimeter, $\chi = 0.26$ m; A is the cross-section of flow, $A = 0.008\ m^2$, then, V_c is 1.89 cm/s, which is obtained by substituting the known parameters into Eq. (7.11). The current velocities measured in all the flume experiments all exceeded that value, so the currents in these experiments can be considered turbulence, and then the critical shear stress for each wave loading condition can be calculated based on the corresponding formulas as follows:

$$\frac{u}{u_*} = 5.5 + 5.75 \log \frac{yu_*}{\upsilon} \tag{7.12}$$

$$\tau_0 = \rho u_*^2 \tag{7.13}$$

where τ_0 is the critical shear stress of the bed, ρ is the seawater density (the value is 1.025 g/cm^3 here), u_* is the friction velocity, u is the threshold velocity here, y is the location of the flow meter probe, the value of which is 0.015 m here, and υ is the kinesis viscous coefficient of the seawater ($\upsilon = 1.792 \times 10^{-6} \exp(-0.042\mathrm{T}^{0.87})$ $m^2/s \approx 1.01$, T = 20 °C).

In addition, wet bulk density, water content, undrained shear stress, and penetration strength were measured to investigate the dynamic variety of the silty sediment's physical and mechanical properties under wave-pressure action, and dry bulk density and pore ratio were calculated based on the wet bulk density, moisture content, and specific gravity. Furthermore, the particle size of the surface sediment at a depth of ~2 cm that was affected by wave-pressure action was analyzed using a sieve and a hydrometer.

7.4.2 Results

(1) ***Laboratory Data***

The data on erodibility (i.e., threshold velocity, erosion rate, and critical shear stress) and strength (i.e., undrained shear strength, residual shear strength, and penetration strength) of the sediment samples under simulated wave loadings under different conditions was obtained in laboratory experiments, and is presented in Table 7.8.

Moreover, the influence of wave loadings of different magnitudes and cycle numbers on sediment strength and erodibility are highlighted in Figs. 7.24, 7.25, 7.26 and 7.27, which indicates a significant role of wave loadings on sediment strength and erodibility in the Yellow River estuary. As shown in Figs. 7.24, 7.25, 7.26 and 7.27, wave loadings significantly reduced the erodibility and strength of the silty sediment samples. Furthermore, erodibility and strength of the silty sediment samples had a different sensitivity to the wave loadings depending on magnitude and cycle number. Specifically, good-fit regression lines can describe the variations in sediment erodibility and strength with magnitude and cycle number of the simulated wave loadings, all of which had a high r^2 value. In the laboratory data, no "stable state" is found due to the precision of the dynamical triaxial device. In this study, the "stable state" means sediment erodibility and strength changes little with the increasing of cycle number

Table 7.8 Laboratory results of the strength tests and flume experiments

Sample number	Magnitude of dynamic loadings (kPa)	Cycle number	Undrained shear strength (kPa)	Residual strength (kPa)	Penetration resistance (N)	Current velocity (ms^{-1})	Threshold velocity (ms^{-1})	Critical shear stress (Pa)	Erosion rate ($g\ cm^{-2}s^{-1}$)
0–1 0–2	0	0	175.8	151.9	12.3	1.37	0.82	1.82	0.038
1–1 1–2	15	30	138.5	127.4	11.4	1.37	0.75	1.55	0.074
2–1 2–2	20	30	135.7	122.6	8.1	1.37	0.62	1.11	0.089
3–1 3–2	25	30	106.6	91.1	7.9	1.37	0.55	0.89	0.092
4–1 4–2	30	30	106.0	98.8	6.1	1.37	0.41	0.53	0.121
5–1 5–2	16	30	169.5	138.3	8.0	1.37	0.75	1.55	0.076
6–1 6–2	16	60	136.3	136.3	7.3	1.37	0.62	1.11	0.094
7–1 7–2	16	90	122.5	107.2	5.5	1.37	0.42	0.55	0.116
8–1 8–2	16	110	114.8	104.2	4.5	1.37	0.25	0.22	0.148

Note Undrained shear strength was determined by the peak value on the curve of deviatoric stress versus axial strain, or the half value of the deviatoric stress as the axial strain reaches 5% when no obvious peak value appeared in the curve. Residual strength was determined by the half value of the average deviatoric stress when the curve of deviatoric stress versus axial strain remained stable. In the laboratory portion of this study, the undrained shear strength and residual strength were tested under a confinning pressure of 50 kPa

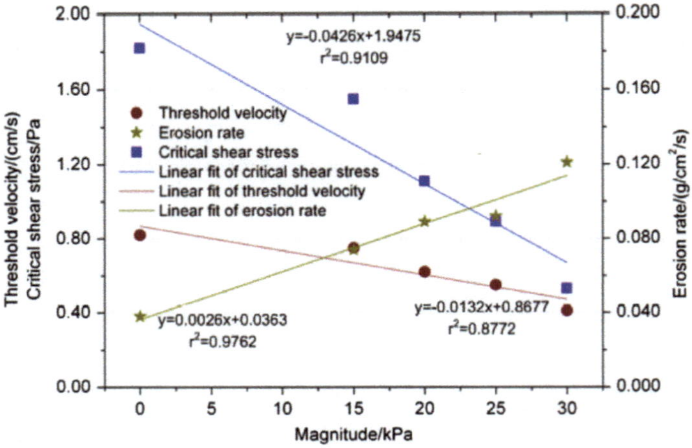

Fig. 7.24 Plots and fitted curves of the variation in threshold velocity (cm/s), critical shear stress (Pa), and erosion rate $\left(g/cm^2/s\right)$ with the magnitude of simulated wave loadings at a cycle number of 30; the current velocity for the erosion rate was 1.37 m/s

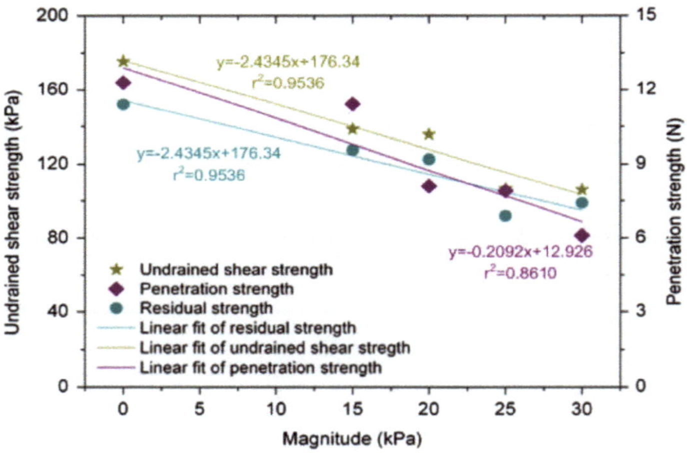

Fig. 7.25 Plots and fitted curves of the variation in undrained shear strength (kPa), residual shear strength (kPa), and penetration strength (N) with the magnitude of simulated wave loadings at a cycle number of 30; the current velocity for the erosion rate was 1.37 m/s

and magnitude of wave loadings. Limited by the designed experimental parameters, the application range of the linear model describing the variation in erodibility and strength with wave conditions could not be determined in this experiment.

Wave loadings with different magnitudes had different capabilities of reducing the erodibility and strength of the silty sediment samples (Figs. 7.24 and 7.25). The threshold velocity and critical shear stress (for both, the higher the value, the higher the erodibility of the sediment) obviously decreased after the wave loading actions,

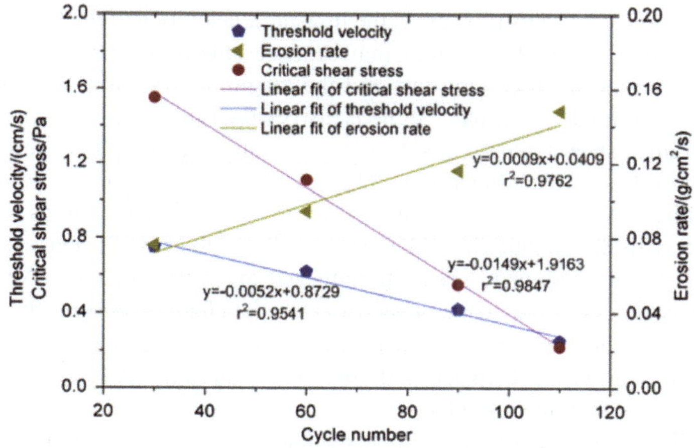

Fig. 7.26 Plots and fitted curves of the variation in threshold velocity (cm/s), critical shear stress (Pa), and erosion rate $\left(g/cm^2/s\right)$ with cycle number of simulated wave loadings with a magnitude of 16 kPa; the current velocity for the erosion rate was 1.37 m/s

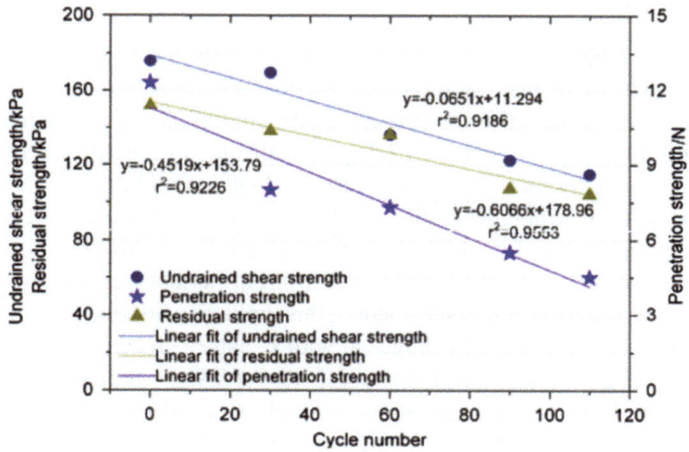

Fig. 7.27 Plots and fitted curves of the variation in undrained shear strength (kPa), residual shear strength (kPa), and penetration strength (N) with cycle number of simulated wave loadings with the magnitude of 30; the current velocity for the erosion rate was 1.37 m/s

and the larger the magnitude of the wave loadings, the sharper the decrease in the threshold velocity and critical shear stress of the sediment samples; the reverse trend appears to be variation in the erosion rate, for which a higher value means lower erodibility, with the magnitude of the wave loadings (Fig. 7.24). In addition, good-fit regression lines are shown between erosion parameters and the magnitude of the wave loadings; all with r^2 values greater than 0.85. Conducting a comparative analysis on the three regression curves, erosion rate had the highest correlation coefficient

(0.9762), which is close to 1, whereas the threshold velocity had the lowest correlation coefficient (0.8772). Therefore, it may be more suitable and accurate to describe the relationship between erosion rate and the magnitude of wave loadings with a linear model. Similarly, a visual decline is also evident in the three strength parameters: undrained shear strength, residual strength, and penetration strength with the magnitude of the wave loadings (Fig. 7.25); moreover, good-fit regression lines are shown, with the highest r^2 value greater than 0.85. Among the variations in the three parameters with the magnitude of the wave loadings, the best-fit regression line is determined by the undrained shear strength and magnitude of the wave loadings with an r^2 value of 0.9536, while penetration strength had the poorest fit regression line with an r^2 value of 0.8610. Hence, it might be more suitable and accurate to describe the relationship between undrained shear strength and the magnitude of wave loadings with a linear model.

Generally, the stability of the sediment declines due to a vertical destructive force. Furthermore, higher magnitude wave loadings had higher destructive effects on the exposed sediment, and a linear regression can be adopted to describe the trend in variation in sediment erodibility and strength with the magnitude of wave loadings, especially for the selection parameters of erosion rate, and undrained shear strength, respectively.

Likewise, compared to the variation in sediment erodibility and strength (Figs. 7.24 and 7.25), the variation in wave loadings at different cycle numbers had different capabilities in terms of reducing the erodibility and strength of the silty sediment samples (Figs. 7.26 and 7.27). Specifically, the threshold velocity and critical shear stress (for both, a higher value refers to higher erodibility of the sediment), declined with the cycle number of the wave loadings; the reverse trend occurred in the variation in erosion rate, for which the higher value refers to lower erodibility, with the magnitude of the wave loadings (Fig. 7.26). Furthermore, a linear relationship exists in the variation of critical shear stress, threshold velocity, and erosion rate with the cycle number of the wave loadings, all with r^2 values greater than 0.95. Critical shear stress had the highest correlation coefficient at 0.9847, with the cycle number of wave loadings, erosion rate had the second highest correlation coefficient, and the threshold velocity had the lowest. Therefore, critical shear stress may be the most suitable parameter to describe the dynamic response of sediment erodibility using the linear model. Similarly, a visual decrease also occurred to the three strength parameters, undrained shear strength, residual strength, and penetration strength, with the cycle number of the wave loadings (Fig. 7.27); moreover, good-fit regression lines are shown with the higher r^2 values more than 0.90. Among the variations of the three parameters with the cycle number of the wave loadings, the best-fit regression line is determined by the residual strength and the cycle number of the wave loadings with an r^2 value of 0.9553, penetration strength is second with a correlation coefficient of 0.9226, while undrained shear strength had the poorest fit regression with an r^2 value of 0.8610. Hence, the linear model may be most practical for the application of the erodibility calculation with the adoption of the residual shear strength as the describing parameter.

Similar to the demonstrations in Figs. 7.24 and 7.25, the stability of the sediment decreased due to the vertical destructive force of wave loadings with a different magnitude; a higher cycle number of wave loadings also had a higher destructive effect on the exposed sediment, and the linear regression can be adopted to describe the trends in variation in sediment erodibility and strength with cycle number of wave loadings. However, the optimum parameters for the linear model to describe the relationship between erodibility, strength, and cycle number of wave loadings were, respectively, critical shear stress and residual shear strength in Figs. 7.26 and 7.27, whereas those are erosion rate and undrained shear strength with the magnitude of wave loadings in Figs. 7.24 and 7.25. In other words, the influence of the magnitude and cycle number of the wave loadings, in a certain application range, can both be estimated using linear regression, but the chosen parameters for the better adoption are not consistent. The inconsistency can be the result of mechanism difference of sediment dynamic response to wave loadings for load effect and time effect, which is still needed to clarified by further studies.

(2) ***Field Data***

To investigate the dynamic response of the surficial sediment to wave loadings at different cycle numbers, data on sediment erodibility, physical and mechanical properties, and particle components were obtained from the field experiments, as well as from the observations of surficial sediment morphology. In contrast to the laboratory experiments, no linear relationship was evident in the variation of the erosion and geotechnique parameters with the cycle number of the wave loadings; the tested parameters of the surficial sediment in the field (i.e., threshold velocity, critical shear stress, undrained shear strength, penetration strength, wet bulk density, dry bulk density, moisture content, average grain diameter, porosity, content of silt, clay and sand) all varied differently (Figs. 7.28, 7.29, 7.30 and 7.31).

A sharp decline occurred in sediment erodibility and strength when the cycle number of the wave loadings was not high. Higher cycle numbers had comparatively little influence, except the unique plot under the wave loadings at cycle number 40 (Fig. 7.28). In another words, sediment erodibility (i.e., threshold velocity and critical shear stress) and strength (undrained shear strength and penetration strength) did not decrease with an increase in the cycle number of the wave loadings in the field experiment, which is different from the linear variation observed in sediment erodibility and strength with the cycle number of wave loadings in the laboratory study, while when the cycle number reaches a certain value, sediment erodibility and strength will be stable with nearly an unchanged threshold velocity and critical shear stress, and undrained shear strength and penetration strength, respectively. On the other hand, similar to the laboratory results, the variation in sediment strength with the cycle number of wave loadings was consistent with sediment erodibility.

The vertical pressure of the wave loadings not only changed the mechanical properties of the surficial sediment (Fig. 7.28), it also altered the physical properties, but the fluctuation range is less obvious than the mechanical properties under wave

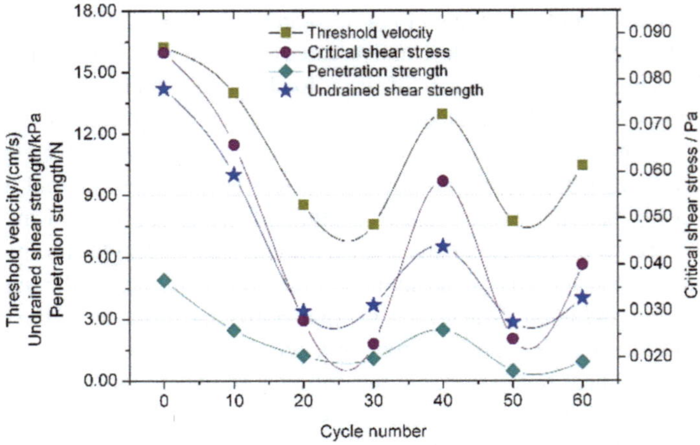

Fig. 7.28 Variations in the threshold velocity (cm/s), critical shear stress (Pa), undrained shear strength (kPa), and penetration strength (N) with the cycle number of wave loadings in field experiments

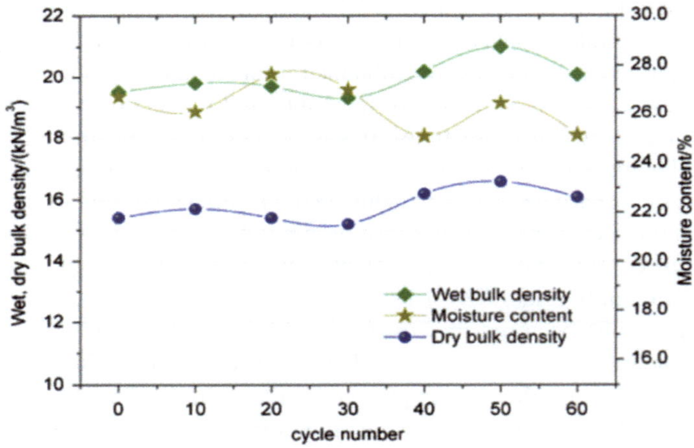

Fig. 7.29 Variation in wet bulk density (kN/m^3), dry bulk density (kN/m^3), and moisture content (%) with the cycle number of wave loading in the field experiments

loadings with different cycle numbers (Fig. 7.29). Furthermore, compared to the variation in sediment strength with different cycle numbers of wave loadings shown in Fig. 7.28, the greatest difference was that when the cycle number was less than 30, no obvious fluctuations were evident in the curve, whereas sediment strength was greatly reduced; when the cycle number exceeded 30, different variations were evident in the relationships between the wet bulk density, dry bulk density, moisture content, and cycle number of wave loadings, respectively, but the overall trend of the three parameters was increasing. As shown in Fig. 7.28, sediment erodibility and strength significantly decrease at cycle number 20, suggesting that fluidization has

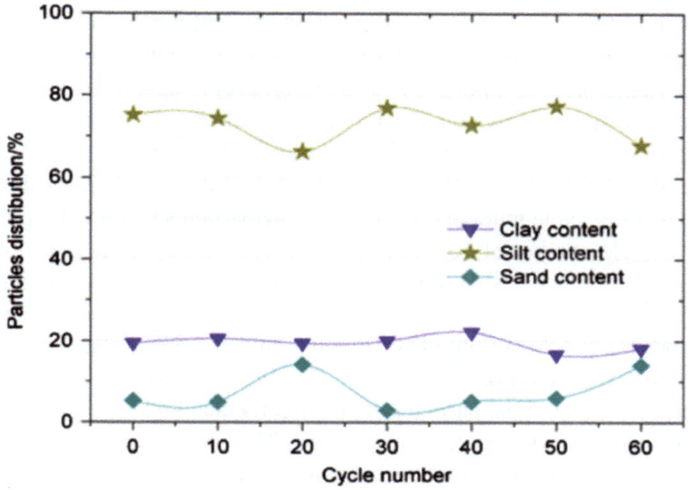

Fig. 7.30 Variation in the particle distributions (%) of clay, silt, and sand content at different cycle numbers for wave loadings in the field experiments

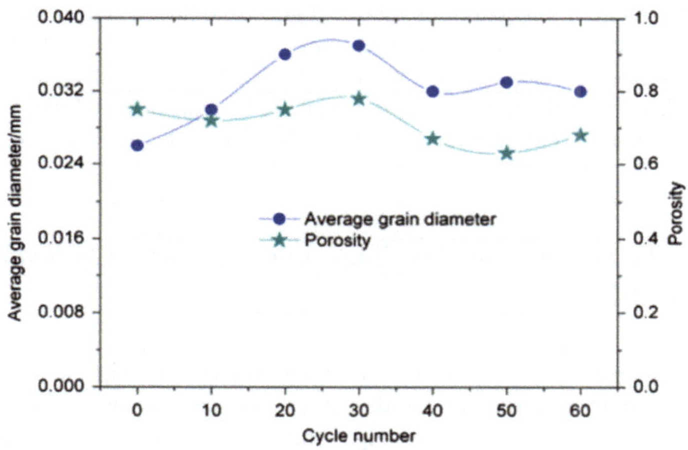

Fig. 7.31 Variation in the average grain diameter (mm) and porosity at different cycle numbers for wave loadings in the field experiments

occurred in the surface sediment. However, the values of sediment erodibility and strength at cycle number 40 are interesting as it is higher than those of the preceding cycle numbers. This can most likely be explained by a certain degree of consolidation accompanied by the dissipation of the excess pore pressure. This phenomenon also suggests that wave-induced fluidization and concurrent sediment densification complicate the response of sediment erosion resistance to hydrodynamic actions.

As shown in Figs. 7.32 and 7.33, wave loadings had the capacity to change the particle distribution of the surficial sediment. After the actions of wave loadings, the particle component and porosity varied to different degrees, as did the average grain diameters, and the wave loadings with different cycle numbers had a different influence on the particle distribution. In the exposed silty sediment under wave loadings, the content of sand particles was low, but it had the most remarkable changes with wave loading cycle numbers; content of clay particles had no obvious changes except for a cycle number of 50; the content of silty particles, the main component of the sediment that was tested, fluctuated with the increase in cycle number of the wave loadings; the lowest value was for cycle number 20 and the highest was for cycle number 50, but the amplitude of fluctuation was minor relative to the overall content of silty particles (Fig. 7.30).

The transformation of the particles could directly change the average grain diameters of the surficial sediment that were tested. Therefore, after the action of wave loadings with different cycle numbers, a change in the response of the average grain diameter also took place, and an analog normal curve was displayed in the variation of the average grain diameter with cycle number of wave loadings, and the two peak values highlighted at cycle numbers of 20 and 30 (Fig. 7.31) in the condition of which, the content of sand particles was the highest and the content of silty particles was the lowest, and the opposite, respectively (Fig. 7.30). In addition, the porosity of the sediment also changed after the wave loading actions with different cycle numbers, but when the cycle number was less than 30, it changed very little, a relatively large reduction and small recovery occurred when the cycle number exceeded 30.

7.4.3 Influence of Wave Loadings on the Variation of Seabed Erodibility

Erosion resistance of fine sediment under waves and currents (i.e., sediment erodibility) is important for research on sediment erosion and transport, which has been demonstrated in a number of studies (e.g., Aberle et al. 2004), and also highlighted in this section (Figs. 7.32, 7.33, 7.34, 7.35). Sediment erodibility is usually assessed by the parameters threshold velocity, critical shear stress, and erosion rate in related research work. Some authors have demonstrated that critical shear stress is some complex function of shear strength, clay content, structure, and other geotechnical properties (Parchure and Mehta 1985), and the erosion rate can be influenced by bulk density, average particle size, particle size distribution, mineralogy, organic content, volume of gas in the sediment, salinity of the pore waters, time after deposition, and so on.

Figures 7.34 and 7.35 show the overall trend, both in the laboratory and field experiments, that the higher the strength of the surficial sediment, the higher the threshold velocity and critical shear stress and the lower the erosion rate. In other words, when the sediment is characterized as having high strength, which could be

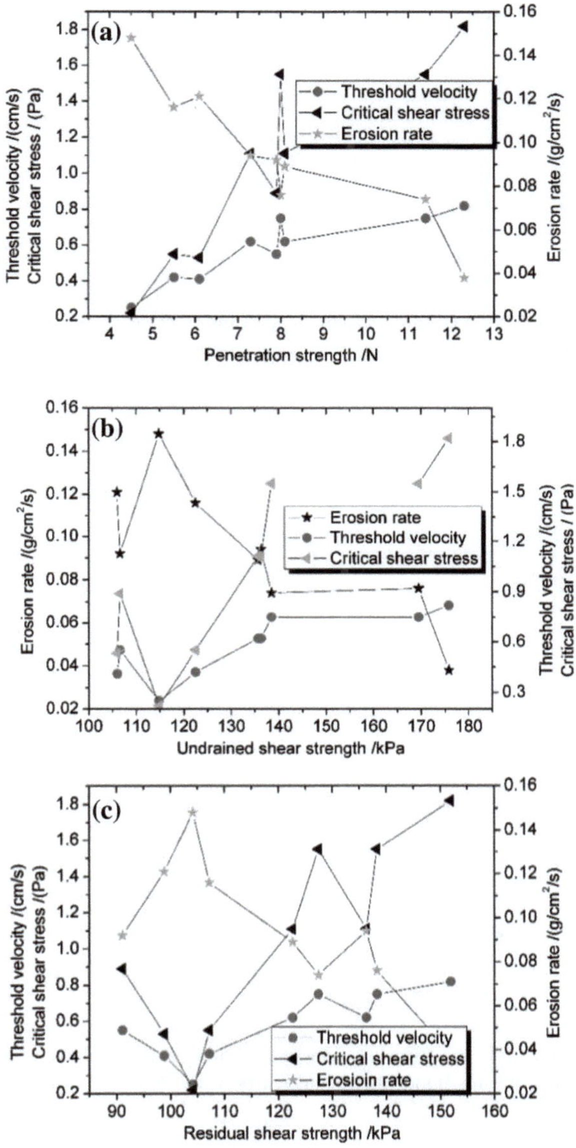

Fig. 7.32 Variation in sediment erodibility of threshold velocity, critical shear stress, and erosion rate with sediment strength of penetration strength (**a**), undrained shear strength, (**b**) and residual shear strength (**c**) in laboratory experiments, where the sediment samples are all well-consolidated and compacted with high strength

penetration strength, undrained shear strength, or residual shear strength, it could not be eroded by the shear stress exerted by the flows. Therefore, with regard to the homogeneous seabed of fine particle sediment in the process of both consolidation and unconsolidation, erodibility can be indicated closely by the strength parameters, which can be more easily measured by the designed devices than sediment erodibility.

Sediment erodibility not only relates to the mechanical properties of the sediment, but also to its physical properties and grain-size distribution (Figs. 7.34 and 7.35).

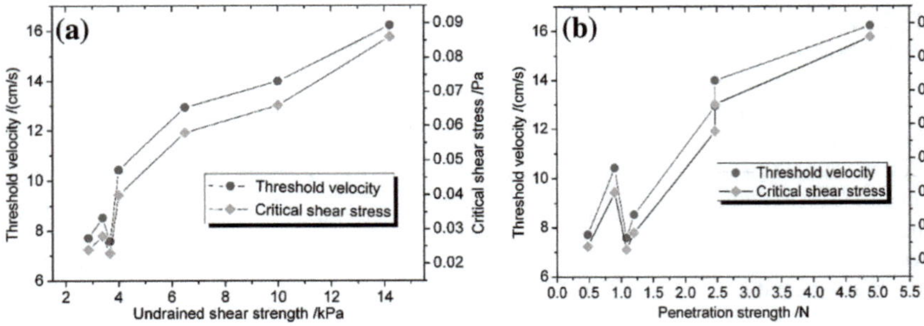

Fig. 7.33 Variation in sediment erodibility of threshold velocity and critical shear stress with sediment strength of **a** undrained shear strength and **b** penetration strength in field experiments, where the sediment samples were flow mud or poorly consolidated

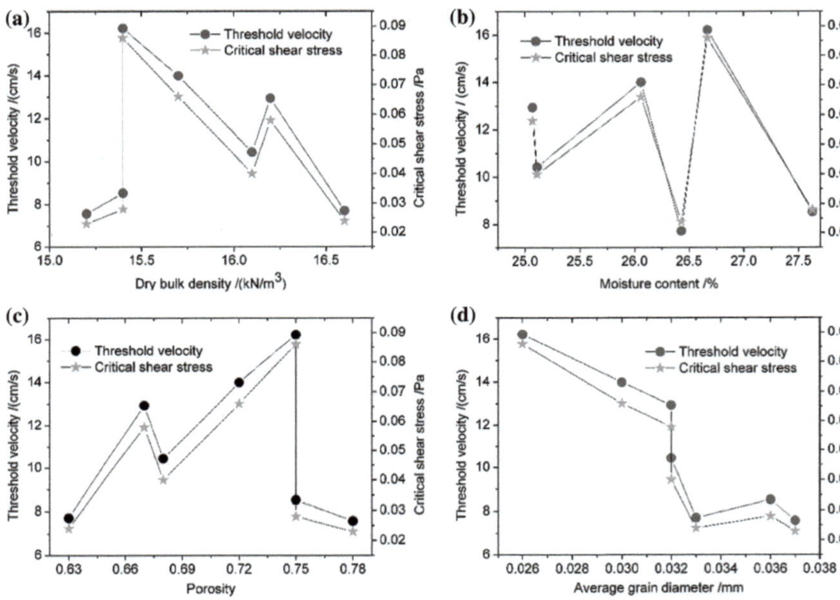

Fig. 7.34 Variation in sediment erodibility of threshold velocity and critical shear stress with **a** dry bulk density, **b** moisture content, **c** porosity, and **d** average grain diameter in field experiments

For each shear stress, erosion rate was related uniquely to the local bulk density determined by the particle size and mineral and that the rapid decrease in the erosion rate occurred as the bulk density increased, which is equivalent to the effect of decreasing moisture content (Aberle et al. 2004). However, as shown in Fig. 7.36a, b, monotonicity does not exist either in the relationship between sediment erodibility and dry bulk density or moisture content.

A number of previous investigations of sediment erodibility pointed out the significant role of surficial porosity in controlling sediment erodibility, sediment with

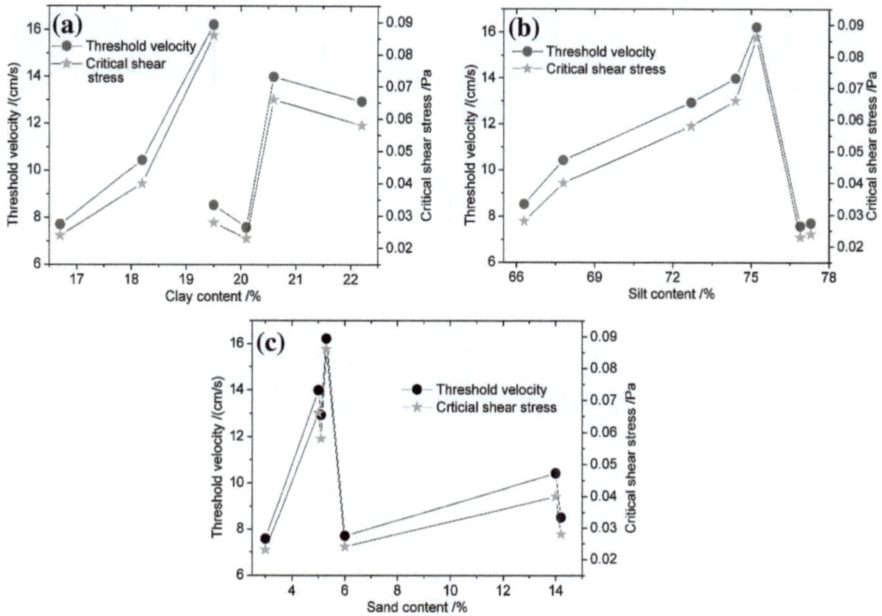

Fig. 7.35 Variation in sediment erodibility of threshold velocity and critical shear stress with **a** clay content, **b** silt content, and **c** sand content in field experiments

low porosity was more easily eroded than sediment with higher porosity. In this study (Fig. 7.34c), there were two stages in the variation of sediment erodibility with porosity, which can be described as follows: first, sediment erodibility rapidly increased with an increase in porosity, reached a maximum, and then decreased at higher porosities. Furthermore, porosity has long been known to depend on particle size; hence, sediment erodibility could change with the average grain diameter (Fig. 7.34d), which was presented in this study as: the greater the average grain diameter of the surficial sediment, the lower the erodibility of the sediment. However, Roberts et al. (1998) found that it increased rapidly for the smaller particles, reached a maximum, and then decreased rapidly for the larger particles.

Some researchers have demonstrated the significant effect of particle size distribution on erosion rate. The capability of a cohesive soil to resist erosion increases with clay content and plasticity index. Aberle et al. (2004) found that erodibility decreased with increasing sand content when other conditions remained unchanged. Sediment erodibility was closely correlated with the silt content determined the bulk density. As shown in Fig. 7.35, within the different ranges of particle content for the clay, silt, and sand, the influence of each on sediment erodibility is different. When sediment is composed by a certain percentage of clay, silt, and sand content, sediment erodibility can reach a maximum value and this percentage of clay, silt, and sand content is defined as the "optimal percentage". For the sediment tested in this study, the optimal percentages for maximum erodibility was a clay content of ~19.4%, a

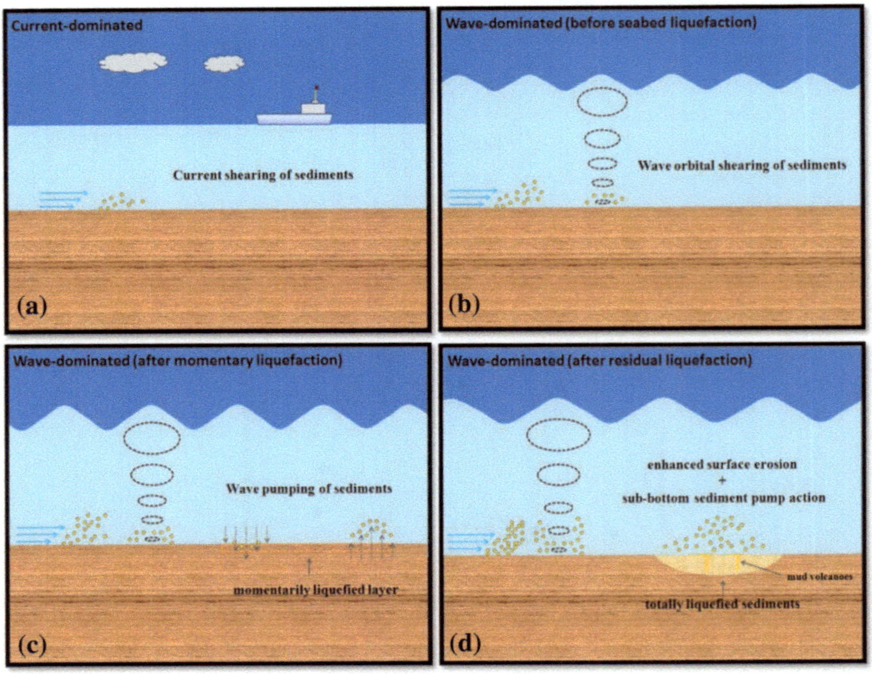

Fig. 7.36 An improved conceptual model for silty sediment resuspension: **a** shear erosion of sediments by current flows, **b** enhanced erosion from wave orbital shearing, **c** wave pumping of sediments, and **d** enhanced shear erosion due to the attenuation of the sediment erodibility (e.g., Clukey et al. 1985) and sub-bottom sediment pumping action in liquefied sediments

silt content of ~75.3%, and a sand content of ~5.3%. When the ratio of three particles content deviates from the optimal percentage value, erodibility can decrease to different degrees, but the content cannot consist of just one type of particle.

However, there are some discrepancies in the literature, which may result from the distinctive nature of the flow sediment under the influence of wave loadings, which is different from the normally consolidated sediment. Further work should be conducted to investigate its unique characteristics. On the other hand, a series of experiments indicated that traditional physical property sampling procedures were not adequate to predict the resuspension potential of surficial sediment. Therefore, the limitation of the test methods adopted in this study may also have contributed to the discrepancy to some extent.

The sediment of the Yellow River Delta can be rapidly consolidated and an over-consolidated hard shell may be formed within the surface sediment 1–2 days after deposition, which led to the shear strength of the seabed sediments generally increasing to several kilopascals. Using numerical simulations, Gao and Jia (2002) concluded that the magnitudes of the shear stresses under the combined wave and current loading were only several pascals at the soil bed surface. From a comparison of the two values, it seemed difficult for the wave/current to erode and re-suspend the seabed sediments. However, the stiff sediment may be quickly liquefied resulting in its strength to rapidly decreasing, even under slight vibrations. Wave-induced pore pressure accumulation in the liquefied cohesive sediments was found to enhance bed erosion and suspension (Aldridge and Rees 1997).

The phenomenon of the wave-induced accumulated pore pressure and the resultant soil bed liquefaction has been studied by numerous researchers. Foda and Tzang (1994) found that silt bed was liquefied after a sudden increase in pore pressure under wave action. Using flume experiments, de Wit and Kranenburg (1997) established the pore pressure threshold at which soft soil liquefaction will occur. Through one-dimensional experimental studies, Zen and Yamazaki (1990) and suggested that the mechanism of the wave-induced seabed liquefaction was dominated by excess pore pressure redistributions, which were caused by wave damping and phase lags. Sumer et al. (2006) concluded that the silty sediment is liquefied when the "accumulated" excess pore pressure reaches its maximum value. Feng (1992) reported that the liquefaction depth was positively correlated to wave height and reversely correlated to consolidation time.

We hypothesize that cyclic wave loading may lead to an increase in accumulative pore pressure and a decrease in effective stress, which may result in a reduction of the seabed's critical shear stress until the sediment is fully liquefied, and subsequently, soil bed sediment resuspends under an altered, liquefied sediment condition. Moreover, among the previous studies, for example, wave dynamic loading was also found to disengage fine-grained sediment from the seabed skeleton (Bennett 1977). The sediment was transported upward under hydrodynamics, which led to the suspension of the silty sediment and further transportation to the surrounding waters. The horizontal wave shear stress and cyclic loads of vertical pressures have been recognized as two key factors dominating each state transition of the silty sediment during wave action. The specific combination of the two forces may cause some variation in the seabed sediment resuspension process, especially in the nearshore seabed.

However, few studies have focused on the internal responses of sediment under wave action with regard to its role in sediment resuspension. Fluidisation and its subsequent upward fluid propagation caused the cohesive sediment particles to separate, and essentially affected the fluvial erosional strength of sediments. Foda and Tzang (1994) initially observed that the phenomenon of sediment transportation and suspension was much more spectacular over fluidised beds than in unfluidised beds under the action of non-breaking waves. Tzang (1998) observed in a laboratory flume that the silt seabed formed a liquid surface layer after a short time and produced suspended materials. More recently, Tzang and Ou (2006) further explored some parameters inside fluidisation beds for the SSC and mechanism of sediment suspension, and found that the SSC occurs several wave cycles after the occurrence of the fluidisation response. Therefore, liquefaction plays an important role in resuspension, the effect of wave-induced liquefaction in silt sediment on the resuspension feature and component is critical to understanding the mechanism of sediment resuspension. However, to date this aspect has not been investigated experimentally in the laboratory, not to say in the field. The objectives of this chapter are to use fine-grained sediment from the Yellow River Delta, exploring the physical mechanisms for how seabed liquefaction will influence the sediment resuspension in the Yellow River Delta.

In estuarine and coastal regions, wave-induced seabed instability can be classified into two major types: shear failure and liquefaction, which could be due to wave-induced shear stress and the pressure gradient of wave loadings, respectively. However, in research on seabed erosion, most studies have focused only on external shear-induced sediment erosion and on the evaluation of the inherent resistance strengths of seabed sediment (e.g., Aberle et al. 2004). The lack of knowledge about the influence of the pressure gradient of wave loadings on sediment erodibility is that for the certain seabed sediment under waves, the properties can be changed by the wave loadings, which implies that erodibility cannot be due to the characteristics of the seabed sediment, but also relates to the pressure gradient of the wave loadings (Figs. 7.26, 7.27, 7.28, 7.29 and 7.30).

A number of studies have shown the complicated dynamic response of seabed sediment to wave loadings, such as nonuniform consolidation, physical and mechanical properties, and granular composition and microstructure. Additionally, the variation in physical and mechanical properties, particle distributions with wave loadings of different magnitudes, and cycle number is presented in Figs. 7.26, 7.27, 7.28, 7.29, 7.30, 7.31, 7.32, and 7.33 for both the laboratory and field experiments. Sediment properties can control the erodibility. Therefore, wave loadings can change sediment erodibility by changing its controlling factors, such as sediment strength, bulk density, moisture content, and so on.

Furthermore, the dynamic response of sediment erodibility of certain seabeds to wave loadings depends on the related parameters of the waves, such as magnitude, cycle number, and period. Sediment erodibility has a different sensitivity to wave loadings in different conditions, as shown in Figs. 7.26, 7.28 and 7.30, which may result from the complex interaction between wave loadings and seabed sediment and interdependence with the characteristics of the seabed sediment, which still needs further research.

7.4.4 *Physical Mechanisms for the Attenuation of Erodibility Under Waves*

In this section, we proposed three modes of detailed mechanism accounting for influence. The first, also the most direct, mechanism is the attenuation or even loss of cohesive force between soil particles. It is well known that cohesive sediments are cemented in the form of aggregates normally, and this will be broken up when the seabed is fluidized. For that fluidization is a process that solids are transformed into fluids due to the increase in pore pressure and the accompanying loss of interparticle friction. In this case, both the volume and mass of the aggregates to be eroded are dispersed into separate particles. Consequently, the original erosion resistance is decreased. The second mechanism was referred to in the previous text that strong upward seepage forces acting on surface sediments provide additional uplifting forces and thus partially offset the particle gravity. This opinion has been commonly accepted and employed to interpret the effect of upward seepage on sediment motions (e.g., Santos et al. 2012). Last but not least, a rarely mentioned mechanism is that the enhanced erodibility of fluidized silts is partially due to a process namely sub-bottom "Sediment Pump Action" (SPA) SPA refers to a process that fine-grained sediments are transported vertically by upwelling seepage flows from the interior to the surface of the fluidized deposits, which is exactly the physical mechanism of "micro mud volcanoes" appeared on the surface. As the pore fluid velocities are sufficient to detach the fines from the soil skeletons, "internal suspension" (Tzang 1998) occurs; further, when the pore fluid velocities are sufficient to transport the detached fines between the soil skeletons, "SPA" occurs. This process will surely add the fine content to the surface sediments and thus enlarge the Shield parameter, finally contribute to the increase of sediment erodibility.

7.5 Summary

This chapter examined the detailed mechanisms for the sediment resuspension process under waves in the Yellow River Delta. We found sediment resuspension can be significantly influenced by the wave–seabed interactions. First, there will be a transient pumping of sediments due to wave-generated transient seepage flows. Second, there will be a residual pumping of sediments due to wave-generated residual seepage flows. Third, there will be an attenuation of sediment erodibility when seabed is liquefied by waves.

For the first mechanism, a specially designed in situ benthic chamber was developed to continuously measure sediment resuspension that is caused by wave-induced oscillatory seepage effects (i.e., the wave pumping of sediments). We found the WPS comprised 20–60% of the total suspension in the continuous presence of normal waves (H_s ~ 1–1.5 m). For the second mechanism, our flume experiment indicated that (1) fine particles, mainly clay and fine silt (<10 μm), are transported upward during wave loading, by the wave-driven pore water flow through the sediment. The residual pore pressure is primarily responsible for the seepage carrying fine particles that results in seabed stratification. To examine the third mechanism, i.e., variability of erodibility of silty sediments under wave loadings, straight flumes and annular flumes were built, tested, and deployed in the laboratory and field, respectively. Results indicated that there is a linear reduction in erodibility with an increasing magnitude and cycle number of wave loadings.

To summarize the historical consensus has been that sediments are largely eroded by bottom shear stress (e.g., Van Rijn 1989), which is related to the roughness of the seabed (Bartholdy et al. 2010) and mean current velocity (e.g., Sanford et al. 2001). Typically, the magnitude of resuspension is characterized by the erosion rate (resuspension flux), which is proportional to the differential, or ratio of, the bottom shear stress and critical shear stress, namely, the excess shear stress. Although many experimental works were conducted to test the role of the effect of wave-induced seepage on sediment erosion or transport rates, few published studies have proposed a summative conceptual model for silt resuspension that fully considers all the aspects of wave effects. According to the findings in this book, we finally propose an improved conceptual model for the resuspension mechanism of silty sediments for the first time (Fig. 7.36).

As shown in Fig. 7.36, sediments are largely eroded from the seabed surface when the natural hydrodynamics are dominated by currents because of the resultant shearing effect. However, larger quantities of sediments are eroded by an enhanced wave orbital shearing effect in wave-dominated environments. The momentary seabed liquefaction of shallow layers may occur as waves become larger or last for a longer duration, so sediments are partially resuspended through the wave pumping of sediments. In extreme circumstances of continuous storm waves, residual seabed liquefaction may occur (Xu et al. 2009) and thus enhance erosion because of the attenuation of the erosion threshold of surface sediments (e.g., Clukey et al. 1985) and the vertical migration of fine-grained sediments from the interior to the surface of a liquefied seabed, which is driven by upward seepage flows, namely, sub-bottom sediment pumping action. This improved conceptual model for silty sediment resuspension may have some implications for numerical works, which have rarely parameterized the influence of seabed fluidization/seepage flows into the erosion/resuspension module.

References

Aldridge JN, Rees JM (1997) Interpreting observations of near-bed sediment concentration and estimation of 'pick-up' function constants, pp 289–303. In: Burt N, Parker R, Watts J (Eds.), Cohesive sediments. Wiley, Chichester, 458 pp

Aberle J, Nikora V, Walters R (2004) Effects of bed material properties on cohesive sediment erosion. Mar Geol 20(7):83–93

Bennett RH (1977) Pore-water pressure measurements: Mississippi delta submarine sediments. Mar Geotechnol 2(1–4):177–189

Baldock TE, Holmes P (1999) Seepage effects on sediment transport by waves and currents. Coast Eng 1998:3601–3614

Bartholdy J, Flemming BW, Ernstsen VB et al (2010) Hydraulic roughness over simple subaqueous dunes. Geo-Mar Lett 30(1):63–76

Clukey EC, Kulhawy FH, Liu PLF et al (1985) The impact of wave loads and pore-water pressure generation on initiation of sediment transport. Geo-Mar Lett 5(3):177–183

Chu ZX, Sun XG, Zhai SK et al (2006) Changing pattern of accretion/erosion of the modern Yellow River (Huanghe) subaerial delta, China: based on remote sensing images. Mar Geol 227(1):13–30

De Wit PJ, Kranenburg C (1997) The wave-induced liquefaction of cohesive sediment beds. Estuar Coast Shelf Sci 45(2):261–271

Feng J (1992) Laboratory experiments on cohesive soil-bed fluidisation by water waves (Thesis, Report No. UFL/COEL-92/005). University of Florida, Gainesville, USA p. 108

Foda MA, Tzang S (1994) Resonant fluidization of silty soil by water waves. J Geophys Res 99:20463–20475

Friedrichs M, Graf G, Van Duren LA (2006) Bio-fluid-dynamics: Exchange processes at the sediment-water interface. J Sea Res 55(1):1–2

Green MO, Coco G (2014) Review of wave-driven sediment resuspension and transport in estuaries. Rev Geophys 52(1):77–117

Gao S, Jia JJ (2002) Modeling suspended sediment distribution in continental shelf upwelling/downwelling settings. Geo-Mar Lett 22(4):218–226

Grabowski RC, Droppo IG, Wharton G (2011) Erodibility of cohesive sediment: The importance of sediment properties. Earth Sci Rev 105:101–120

Guillén J, Jiménez JA, Palanques A, Gracia V, Puig P, Sánchez-Arcilla A (2002) Sediment resuspension across a microtidal, low-energy inner shelf. Cont Shelf Res 22(2):305–325

Hooshmand A, Horner-Devine AR, Lamb MP (2015) Structure of turbulence and sediment stratification in wave-supported mud layers. J Geophys Res Oceans 120:2430–2448

Jia Y, Zhang L, Zheng J, Liu X, Jeng DS, Shan H (2014) Effects of wave-induced seabed liquefaction on sediment re-suspension in the Yellow River Delta. Ocean Engineering, 89:146–156

Jeng DS, Zhao HY (2015) Twodimensional model for pore pressure accumulations in marine sediments. J Waterway Port Coast Ocean Engineering, ASCE 141:04014042

Lesht BM, Clarke TL, Young RA, Swift DJP, Freeland GL (1980) An empirical relationship between the concentration of resuspended sediment and near bottom wave-orbital velocity. Geophys Res Lett 7(12):1049–1052

Myrhaug D, Holmedal LE et al (2014) A rational approach to seepage flow effects on bottom friction beneath random waves. Appl Ocean Res 47:322–328

Mörz T, Karlik EA, Kreiter S, Kopf A (2007) An experimental setup for fluid venting in unconsolidated sediments: New insights to fluid mechanics and structures. Sed Geol 196:251–267

Maa PY, Wright LD, Shannon TW (1993) VIMS Sea Carousel: A field instrument for studying sediment transport. Mar Geol 115(3):271–287

Neumeier U, Lucas CH, Collins M (2006) Erodibility and erosion patterns of mudflat sediments investigated using an annular flume. Aquat Ecol 40:543–554

Nichols RJ, Sparks RSJ, Wilson CJN (1994) Experimental studies of the fluidization of layered sediments and the formation of fluid escape structures. Sedimentology 41:233–253

Paphitis D, Collins MB (2005) Sediment resuspension events within the (microtidal) coastal waters of Thermaikos Gulf, northern Greece. Cont Shelf Res 25(19):2350–2365

Precht E, Huettel M (2004) Rapid wave-driven advective pore water exchange in a permeable coastal sediment. J Sea Res 51(2):93–107

Parchure TM, Mehta AJ (1985) Erosion of soft cohesive sediment deposits. J Hydraul Eng 111(10):1308–1326

Prior DB, Suhayda JN, Lu NZ, Bornhold BD, Keller GH, Wiseman WJ, Wright LD, Yang ZS (1989) Storm wave reactivation of a submarine land-slide. Nature 341(6237):47–50

Roberts J, Jepsen R, Gotthard D, Lick W (1998) Effects of particle size and bulk density on erosion of quartz particles. J Hydraul Eng 124(12):1261–1268

Simon A, Collison AJC (2001) Pore-water pressure effects on the detachment of cohesive streambeds: seepage forces and matric suction. Earth Surf Proc Land 26:1421–1442

Sumer B.M., Fredsøe J. (2002) The mechanics of scour in the marine environment. World Scientif 552

Sanford L, Maa JPY (2001) A unified erosion formulation for fine sediments. Mar Geol 179:9–23

Seed HB, Rahman MS (1978) Wave-induced pore pressure in relation to ocean floor stability of cohesionless soils. Mar Georesour Geotechnol 3(2):123–150

Sassa S, Sekiguchi H (2001) Analysis of wave-induced liquefaction of sand beds. Geotechnique 51:115–126

Santos IR, Eyre BD, Huettel M (2012) The driving forces of porewater and groundwater flow in permeable coastal sediments: A review. Estuar Coast Shelf Sci 98:1–15

Schaaff E, Grenz C, Pinazo C, Lansard B (2006) Field and laboratory measurements of sediment erodibility: A comparison. J Sea Res 55(1):30–42

Sumer BM, Hatipoglu F, Fredsøe J, Sumer SK (2006) The sequence of sediment behaviour during wave-induced liquefaction. Sedimentology 53:611–629

Smith AJ, Herne DE, Turner JV (2009) Wave effects on submarine groundwater seepage measurement. Adv Water Resour 32(6):820–833

Tzang SY (1998) Unfluidized soil responses of a silty seabed to monochromatic waves. Coast Eng 35:283–301

Tzang S, Ou S (2006) Laboratory flume studies on monochromatic wave-fine sandy bed interactions: Part 1. Soil fluidization. Coastal Engineering 53:956–982

Thompson CEL, Couceiro F, Fones GR et al (2011) In situ flume measurements of resuspension in the North Sea. Estuar Coast Shelf Sci 94(1):77–88

Van Rijn L C (1989) Sediment transport by waves and currents. Report H461. Delft Hydraulics

Willets BB, Drossos ME (1975) Local erosion caused by rapid infiltration. J. Hyd. Div. ASCE 101:1477–1488

Wolanski E, Spagnol S (2003) Dynamics of the turbidity maximum in King Sound, tropical Western Australia. Estuar Coast Shelf Sci 56:877–890

Wright LD, Boon JD, Xu JP et al (1992) The bottom boundary layer of the bay stem plains environment of lower Chesapeake Bay. Estuar Coast Shelf Sci 35(1):17–36

Widdows J, Brinsley MD, Salkeld PN et al (1998) Use of annular flumes to determine the influence of current velocity and bivalves on material flux at the sediment-water interface. Estuaries Coasts 21(4):552–559

Wang HJ, Yang ZS, Li YH et al (2007) Dispersal pattern of suspended sediment in the shear frontal zone off the Huanghe (Yellow River) mouth. Cont Shelf Res 27:854–871

Xu G, Sun Y, Wang X et al (2009) Wave-induced shallow slides and their features on the subaqueous Yellow River delta. Can Geotech J 46(12):1406–1417

Yamada F, Tsujimoto G, Long B (2011) Internal sediment density structures around bars due to beach deformation using X-ray CT. J Coastal Res 64:716–720

Zen K, Yamazaki H (1990) Mechanism of wave-induced liquefaction and densification in seabed. Soils Found 30(4):90–104

Chapter 8
Theoretical Prediction of Wave-Induced Sediment Resuspension

8.1 Overview

As the wave loading continues, pore pressure builds up and seepage flows generates, seabed sediments gradually transfer from solid state into liquid state. In the field environment, sediments can be totally liquefied or partially fluidized as wave loading time varies. Partially fluidized also called as less-consolidated sediments refer to sediments that are not in normally consolidated state but with excess pore pressure in them. Because fluidization and consolidation are two reversible processes. Partially Fluidized Sediments (PFS) distribute extensively on the shelf seafloors or along the banks of estuaries (Saffer et al. 2000). The subaqueous Yellow River Delta (YRD) is also widely covered by PFS, which are mainly formed from two processes: (1) fast sedimentation of the Yellow River sediment discharge, in which the excess pore pressure could not be squeezed out in time (Li et al. 2000); (2) the normally consolidated seabed transferred into PFS by frequent wave events via causing active slope failures, seabed fluidizations (Clukey et al. 1985), sediment collapses (Prior et al. 1989), silt flows (Prior et al. 1986), and gravity underflows (Wright et al. 1988).

It is intuitively believed that PFS is eroded more easily than those well-consolidated sediments under the action of surface waves and tidal currents. That is, sediments with various excess pore pressure may have different erodibilities. However, sediment erodibility is particularly important for the prediction of local scour and sediment transport. Critical shear stress for erosion (τ_{cr}) and erosion rate (E_r) are the two most important parameters to characterize erodibility. In general, erosion rate is assumed to be proportional to the excess of bottom shear stress (τ_b) above the critical shear stress (τ_{cr}) (van Rijn 1989; Neumeier et al. 2008). Previous research indicated that τ_{cr} is related to site-specific characteristics of the sediments, including the particle grain size, density, cohesiveness, water content, and biological binding (Sanford and Maa 2001). In fact, the cohesiveness, water content and biological binding factors will be significantly changed when a consolidated seabed is transferred into a fluidized seabed or in the reverse process. In other words, erodibility of a specific site also can be time-varying (Maa and Lee 1997). However, the existing

Y. Jia et al., *Wave-Forced Sediment Erosion and Resuspension in the Yellow River Delta*, Springer Oceanography, https://doi.org/10.1007/978-981-13-7032-8_8

sediment transport models usually set τ_{cr} as a constant value for a scope of area or a period of time (Qiao et al. 2016), making the τ_{cr} in the models a representive parameter rather than an accurate physical parameter. This simplification is a benefit for the efficiency of mathematical operations rather than the physical meanings. Therefore, more parameterization works are still required to better consider the temporal changing of erodibility due to the fluidization in sediment transport models to date.

In order to overcome the shortcomings of the existing models, numerous studies have been conducted to include the influence of fluidization in the erosion process of sediments. Dade et al. (1992) early reported that critical flow speed (V_{cr}) for sediment entrainment may go down by 10 times as the water content varies from 50% (consolidated) to 90% (unconsolidated). Nielsen (1997) parameterized the seepage effects on sediment mobility using the modified Shield parameter. However, his research mainly focused on seepage effect in non-cohesive sands. Foda and Huang (2001), Rodriguez and Mehta (2000), Wolanski and Spagnol (2003) considered the effects of seabed fluidization on sediment transport by introducing wave parameters into traditional erosion formulations. However, their parameterizations are implicit without clear physical meanings, because wave parameters are indirect indicators for fluidization, the most direct parameter is the pore-water pressure. Therefore, unlike the conventional erosion item that is caused by the shear stress, Lambrechts et al. (2010) considered the influence by adding a new erosion source item that is caused by fluidization. However, it is a pity that the pore pressure, in their study, is estimated from an empirical formula, rather than real measurements. In the chapter, the present book proposed an attenuation curve to describe the relationship between critical shear stress (τ_{cr}), critical flow speed for entrainment (V_{cr}) and the fluidization degree of YRD sediments. However, wave cycles were employed to indirectly indicate the various fluidization degree again, thus no parameterization formulations that are convenient for engineering calculation or numerical modeling was reported in their literature. Recently, a more explicit parameterization on this issue was performed by Zhang et al. (2017). They proposed a modified equation for erosion rates by introducing the yield stress, which is a rheological parameter of sediments, into the Partheniades (1962) erosion formulation. This provides a better calculation of the erosion rate of fluidized cohesive sediments from the Hangzhou Bay of China.

Considering the rheological parameter employed by Zhang et al. (2017) is inconvenient and almost impossible to measure in the field, in this chapter we tried to parameterize pore pressure, an easily measured also the most direct parameter to describe fluidization, into conventional erosion formulation.

8.2 Modification of Sediment Resuspension Model Considering Wave Liquefaction

The first objective of this chapter is also to explicitly parameterize the influence of sediment fluidization degree on the erosion process of silty sediments in the YRD, where PFS widely distributed. Therefore, we tried to construct a modified erosion model that considers the effect of wave-induced pore pressure responses for the Yellow River silts through a series of controlled experiments. To achieve this goal, a more convenient annular flume than that of Zhang et al. (2017) was developed.

8.2.1 Methodology

(1) *Annular Flow Flume*

The experiments were conducted in a self-designed annular flume, which had overall dimensions of 1.8 m in length and 1.1 m in width, creating a 0.3 m wide and 0.2 m high channel for water flow, the experimental water depth was 0.1 m (Fig. 8.1). The shape design and size referenced the advanced laboratory (Stone et al. 2008) or in situ flumes for erosion research (e.g., Voyager II) (Thompson et al. 2011), thus the scaling effect is acceptable. A soil tank with a total volume of 0.4 (L) × 0.3 (W) × 0.3 (D) $= 0.36\ \text{m}^3$ was designed to locate at the end of the flow route, where water flow near the seabed has become stable (Sumer and Fredsøe 2002). Five stepwise increased flow speeds from 16 to 55 cm s^{-1} were generated by a rotating cylinder with paddles that was driven by an engine. The flow channel before arriving the tank is designed as rectangular to reduce secondary flows, to generate horizontal uniform flow speeds on the seabed, although the flow structure near the boundary of flume wall will be slightly influenced inevitably. However, as the present study specially concerns with

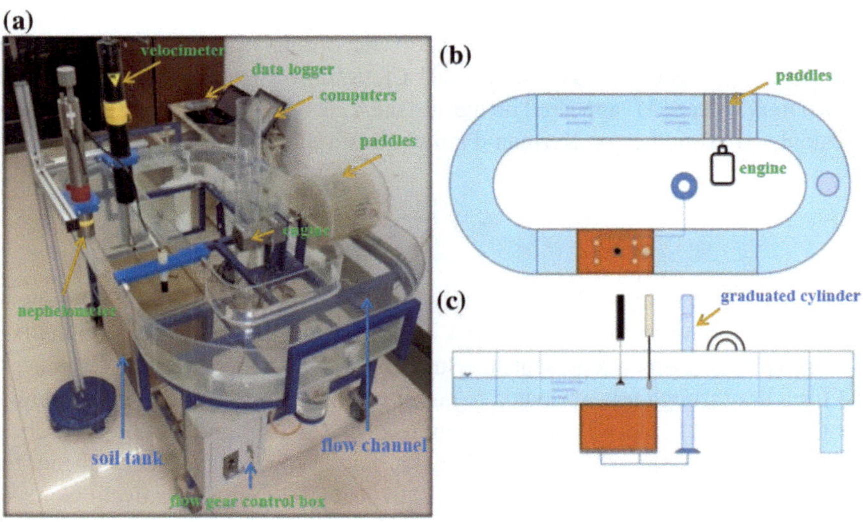

Fig. 8.1 Experimental system **a** the annular flume, velocimeter, nephelometer, data logger for pore pressure sensor and associated recording computers; **b** schematic top view of the setup; and **c** schematic side view of the setup. Note: the graduated cylinder in the central of the flume is designed for simulating seepage gradients into the sediment, this function is not used in this study

the sediment entrainment and seepage flows at the water-seabed interface, this effect could be ignored.

(2) ***Instrumentation***

A highprecision pore pressure sensor (0.001 Pa) was buried 10 cm below the surface of the sediment to record the pore pressure evolution at a frequency of 1 Hz. Flow speed profiles were measured using a portable velocimeter, with its propeller sensor mounting in the mid-channel above the sediments. In the pretests before the normal experiments, total flow profiles were measured for the estimation of the bottom shear stresses. Measured flow speeds are shown in Fig. 8.2.

It is clear that flow speed overall increased from top to the bottom, locally a decrease was observed when the water flow was affected by the underlying seabed (Dyer 1980), forming the bottom boundary layer (BBL). The local flow speed profile (below 3.5 cm) fit the log-profile well (Fig. 8.2b), indicating the presence of a constant shear stress layer. Therefore, the bottom shear stress could be estimated using the measured speed at any depth within that layer. For the present study, the bottommost speeds at 1.5 cm above the bottom were employed.

With low seabed gradient and no obstructions causing modifications to near-bed flow structure, bed shear stresses (τ_b) induced by unidirectional flows is commonly calculated by

$$\tau_b = C_D \rho_w u_*^2, \tag{8.1}$$

where $\rho_w = 1.0 \times 10^3\,\text{kg/m}^3$ is the density of water; $C_D = \alpha(z_0/z)^\beta$ is the drag coefficient. In the Manning–Strickler law, $\alpha = 0.0474$ and $\beta = 0.333$, while Dawson (1983) suggested $\alpha = 0.0190$ and $\beta = 0.208$. Accordingly, the calculated C_D were 0.02 and 0.01, respectively. Herein, a larger C_D (0.02) was employed because of the "micro mud volcanoes" appeared on the surface.u_* is the friction velocity which can be calculated using the von Karman–Prandtl equation:

$$u_* = \frac{u_z \cdot \text{k}}{\ln \frac{z}{z_0}}, \tag{8.2}$$

where u_z is the measured flow velocity within the constant BBL, in this study, at $z = 0.015$ m above the seabed, k = 0.4 is the von Karman constant, z_o is the bed roughness length, which represents the sum of the skin friction (particle roughness) and form drag (roughness attributed to larger elements such as bedforms) (Paphitis and Collins 2005), depends upon the nature of the seabed and its characteristics. According to Drake and Cacchione (1986), z_o was employed as 0.9 and 0.7 mm for cohesiveless silts with mean diameters of 0.07 and 0.02 mm, respectively. Soulsby (1983) suggested $z_o = 0.2$ mm for mud and 0.7 mm for a mixture of mud and sand. Considering the well-mixed flat silty seabed (with mean diameter of 0.05 mm) in this experiment, we employed a relatively larger value of $z_o = 0.75$ mm to account for the influence of "micro mud volcanoes". The calculated results are summarized in Table 8.1.

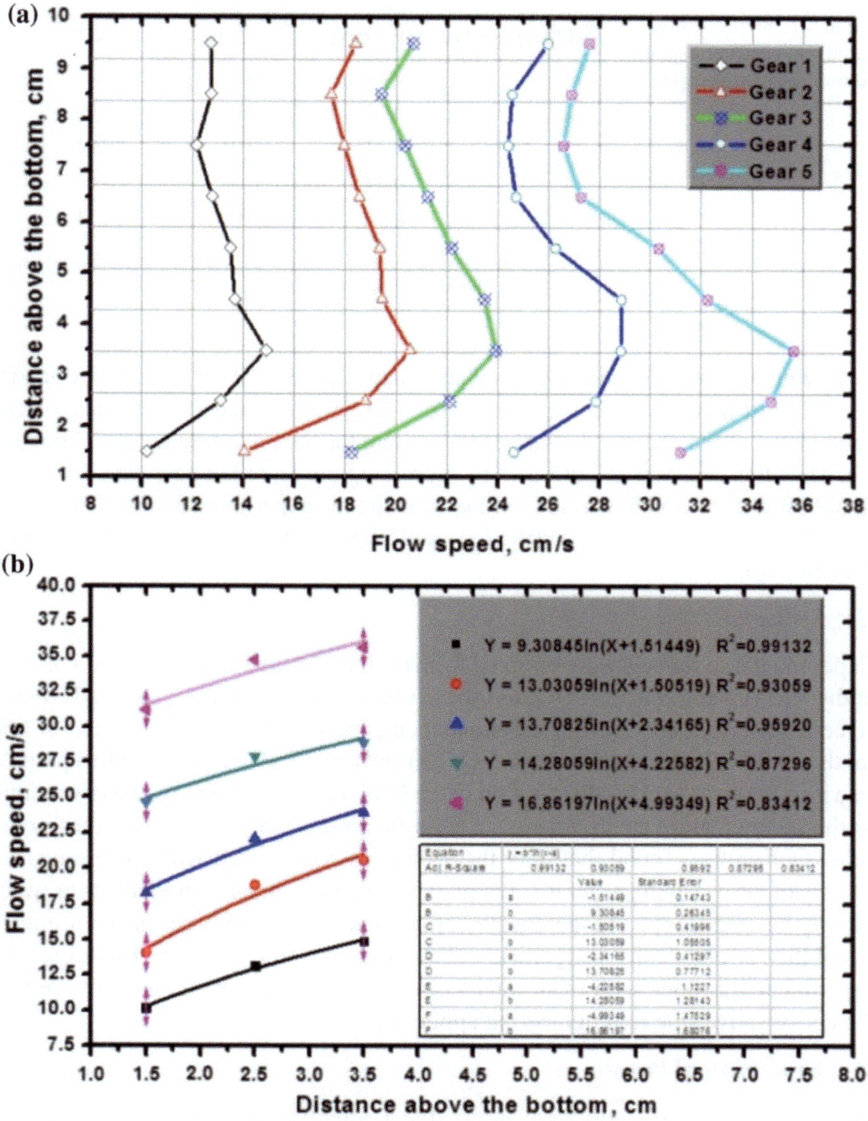

Fig. 8.2 Flow speed profiles at each flow speed (gear) **a** total profiles in the annular flume and **b** logarithmic fits of the local profiles within the constant stress layer. *Note* in figure b, Y = the flow speed and X is the distance above the bottom

Table 8.1 Stepwise flow speeds (F_s) and corresponding bottom shear stresses (τ_b)

Flow gear (G_m)	F_s (m/s)	u_*(m/s)	τ_b(Pa)
1	0.16	0.02	0.01
2	0.24	0.03	0.02
3	0.32	0.04	0.04
4	0.44	0.06	0.07
5	0.55	0.08	0.11

*G_m is the driving gear of the flow generator, F_s is the flow speed at 1.5 cm above the bottom

During the formal erosion experiments, a nephelometer (RBR Concerto, Canada) with a built-in optical backscatter sensor (OBS) was fixed downstream of the velocimeter at 9 cm above the seabed to measure the SSC at a frequency of once per 3 s. The OBS sensor was calibrated by regressing the recorded data with the turbid liquid samples (with known SSC) prepared using the experimental silts. Conversion formula was derived as $C = 0.0047\,T + 0.4457$, $R^2 = 0.97$, where C is the SSC and T is the turbidity.

(3) ***Sediments***

Experimental sediments were sampled from the bank of the abandoned Diaokou lobe of the YRD (Wang et al. 2014). This lobe has been exposed to serious seabed erosion since it was abandoned in 1976. Sediments belong to sandy silts consisting of fine sands (29.28–29.57%), silts (62.87–62.97%) and clay (7.46–7.85%), median particle size is 0.044 mm. Salinity distribution in the YRD is temporal and spatial dependent, so the sedimentation environment can be in freshwater, saltwater or the transitional zone. Our pretests found that after the sediments were mixed with freshwater, the salinity of the overlying water is −12 ppt. It has been previously reported that as water salinity beyond 10 ppt, the incremental effect of salinity on soft cohesive sediments erodibility is practically insignificant (Parchure and Mehta 1985). Therefore, the experiment was done with freshwater to ensure the parameterization results are on the conservative side for engineering applications.

(4) ***Experimental Process***

The field-derived sediments were air dried first, after which many cemented clusters of 1–2 cm in diameter formed. Then, the clusters were pulverized into dried individual particles or micro-aggregates of less than 1 mm in diameter. In this case, the pulverized soils are more appropriate for simulating the deposition of sediments from suspension. Pulverized dry sediments were then mixed with fresh water according to a constant weight proportion of 3:1 to prepare a slurry with a saturated water content of 25%. Subsequently, the slurry was backfilled into the tank to simulate homogeneous seabed of which the soil structure is quite similar to the large volume of fluidized sediments that formed from fast sedimentation in the field. The thickness of simulated seabed is slightly larger than 0.3 m including a reserved space for settlement due to consolidation, to make the seabed surface flush with the floor of the water channel. Subsequently, water was filled into the overlying flow channel to a depth of 10 cm for consolidation under static water pressure.

The single variable of the controlled experiments is the fluidization degree of the sediment bed, here fluidization degree (f_d) refers to the ratio of excess pore pressure to the final normal effective stress.

$$f_d = p_{exc}/\sigma_v^{'}, \tag{8.3}$$

where the excess pore pressure (p_{exc}) was calculated by removing the overlying static water pressure (p_{sta}) from the total pressure (p_{tot}) measured at $Z = 10$ cm subsurface:

$$p_{exc} = p_{tot} - p_{sta}, \tag{8.4}$$

In Eq. (8.3), $\sigma_v^{'}$ is the final normal effective stress, i.e., effective stress of the YRD sediments that are in normally consolidated state.

$$\sigma_v^{'} = \gamma' Z, \tag{8.5}$$

where $\gamma' = 10\text{kN/m}^3$ is the submerged unit weight of normally-consolidated YRD sediments obtained from testing of soil samples; $Z = 0.1$ m is the vertical distance from the seabed surface to the measurement of pore pressure.

In this case, since the final normal effective stress is constant, excess pore pressure can also be considered as a single variable. In this experiment, various excess pore pressure in the sediment bed (i.e., f_d) was obtained via controlling the consolidation time of the initially deposited saturated water-sediment slurry.

Evolution of excess pore pressure was derived from a preexperiment for consolidation in the tank (Fig. 8.3). It is clear that excess pore pressure decreased from ~0.96 kPa to a final constant value of ~0.03 kPa within 10 h, verifying the fast-consolidation feature of the Yellow River silts (Prior et al. 1989). Therefore, any fluidization degree $f_d = p_{ex}/\sigma_v^{'} \in 0.03\text{–}0.96$ $\left(\sigma_v^{'} = \gamma' Z = 10\,\text{kN/m}^3 \times 0.1\text{m} = 1\,\text{kPa}\right)$ can be simulated in this tank by controlling the consolidation time. In the present study, six fluidization degrees were simulated corresponding to six different consolidation time (Table 8.2).

The formal erosion experiment consists of six rounds. Experimental procedure in each round is exactly the same, except for the consolidation time of slurry. Take the first round as an example. After 1 h of consolidation, bottom sediments were exposed to five stepwise increased bottom shear stresses for flow scouring. Each bottom shear stress lasted for 600 s, thus the total period of one experimental round is 3000 s. SSC in the flume and flow speed at 1.5 cm above the seabed were measured continuously and synchronously. After all the five stepwise flow speeds were finished, the overlying turbid water was drained and the sediments were dug out. Then the flume was cleaned up to avoid residual sediments mixing to the next experimental round. In total, six rounds were conducted for comparison to investigate the influence of various fluidization degrees on the erodibility of the YRD sediments.

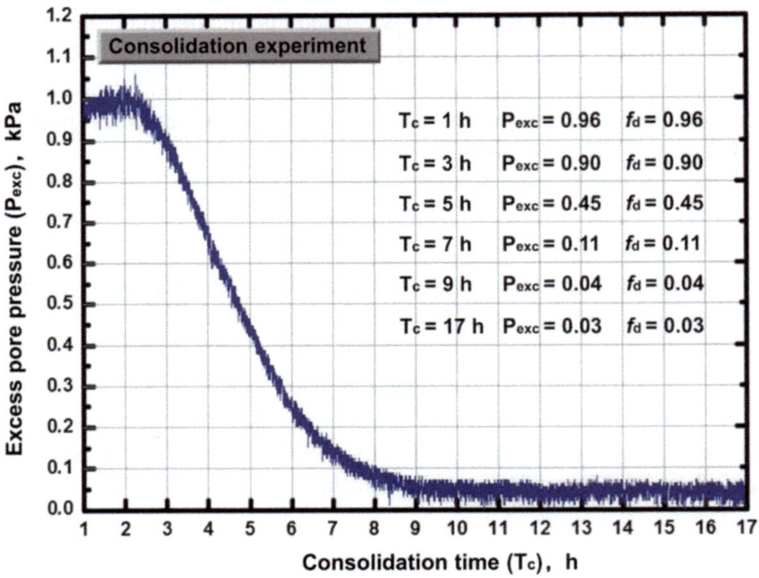

Fig. 8.3 Evolution of excess pore pressure at subsurface 10 cm of the sediments during the consolidation of the backfilled slurry. *Note* The consolidation time in this plot starts after 1h of initial deposition because adding water lasted for 1 h

Table 8.2 Seabed parameters in each experimental round

Round no.	T_c/h	p_{exc}/kpa	f_d
1	1	0.96	0.96
2	3	0.90	0.90
3	5	0.45	0.45
4	7	0.11	0.11
5	9	0.04	0.04
6	17	0.03	0.03

T_c is consolidation time, p_{exc} is excess pore pressure, and f_d is seabed fluidization degree

8.2.2 *Results*

Time evolution of the SSCs corresponding to stepwise bed shear stresses is presented in Fig. 8.4. We can find that the SSC gradually increased from ~0.45 g/l to a final level of ~0.55 g/l well corresponding to the changing of bed shear stresses. Exposed to each shear stress for 600 s, all the stepwise increases in SSC reached equilibrium states.

It is strange from Fig. 8.4 that the influence of fluidization on erosion was not monotonic. For highly fluidized sediments (e.g., $T_c = 1$ h), the entrainment of sediments seemed to be even more difficult than the other rounds especially at low flow

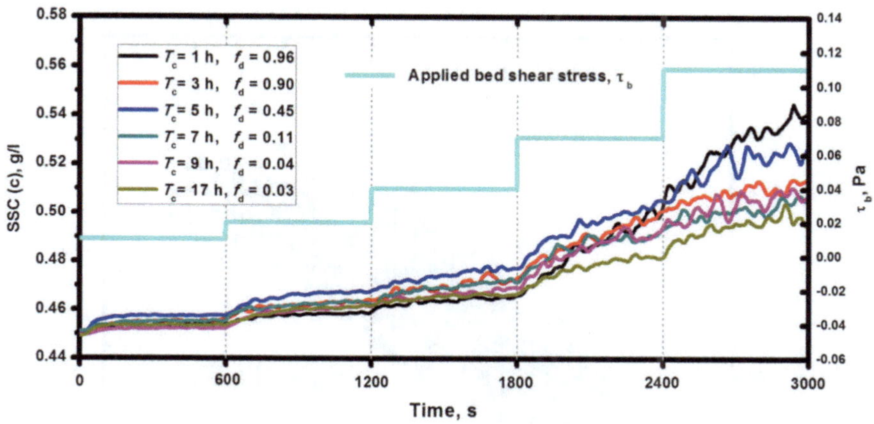

Fig. 8.4 Time series of the SSCs and applied bed shear stresses (τ_b) in all the six rounds of experiment with sediments in various fluidization degrees

speeds. However, as the speed exceeded a threshold ($\tau_b = 0.07$Pa), the SSC level surpassed all the other rounds, eventually showing a feature of high erodibility (see the black line in Fig. 8.4). Similarly, the SSC levels of round $T_c = 3$ h were also lower than those of round $T_c = 5$ h, supporting the trend mentioned above. These findings do not conform to a perception that sediment erodibility increases with fluidization degrees.

For fear of such a strange result is caused by unknown errors; repeatability experiment for the first three rounds ($T_c = 1$, 3, 5 h) was undertaken to confirm this phenomenon. The R^2 values between the data of first and repeated experiments were high to 0.91–0.95 (Fig. 8.5), proving the experimental data were sufficiently convincing. Highly fluidized sediments ($f_d = 0.90$–0.96) were indeed more resistant to entrainment than the moderately fluidized ones ($f_d = 0.44$) in this experiment. However, this apparent phenomenon is not all the truth and further analysis is conducted in a later section. Herein, after T_c exceeded 5 h, the SSC levels became monotonously decrease with the fluidization degrees.

8.2.3 Parameterization Equation Construction Between Liquefaction Degree and Erodibility

(1) ***Erosion Models***

The most important parameters for charactering sediment erosion are the critical shear stress (τ_{cr}) and the τ_{cr} dependent erosion rates (E_r), rather than the SSC. Therefore, the present paper made further analysis on the two parameters using the popular erosion formulations. Two forms of erosion formulation are commonly employed in

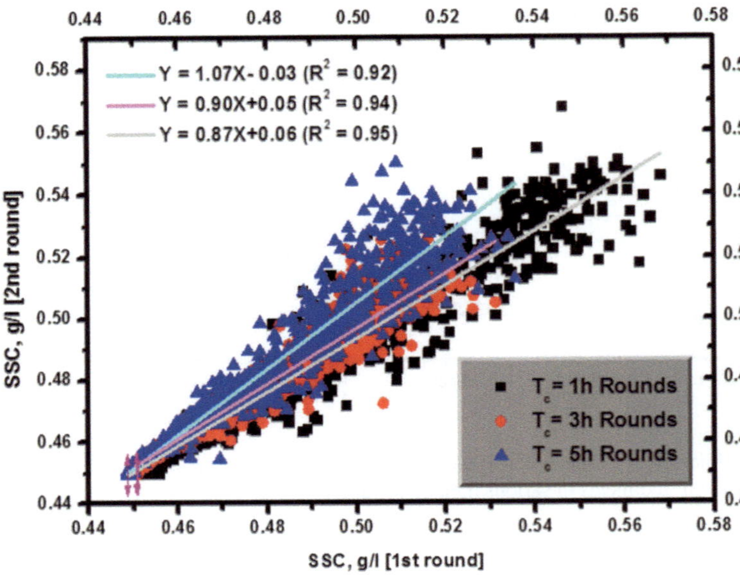

Fig. 8.5 Confirming the data reliability in rounds $T_c = 1$, 3, and 5 h by linear regression. The X axis and Y axis refer to the first measured data and repeated data for verification, respectively. Slopes of $a \approx 1$, intercepts of $b \approx 0$, and $R^2 \approx 1$ demonstrate a good consistency and high data reliability

literatures. One is exponential erosion formulation which is usually employed to describe Type I (depth-limited) erosion (Parchure and Mehta 1985; Amos et al. 1992):

$$E_r = M_e \exp[\epsilon(\tau_b - \tau_{cr}(z))^{\varphi}] \tag{8.6}$$

where E_r is erosion rate, M_e is erosion coefficient, τ_b is bed shear stress, τ_{cr} is critical entrainment stress, ϵ and φ are empirical coefficients, z is the erosion depth, for Type I erosion τ_{cr} increases with depth into the sediments.

The other form is linear erosion formulation to describe Type II (unlimited) erosion (e.g., Mei et al. 1997), with a single, constant τ_{cr} that does not change with depth into the sediments.

$$E_r = M_e(\tau_b - \tau_{cr}) \tag{8.7}$$

As the newly deposited sediments always consolidated via upward drainage, it is well accepted that the τ_{cr} of consolidating mud usually increases with depth and experience Type I erosion (Parchure and Mehta 1985). While consolidated beds almost always behave Type II features due to the insignificant variations in consolidation degree with depth (Le Hir et al. 2007). Based on this idea, most of the erosion cases in the present experiment belongs to Type I erosion, because the sediments are consolidating muds. But the erosion type is gradually transferring from I to II as the consolidation time increases.

According to the existing erosion formulations and their applicable conditions above, it seems that different erosion formulations should be employed when dealing with sediments in various consolidation/fluidization states. In fact, Piedra-Cueva and Mory (2001) reported an important observation that point to the possibility that Eq. 8.7, the simplest and least parameterized erosion formulation, may be used to describe Type I erosion as well by allowing τ_{cr} to increase with depth. Sanford and Maa (2001) further made a theoretical derivation based on Eq. 8.7 that supports and extends the finding of Piedra-Cueva and Mory (2001), and argued that a linear erosion formulation with depth-varying τ_{cr} can be used under general time-varying conditions, for either Type I or Type II erosion. In the present paper, we further support this unified erosion formulation using our experimental data. We found the liner erosion formulation could indeed well describe Type I erosion, on the condition of allowing the τ_{cr} to increase with depth. More importantly, we found the reason is the influence of fluidization on erodibility is essentially parameterized into the erosion coefficient.

Therefore, linear erosion formulation is employed to further analyze the experimental data. In this case, once a series of erosion rates (E_r) and corresponding excess shear stresses ($\tau_{exc} = \tau_b - \tau_{cr}$) were derived from the experimental data, erosion coefficient (M_e) of each round can be obtained via linear regressions. Since the single variable between each round is the fluidization degree, a parameterization equation between fluidization degree (f_d) and erosion coefficient M_e can be derived to quantificationally consider the influence of fluidization in the liner erosion formulation.

(2) ***Erosion Rates***

During each applied shear stress period, time series of measured SSC scatters are fitted with polynomials first:

$$C = \alpha t^k + \beta t^{k-1} + \gamma t^{k-2} + \varepsilon t + \eta, \tag{8.8}$$

where C is SSC, t is time in seconds and α, β, γ, ε and η are all fitted coefficients.

Erosion rate (E_r) equals to the changing rate of SSC multiply by water depth (Parchure and Mehta 1985). In this paper, we further defined an initial erosion rate (E_i) as the erosion rate when a new bed shear stress is applied, which can be calculated as

$$E_i = \mathrm{h} \lim_{t \to 1} \frac{\partial c}{\partial t} = \mathrm{h}\left[\alpha n t^{k-1} + \beta(n-1)t^{k-2} + \gamma(n-2)t^{k-3} + \varepsilon\right], \tag{8.9}$$

Since in Eq. (8.8) the SSC scatters are fitted against time in seconds, all the E_i can be obtained by inputting t = 1 s into the derivative Eq. (8.9).

As an example, Fig. 8.6 shows a series of polynomial fittings and thereby derived E_1 values in all the six rounds (where E_1 refers to the initial erosion rate under the

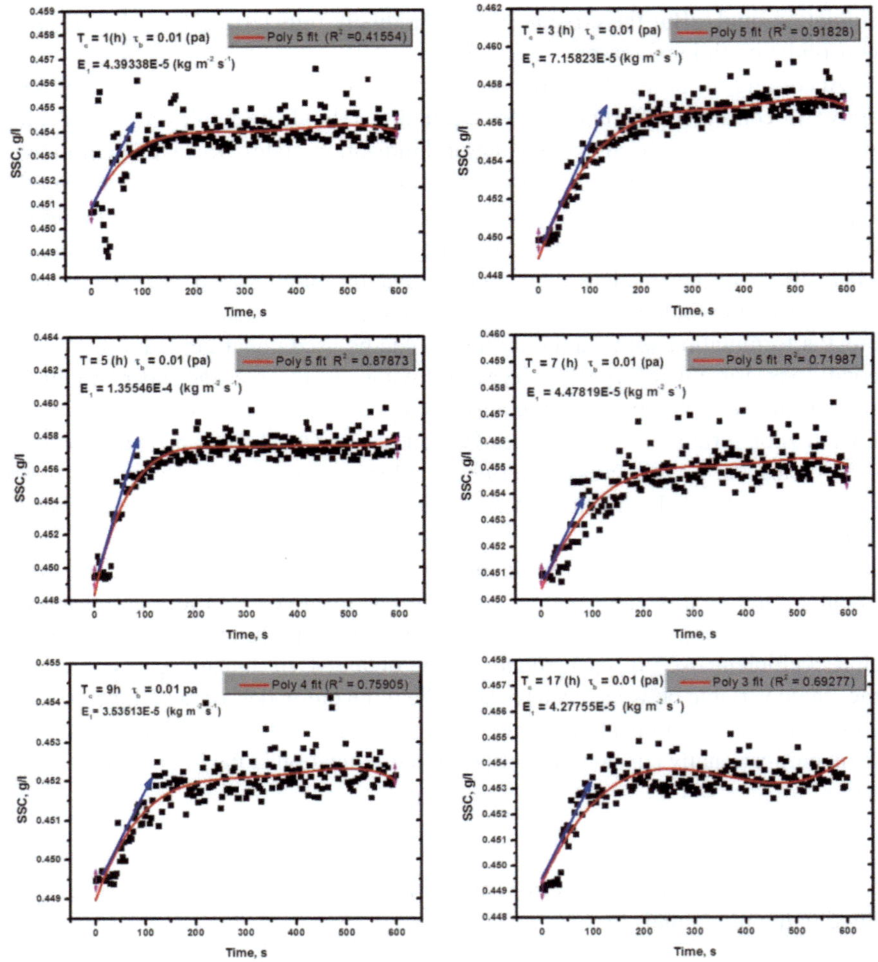

Fig. 8.6 Time series of the SSC scatters under bed shear stress ($\tau_b = 0.01$ Pa) in all the 6 experimental rounds, marked with the optimal regression fits (red curve) and derived initial erosion rates (blue arrow)

first-bed shear stress, i.e., $\tau_b = 0.01$Pa). The remaining results of E_i under other bed shear stresses and experimental rounds are summarized in Table 8.3.

(3) ***Excess Shear Stress***

Excess shear stress (τ_{exc}) refers to the difference between τ_b and τ_{cr}. In the present study, τ_b is the stepwise increased bed shear stress that is applied on the surface and τ_{cr} increased with depth into the sediments (Type I erosion). We argue that in the annular flume the τ_{cr} of the sediment layer exposed to scour during τ_b period (n) is just the applied bed shear stress of the former period (n − 1), this can be expressed as

Table 8.3 Summary of the measured and calculated parameters in the experiments

T_c	f_d (%)	G_m	τ_b(Pa)	τ_{exc}(Pa)	E_r	S_e	M_e
T = 1	96	1	0.01		4.39 E-05	1.17 E-05	2.22 E-04
		2	0.02	0.01	5.91 E-05	1.57 E-05	
		3	0.04	0.02	5.92 E-05	4.05 E-05	
		4	0.07	0.03	5.58 E-05	8.64 E-05	
		5	0.11	0.04	1.25 E-04	1.13 E-04	
T = 3	90	1	0.01		7.16 E-05	8.31 E-06	2.90 E-04
		2	0.02	0.01	1.79 E-05	1.25 E-05	
		3	0.04	0.02	3.34 E-05	1.26 E-05	
		4	0.07	0.03	8.69 E-05	4.04 E-05	
		5	0.11	0.04	1.29 E-04	7.51 E-05	
T = 5	45	1	0.01		1.36 E-04	9.80 E-06	5.44 E-04
		2	0.02	0.01	1.04 E-04	1.60 E-05	
		3	0.04	0.02	6.23 E-05	3.19 E-05	
		4	0.07	0.03	2.22 E-05	6.56 E-05	
		5	0.11	0.04	1.99 E-04	1.15 E-04	
T = 7	11	1	0.01		4.48 E-05	1.01 E-05	3.25 E-04
		2	0.02	0.01	7.08 E-05	2.64 E-05	
		3	0.04	0.02	9.22 E-05	3.85 E-05	
		4	0.07	0.03	9.84 E-05	6.16 E-05	
		5	0.11	0.04	1.77 E-04	9.07 E-05	
T = 9	4	1	0.01		3.54 E-05	3.42 E-06	1.03 E-04
		2	0.02	0.01	3.01 E-05	2.07 E-06	
		3	0.04	0.02	3.13 E-05	2.01 E-05	
		4	0.07	0.03	3.51 E-05	2.37 E-06	
		5	0.11	0.04	6.32 E-05	3.40 E-05	
T = 17	3	1	0.01		4.28 E-05	2.72 E-06	8.27 E-05
		2	0.02	0.01	3.16 E-05	1.42 E-05	
		3	0.04	0.02	4.51 E-05	1.99 E-05	
		4	0.07	0.03	7.43 E-05	3.96 E-05	
		5	0.11	0.04	5.54 E-05	7.08 E-05	

T_c is the consolidation time in hours, f_d is the fluidization degree, G_m is the gear of flow generator, τ_b is the bottom shear stress in Pa, τ_{exc} is the excess pore pressure in Pa, E_r is the erosion rate in kg m^{-2} s^{-1}, S_e is the standard error of the calculated E_r, and M_e is the regressed erosion coefficient

$$\tau_{\text{cr n}} = \tau_{\text{b n}-1}, (\text{n} \geq 2). \tag{8.10}$$

The necessary condition is that the increase of SSC during period (n − 1) reached an equilibrium state finally ($\partial c/\partial t = 0$). Indeed, final equilibrium indicates two possible cases: (a) all the sediments that could be eroded were resuspended, i.e., the erodible material was exhausted (Type I), or (b) the entrainment and settling rates were in balance (Type II), this can be expressed as

$$\frac{\partial c}{\partial t} = E - D, \tag{8.11}$$

where E is the actual rate of entrainment and D is the rate of deposition of the entrained materials. $\partial c/\partial t$ will approach zero provided: (a) $D \to 0$ and $E \to 0$, or (b) $D \to E$. In fact, for the soft (partially fluidized) cohesive sediments, this question has been investigated thoroughly (see Parchure and Mehta 1985) and the answer was determined as the first one ($D \to 0$ and $E \to 0$). Sediments in the present study are soft cohesive deposits that are quite similar to the subject investigated by Parchure and Mehta (1985). More directly, the experimental phenomenon is that no significant surface sediment movement can be detected at the end of each τ_{b} period. Therefore, we concluded the erosion type of the present study as Type I erosion, the τ_{cr} of the sediment increases with depth. Herein, the annular flume has the advantage of giving this increment. This is because the exhaustion of erodible material under the action of $\tau_{\text{b n}-1}$, indicates a final $E_r = 0$, thus $\tau_{\text{b}} = \tau_{\text{cr}}$. Subsequently, when the next larger $\tau_{\text{b n}}$ was applied onto the seabed, sediments start to be entrained again, because the $\tau_{\text{exc}} \neq 0$ any more, at this moment, Eq. 8.11 holds (i.e., $\tau_{\text{cr n}} = \tau_{\text{b n}-1}$).

(4) ***Erosion Coefficient***

With a series of initial erosion rates (E_i) and corresponding excess shear stresses (τ_{exc}) derived from the experimental data, erosion coefficients (M_{e}) were further obtained via regressions using the linear erosion formulation (Eq. 8.12). Therefore, erosion coefficients of sediments with various fluidization degrees were obtained, respectively (Fig. 8.7).

$$E_i = \text{h} \lim_{\text{t} \to 1} \partial c/\partial t = M_{\text{e}}(\tau_{\text{b n}} - \tau_{\text{b n}-1}) \quad (\text{n} \geq 2) \tag{8.12}$$

It is clear from Fig. 8.7 that erosion coefficients of the same YRD silts should decrease by 6–7 times (from 8.27×10^{-5} to 5.44×10^{-4} kg m^{-2} s^{-1}) positively corresponding to seabed fluidization degrees of 0.03–0.96. The experimental results support that of Dade et al. (1992), i.e., critical flow speed (V_{cr}) for sediment entrainment may change by 10 times as the water content varies from 50% (consolidated) to 90% (unconsolidated). The erosion coefficients also vary in the normal ranges of 10^{-5}–10^{-4} and 1.5×10^{-5}–3×10^{-4} kg m^{-2} s^{-1} for silty flat bottoms, although higher and lower values have also been reported (e.g., Bedford and Lee 1994).

8.2.4 Modification of Linear Erosion Model by Integrating the Parameterization Equation

Analysis of the experimental data shows that erosion coefficients (M_e) are significantly influenced by the fluidization degree of deposits (Fig. 8.7). For the silty sediments in the YRD, the overall influence curve is furtherly shown in Fig. 8.8. M_e are found to increase with fluidization degrees except for the first two experimental rounds ($T_c = 1$ and 3 h), during which the M_e is even smaller than that of the round $T_c = 5$ h. This strange phenomenon has been referenced when describing the SSC level results. It seems that the erodibility of highly fluidized sediments is even smaller that of the moderately fluidized; while moderately fluidized is regularly easier to be eroded than the slightly fluidized.

To identify the mechanisms for the inhibited and enhanced erodibility respectively, a qualitative correlation analysis was made between M_e with the corresponding excess pore pressure and seepage rates (Fig. 8.8). Here, averaged seepage rates $\left(\bar{S}_r\right)$ were employed which were calculated as follows:

$$\bar{S}_{r\,[t_n]} = \frac{\partial p}{\partial t} = \frac{p_{exc\,[t_{n-1}]} - p_{exc\,[t_{n+1}]}}{2\Delta t} \tag{8.13}$$

where $p_{exc\,[t_{n-1}]}$ is the excess pore pressure at the $T_c = t_{n-1}$, $p_{exc\,[t_{n+1}]}$ is the excess pore pressure at the $T_c = t_{n+1}$. $\bar{S}_{r\,[t_n]}$ is the averaged seepage rate from t_{n-1} to t_{n+1}.

From Fig. 8.8, we can also infer that although adding water has finished at $T_c =$ 1 h, the p_{exc} was still increasing until $T_c = 2$ h, indicating that the overlying water was infiltrating downward rather than dewatering upward, thus caused negative seepage

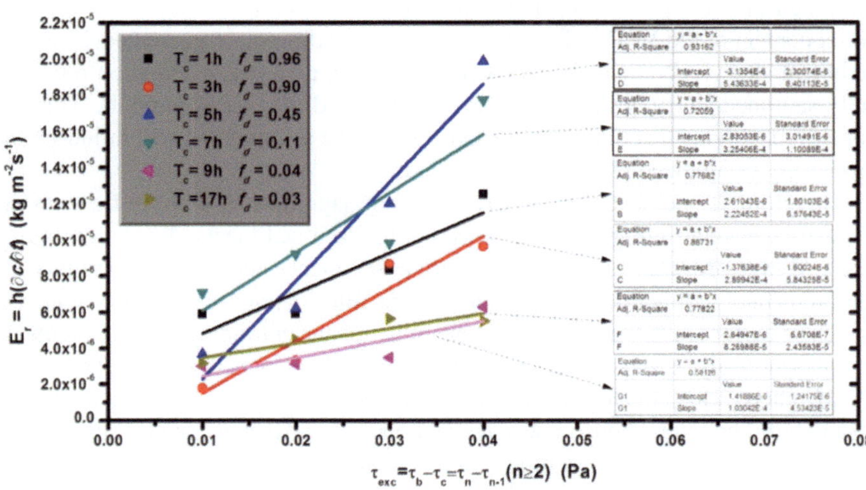

Fig. 8.7 Erosion coefficients for sediments with various fluidization degrees derived from linear regressions between measured E_i and τ_{exc}

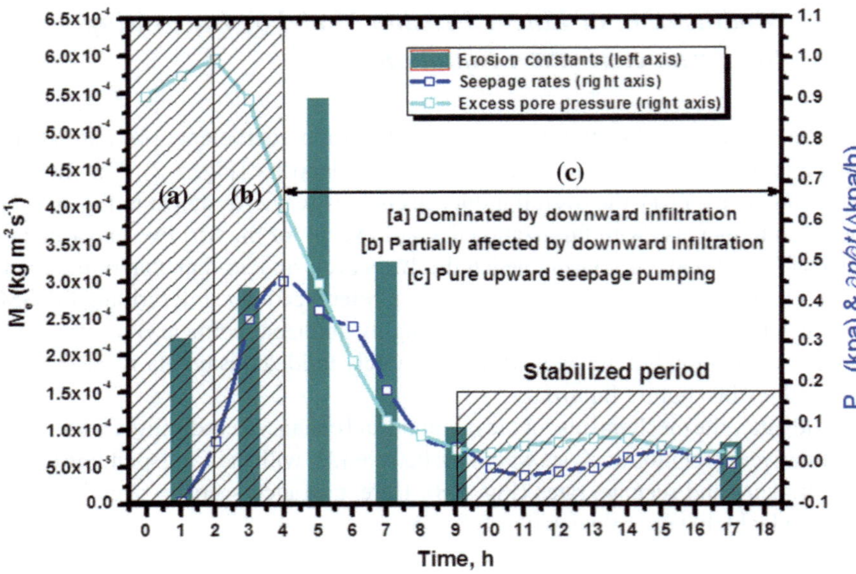

Fig. 8.8 Variation of erosion coefficient (M_e), excess pore pressure (p_{exc}) and seepage rate $(\bar{S}_r)$ with increasing consolidation time (i.e., decreasing fluidization degree). Note that adding water into the flume was finished at $T_c = 1$ h

rates. Similarly, when $T_c = 3$ h, the $\bar{S}_r$ has begun to increase but not reached the maximum; a larger $\bar{S}_r$ was detected when $T_c = 5$ h than that of $T_c = 3$ h. After $T_c = 5$ h, the M_e decreased monotonously with the attenuation of $\bar{S}_r$. The existence of downward seepage in the present study is resulted from the following reasons. Although the slurry was prepared according to the previously reported saturated water content of the YRD sediment. In practice, it is difficult to guarantee the water-sediment slurry is completely saturated. Therefore, as water is added onto the not fully saturated slurry, a short period with downward seepage is possible. This has some implications that tidal level variations or wave set up/down will influence the erodibility of fluidized sediments in the field.

Therefore, it is essentially the direction and magnitude of seepage flows in the deposits that controlling the erodibility of fluidized sediments. Water infiltration exerts downward drag forces on the surface sediments and therefore inhibits the entrainment of sediments, while upwelling percolation of pore-water applies uplifting forces on surface sediments and therefore enhances the entrainment of sediments. In the present study, it was the maximum upward seepage rates leading to the peak in the M_e curve. This conclusion agrees with the previous findings of some hydraulic research on the interactions of seepage and open-channel flows, which also argued that downward seepage flows might inhibit sediment transport rates (e.g., Lu et al. 2008; Cao and Chiew 2014) by a "suction effect" that acting on surface sediments resulting from down-directed seepage flows. On the other hand, upward seepage flows might enhance sediment transport rates (e.g., Cao et al. 2015) by an "injection

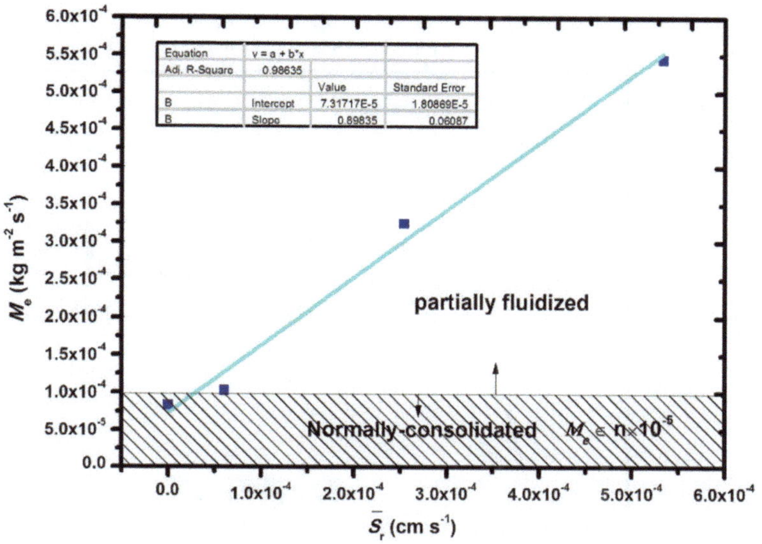

Fig. 8.9 Erosion coefficient (M_e) against upward seepage rates ($\bar{S}_r$) in the fluidized YRD sediments. The shallow area refers to sediments that are in normally consolidated state, of which the M_e has an order of magnitude of 10^{-5} kg m^{-2} s^{-1}, whereas the M_e of partially fluidized silts has an order of magnitude of 10^{-4} kg m^{-2} s^{-1}

effect" that acting on surface sediment particles resulting from upwelling seepage flows.

In order to specifically see the quantitative relationship between sediment erodibility (reflected in M_e) and upward seepage rates. The periods with downward seepage effect are excluded. Thereby derived relationship is shown in Fig. 8.9. It is clear that upward seepage rates have strong a linear correlation with erosion coefficients, demonstrating a dominated role of upward seepage rates in controlling the erodibility of fluidized silts. Furthermore, a power law attenuation in M_e with the increasing of fluidization degree is identified, the mathematical formulation can be expressed by $M_e = 8.74 \times 10^{-4} f_d^{0.56}$. Correspondingly, a power law increasing in M_e with consolidation time is determined, the mathematical formulation can be expressed as $M_e = 1.43 \times 10^{-2} T_c^{-2.02}$ (Fig. 8.10). We suggest that similar influence curves for sediments from other locations can be obtained with the same method. Such parameterization equations are important because it indicates that the employment of erosion coefficients to date has been somewhat arbitrary, at least when applied to areas covered with fluidized sediments. In the field environments of the YRD, we suggest that the calculation of erosion rates should consider the influence of sediment fluidization degree, which can be modeled as follows:

$$E_r = 8.74 \times 10^{-4} p_{ex}/\sigma_v'^{0.56}(\tau_b - \tau_{cr}), \quad \left[R^2 = 0.91\right] \tag{8.14}$$

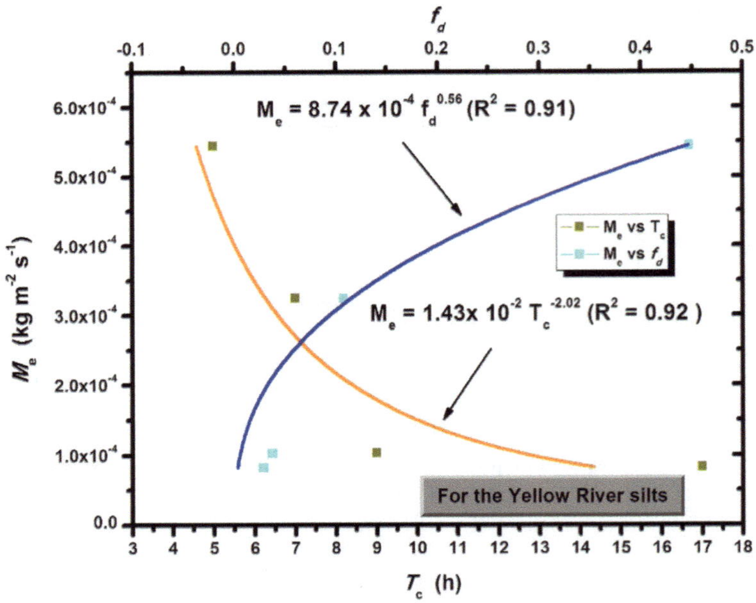

Fig. 8.10 Erosion coefficients as functions of fluidization degree and consolidation time for the Yellow River silts. The blue line is plotted against the top X axis, whereas the yellow line is plotted against the bottom X axis. Both share the same Y axis

where p_{ex} can be measured in situ with pore pressure sensors and $\sigma_v^{'}$ can be tested in the laboratory using sediment cores sampled from the field.

It is worth noting that the significant influence exists particularly in post-fluidized sediments because the permeability of sediments is significantly increased. With respect to pre-fluidized sediments (i.e., well-consolidated or with progressive building up of excess pore pressure before fluidization), this effect should be less important and needs further investigation. That is why some previous research found negligible influences (Baldock and Holmes 1999) of seepage flows on sediment entrainment. For example, Carstens et al. (1976) found that the influence of excess pore pressure on the initiation of sediment transport was negligible until pressure gradients approached liquefaction for both steady and oscillatory conditions. At a vertical pressure gradient of approximately 80% of liquefaction, the threshold velocity required to initiate transport was reduced by only 10%. However, this does not mean that seepage flows are not important in affecting sediment erosion. The present paper argues that the influence is quite significant for fluidized silts. Considering the wide distribution of fluidized sediments on the seafloor of subaqueous YRD, the macroscopic contribution of seepage effect to sediment dynamics merits attention. Moreover, even the consolidation process seems to run to completion in a few hours, in fact, the active also repeated transfer of sediment state between consolidation and fluidization in the YRD due to frequent waves (Prior et al. 1986) make a persistent existence of fluidized silts.

8.3 Validation of the Modified Sediment Resuspension Model

Previous parameterization equations are further summarized into the traditional linear erosion formulation to form a newly modified erosion model in this section. Moreover, the prediction effect was verified via inputting field-measured hydrodynamic data into the new model and thereafter comparing the modeled results with field measurement, respectively. Therefore, an in situ observations of hydrodynamics and seabed erosion process is conducted in the subaqueous Yellow River delta.

8.3.1 Month-Long Field Observation

(1) ***Study Area and Observational Site***

The study area (YRD) locates in the southern part of a large semi-enclosed Bohai Bay, China (see more details in Fig. 7.3), where swelling waves from external ocean are mostly excluded by the Changshan Archipelago near the Bohai Strait. Therefore, waves in the bay are mostly wind-driven short waves which generated locally and died out quickly, long-period (>10 s) ocean waves rarely penetrated into the bay. Strong wave direction is NNE-ENE, and moderate-strong wind direction is NNW. North-directed waves prevail during winter while south-directed waves prevail in the summer. Normal wave height is smaller than 1.5 m, although prevalent northwest winds may generate 4–7 m waves in spring or winter. Observation time was selected in winter when sediments are more significantly resuspended and transported than in the summer.

The observation site locates in the northern part of modern YRD, on the subaqueous Diaokou lobe where the mouth of the Yellow River located from 1964 to 1976. Now the river mouth has changed its course to the sea southwardly to the Qingshuigou channel since 1976, thus northern abandoned lobes have little river-laden sediment supply. In this case, the variations of suspended sediment concentration in this area have mostly resulted from local sediment erosion and resuspension. Historical reports have indicated that seabed here has always been suffering from intense scour and erosion. Therefore, the observation site was selected here to observe sediment erosion and resuspension in the subaqueous YRD.

Surface sediments at the observation site belong to cohesive sandy silts that contain substantial fine-grained fraction (clay 4.72%; silt 52.66%; fine sand 42.62%). Mean grain size is 0.053 mm and median grain size is 0.057 mm. Sediments are poorly sorted ($\delta_i = 1.26$) with very positive skewness ($S_{ki} = 0.32$) and very leptokurtic kurtosis $\left(K_g = 1.56\right)$. The grain size distribution is shown in Fig. 8.11.

(2) ***Field Campaign***

A tripod platform instrumented with one wave gauge (RBR virtuoso, Canada), one electromagnetic current meter (JFE Infinity-EM, Alec, Japan), and one nephelometer

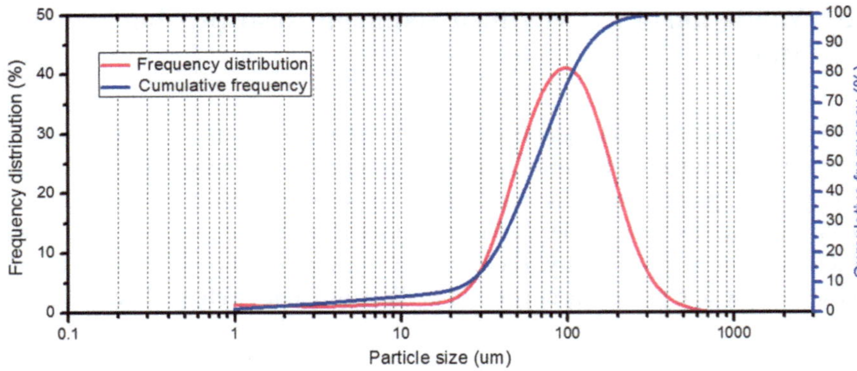

Fig. 8.11 Grain particle size distribution of the bottom sediments

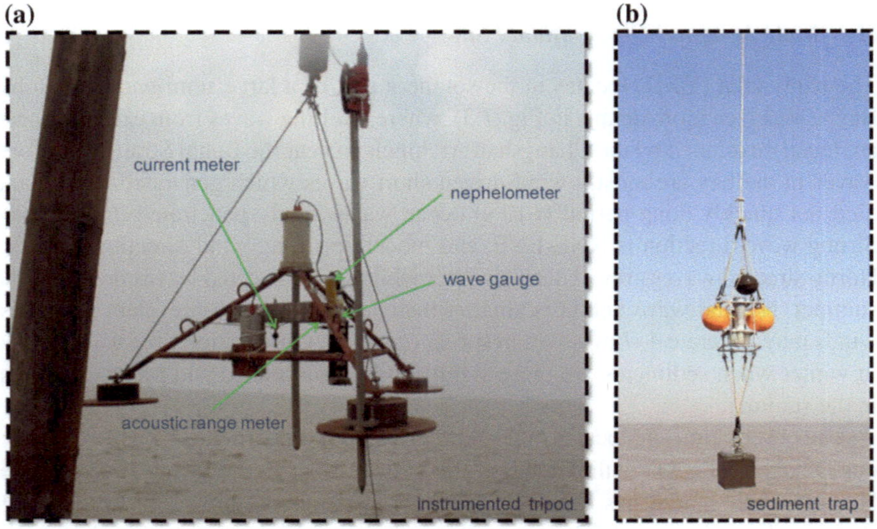

Fig. 8.12 The observation system: **a** instrumented tripod **b** sediment trap mooring

(RBR CTD XR-620, Canada) with built-in OBS (optical backscatter) was deployed at the study site to synchronously measure the waves, mean currents and suspended sediment concentrations (SSC) in the bottom boundary layer from 09/12/2014 to 22/04/2015 (Fig. 8.12a). A time series sediment trap mooring was deployed at the same site for a two-day short-period observation after the retrieve of the tripod platform on 22/04/2015, for the collection of local suspended sediment materials (Fig. 8.12b). Turbidity readings from the nephelometer are converted into SSCs using a pre-calibrated relationship derived from laboratory calibration tests using local silts. Instrument models, position on the tripod and the sampling settings are summarized in Table 8.4.

Table 8.4 Observational apparatus and sampling settings

Parameter	Manufacturer	Settings	Frequency	Period	Duration
Wave	RBR. Canada	5 min/1 h	6 Hz	Dec. 10–Feb. 06	> 27 d
Current	Alec. Japan	1 min/30 min	6 Hz	Dec. 10–Feb. 09	> 27 d
SSC	RBR. Canada	Once/10 min	1 Hz	Dec. 10–Mar. 15	> 27 d
Sediments	NiGK. Japan	Bottle/5 h	*	Apr. 22–Apr. 24	30 h

In the "period" column, Dec. refers to December 2014 while other months are in 2015

8.3.2 Field Data

The synchronous record of all the parameters from December 12, 2014 to January 8, 2015 (27 days) was selected in this paper for analysis because some of the instruments stopped working due to the limited battery, so it is the period with the most complete dataset. As a general setting, time series of the tidal level (L_t), significant wave height (H_s), significant wave period (T_w), wind speed (S_w), mean current velocity (V_c), suspended sediment concentration (SSC) are plotted in Fig. 8.14.

(1) ***Tides***

Tidal level ranges from 5 m in neap tides to 7.5 m in spring tides (Fig. 8.13a). The mean water depth $\bar{D}$ averaged over the deployment period of 27 days is 6.326 m. Tidal ranges relative to $\bar{D}$ are computed as $\left(L_t = h - \bar{d}\right)$. The spring tidal range is >2 m and the neap tidal range <1 m, indicating the YRD belongs to microtidal coast with predominantly unequal semi-diurnal tides.

(2) ***Waves***

Significant wave height ranges from 0.03 to 2.36 m (Fig. 8.13b), the observation site experiences several small storm waves ($H_s > 1.5$ m) during the observation period, compared to the normal waves ($H_s < 1.5$ m) in this area. Measured significant wave period ranges from 4.7 to 7.4 s (Fig. 8.13c), supporting that the waves are predominantly wind-generated, not penetrated ocean swell.

(3) ***Winds***

Wind speed (V_w) ranges from 0.1 to 21 m/s (Fig. 8.13d) which well corresponds to the variation of wave parameters, indicating a direct role of winds in generating local short period waves. Force 6 (>10.8 m/s) and 8 (>17.2 m/s) winds were captured.

(4) ***Currents***

The measured maximum current speed consists of tidal and residual currents. The measured mean current speeds are shown in Fig. 8.13e which are found to be generally less than 40 cm/s, while larger mean speed (40–66 cm/s) is recorded during wavy periods. The measured current speed includes waves-generated current component. A previous report states that the dominating driving force for residual current in the YRD is wind-generated currents. During calm periods (weak waves), tidal

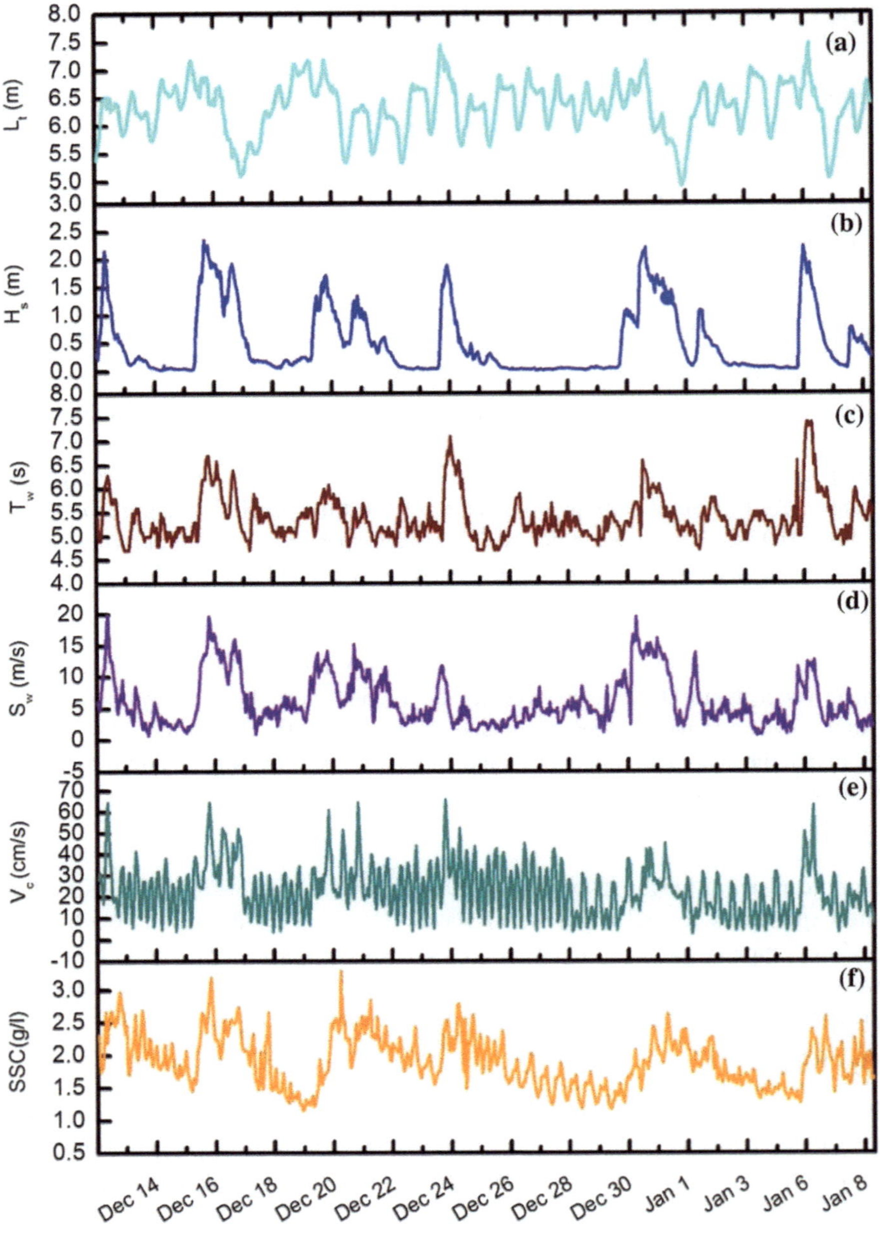

Fig. 8.13 Measured bottom boundary layer parameters from Dec. 12, 2014 to Jan. 8, 2015

currents dominate the total mean current because it was found that two-speed spikes (Fig. 8.13e) occur in each tidal cycle (Fig. 8.13a), which well matched the speed variation during flood and ebb tides respectively.

Previous reports indicated that the current here belongs to reciprocating flow with K = 0.01–0.32 for the M_2 tidal current component in this area. Influenced by the topography, the tidal current is roughly parallel to the coast, the direction of the tidal current is ESE (100–115°) and WNW (280–295°).

(5) ***Suspended Sediment Concentrations***

Generally, suspended sediment concentration ranges from 1.1 to 3.3 g/l, compared with other regions, it is quite high. Although the background concentration is high to >1 g/l, no more sediments settle out under low energy, successfully sustaining a distribution of high-turbidity water mass in this region compared with the offshore seawater (Fig. 7.3).

Specifically, six large sediment resuspension events (significant increase of SSC from the background level) were recorded during the wavy periods (Fig. 8.13f). Each one lasted for 40–60 h from the initial increase from background SSC to the end of decreasing back to the background level. Except for these large events, some relatively small spikes of the SSC also appeared periodically during calm periods, seemingly at the same frequency with simultaneous current speeds.

8.3.3 The Modified Erosion Model

The present paper discusses the situation that liquefied sediments would not be swept away directly under the large shear stress of storm waves, but stay where they were, i.e., wave-induced shear stress amplitude (τ_{max}) in seabed never reached the critical shear strength (S_r) of liquefied sediments, represented by the Coulomb equation

$$\tau\text{max} \leq S_r = c' + (\delta - u)\tan\varphi', \qquad (8.15)$$

where c' is the effective cohesion, $\delta = W\cos\beta$ is total normal stress in which W is the weight of failure block and β is the angle of failure plane, u is pore-water pressure, φ' is the effective friction angle.

The resuspension process (reflected in suspended sediment concentrations) after liquefaction will be significantly changed, in Chap. 5, we constructed a parametric equation between the hydrodynamic condition (relative wave height) and the contribution of liquefaction to resuspension

$$\frac{C_f}{C} = 108.24 + 26.94\ln\frac{H_w}{D} - 0.08, \qquad (8.16)$$

where C_f is the SSC caused by liquefaction, C is the total SSC, H_w is the wave height, and D is the depth of water.

In Chap. 7, this book furtherly established the parameterization equation between the seabed liquefaction degree and the erosion coefficient in the Yellow River Delta, thus modified the traditional erosion model by introducing a liquefaction parameter.

$$E_r = 8.74 \times 10^{-4} \frac{p_{ex}}{\sigma_v'}^{0.56} (\tau_{cw} - \tau_{cr}), \tag{8.17}$$

where p_{ex} is the excess pore pressure which can be measured with piezometers, σ_v' is the normal effective stress of seabed which can be measured from sediment cores. When the pore pressure data is missing, p_{ex} can be estimated from the empirical formula where the excess pore pressure is assumed to be proportional to the bottom wave pressure (P) (Kuo and Chiu 1994; Tsai et al. 2005).

$$p = \frac{\gamma_w H}{2\cosh(\lambda h)} \cos(\lambda x - \omega t), \tag{8.18}$$

where γ_w is the seawater density, H is the wave height, wave number $\lambda = 2\pi/L$, h is the water depth, angular frequency $\omega = 2\pi/T$.

Specifically, the coupled bottom shear stress (τ_{cw}) consists of wave-orbital shear stress (τ_w) and unidirectional current-induced shear stress (τ_c):

$$\tau_{cw} = \tau_c + \tau_w = \frac{1}{2} f_w \rho u_w^2 + \rho u_*^2, \tag{8.19}$$

where u_w is the maximum over-the-wave-cycle horizontal wave-orbital speed at the bed, u_* is the friction shear velocity, f_w is the wave friction factor:

$$f_w = \exp\left[5.213(k_b/A_b)^{0.194} - 5.977\right], \tag{8.20}$$

where $k_b = 2.5D_{50}$ is the bed roughness (Nielsen et al. 2001), $A_b = u_w T/2\pi$ is the wave-orbital semiexcursion. For linear waves

$$u_w = \frac{\pi H}{T \sinh kh}, \tag{8.21}$$

where H is wave height, T is wave period, wave number $k = 2\pi/L$, h is water depth.

To examine the potential of the mean current flow to induce bottom resuspension, the velocity measurement at 0.5 m above the bed (u_z) has been converted into bed shear stress. The current-induced bed shear stresses (τ_c) were computed assuming a logarithmic velocity gradient in the bottom boundary layer (low seabed gradient and no obstructions causing modifications to the near-bed flow structure) (Paphitis and Collins 2005):

$$u_* = \frac{u_z k}{\ln\left(z/z_0\right)}, \tag{8.22}$$

where k is the von Karman constant (0.4, in seawater), z is measuring the height of currents, z_0 is the bed roughness length.

8.3.4 *Prediction Effect of Traditional and the Modified Models*

The concentration of particles in the water column is determined by advection, deposition, resuspension, vertical mixing, and internal sources and sinks:

$$\frac{\partial c}{\partial t} + u\frac{\partial c}{\partial x} + v\frac{\partial c}{\partial y} = \mathrm{E_r} - (w - \omega_s)\frac{\partial c}{\partial z} + \frac{\partial c}{\partial z}\left(K\frac{\partial c}{\partial z}\right), \tag{8.23}$$

where c is SSC, t is time, d is water depth, u, v, w, refers to the 3D velocity component, $\mathrm{E_r}$ is erosion rate, ω is settling velocity, K is vertical eddy diffusivity. Note: c should be the depth-averaged SSC (Parchure and Mehta 1985) but here the local concentrations measured at 0.5 MAB were employed, because full-depth SSC profiles are difficult to observe in situ for a long period of time, moreover, SSC profiles in the YRD have been found to be well mixed (relatively uniform) by the frequent also energetic wave actions in winter–spring seasons, moreover, horizontal advection flux is assumed to be small in the study area, according to the principles indicated by Wang (2003): the near-surface sediment distribution is rather uniform over the tidal excursion distance, besides, fluctuations in suspended sediment concentration are similar between the flood and the ebb (Fig. 8.14). In this case, similar to the analysis of Wang (2003), time series of erosion/settling rates are computed from the SSC time series (Fig. 8.14a) using the vertical mixing model. The sediment settling velocity is 0.1 mm/s according to Qiao et al. (2016). The vertical eddy diffusivity is estimated from $\mathrm{K} = 0.16\ (C_\mathrm{d} \times \mathrm{V})^{0.5}$.

The physical meaning of Eq. 23 is clear that the changing rate of SSC with time can be taken as an indicator for erosion or settling rates in scope of seas, i.e., when the SSC is increasing, sediments around the observation site are resuspending, on the contrary, sediments are settling when the SSC is decreasing.

(1) ***Prediction Effect of Traditional Model on the Total SSC***

Excluding the settling rates (i.e., negative values of calculated E_r), time series of erosion rates deriving from field-measured SSC data were obtained (Fig. 8.14b). As these $\mathrm{E_r}$ time series are directly deduced from measured data according to the definition, they are referred to as derived erosion rates in the present paper. On the other hand, erosion rates were predicted by inputting the field-measured hydrodynamic data (shown in Fig. 8.13) into the traditional erosion model (Eq. 8.7), predicted results were shown in Fig. 8.15c.

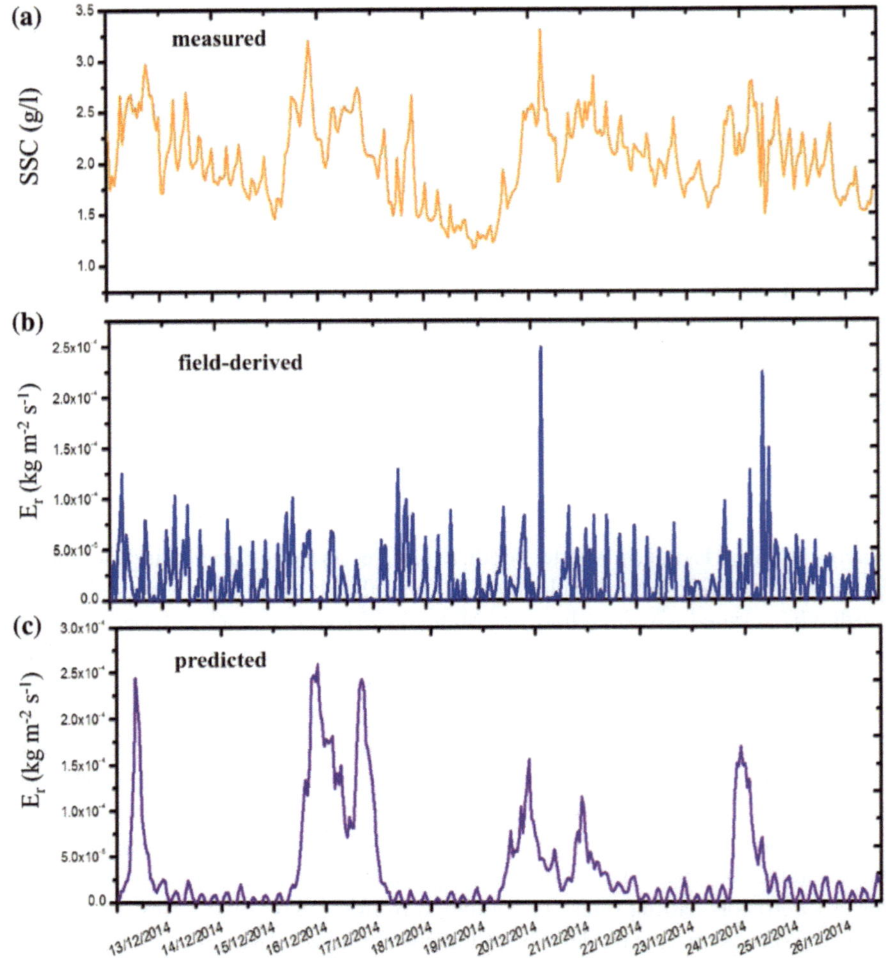

Fig. 8.14 Time series of **a** the measured total SSC **b** derived total erosion rates **c** predicted total erosion rates with the traditional model

It was found that the field-derived erosion rate (Fig. 8.14b) has a maximum value on December 20 and 24, respectively, while the predicted result has a maximum value on December 12 and 15–16. Neither of the peak value and variation trend is consistent between the field-derived and predicted erosion rates (Fig. 8.14).

(2) ***Prediction Effect of Traditional Model on the SSC without Liquefaction***

According to the parameterization equation (Eq. 8.16) constructed in Chap. 5, measured total SSC (Fig. 8.13f) can be separated into SSC that are caused merely by shear stress (C_s) (Fig. 8.15a) and SSC that are caused by liquefaction (C_L) (Fig. 8.16a).

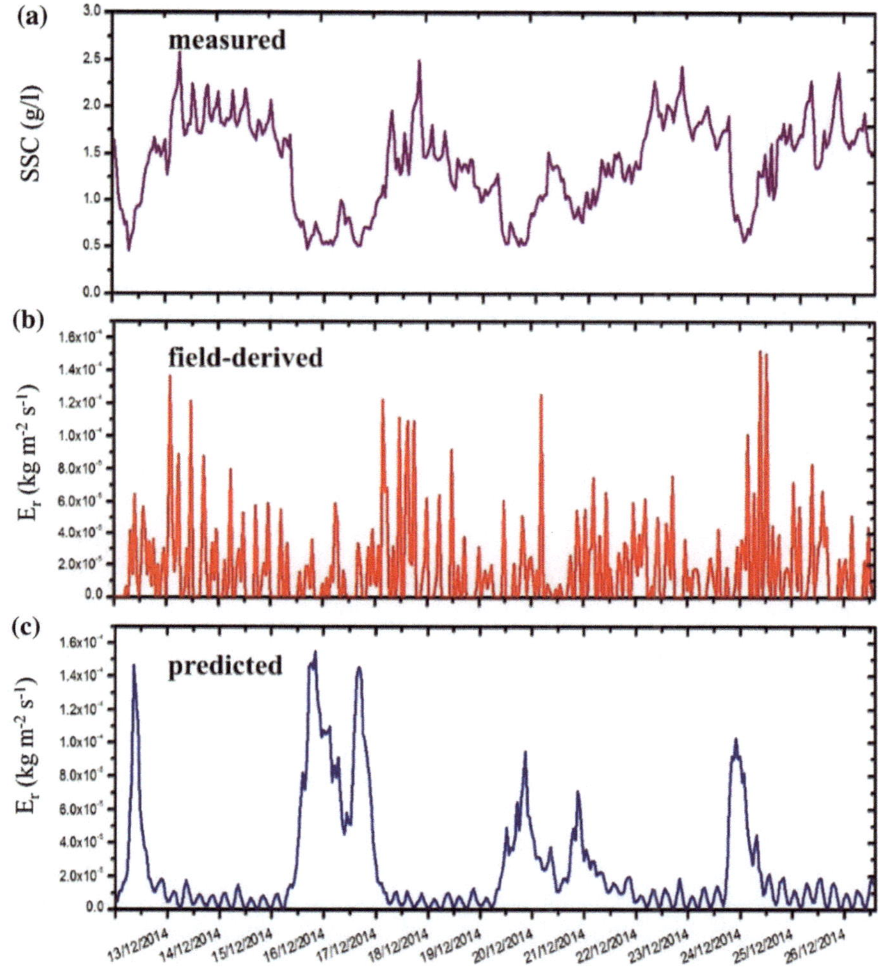

Fig. 8.15 Time series of **a** the separated SSC caused by normal shear **b** derived erosion rates caused by normal shear **c** predicted erosion rates with the traditional model

Similarly, time series of field-derived shear erosion rate (E_s) (Fig. 8.15b) was deduced from the measured C_s data (Fig. 8.15a). Meanwhile, shear erosion rates were predicted by inputting the field-measured hydrodynamic data into the traditional erosion model, predicted results were shown in Fig. 8.15c.

By comparing the field-derived E_s with that predicted by traditional erosion model, it was found that the reproducing effect of the 4 resuspension events was improved, both in the aspect of peak value and variation trend (Fig. 8.15). Therefore, the mathematical method proposed by in Chap. 5 is proved to be effective in separating the SSC caused by shearing and liquefaction, respectively. Traditional model can effectively predict the erosion rates that are not affected by liquefaction. However, if the

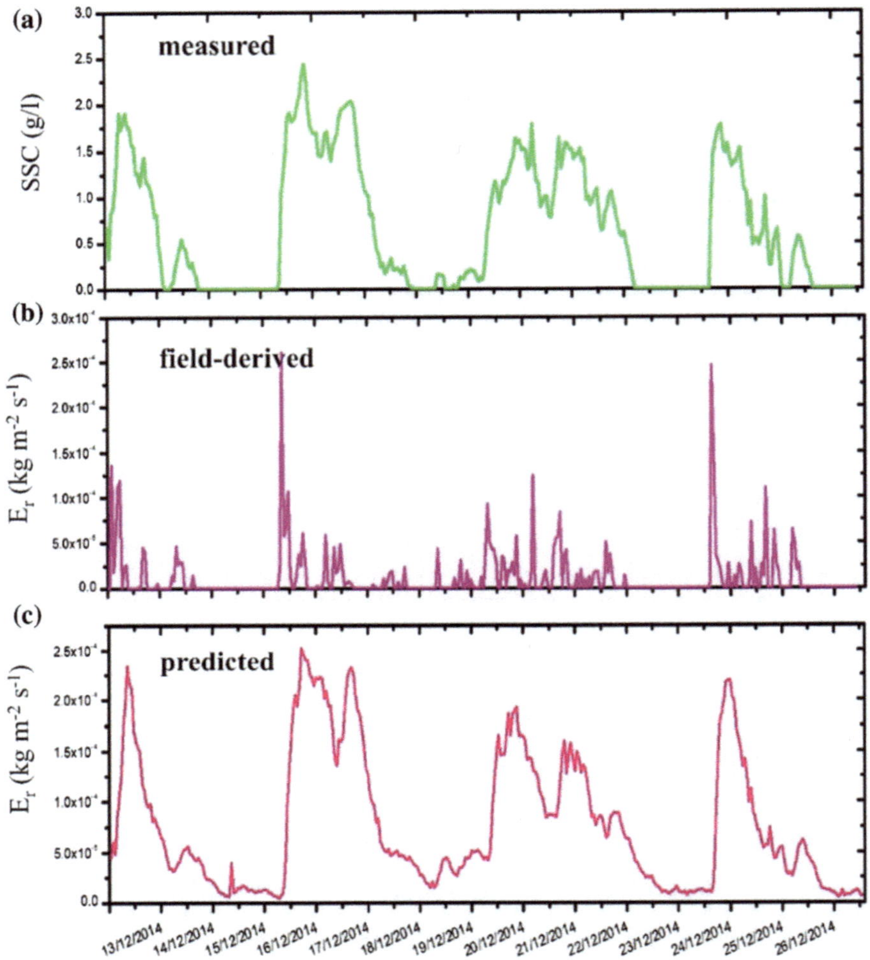

Fig. 8.16 Time series of **a** the separated SSC caused by liquefaction **b** derived erosion rates caused by liquefaction **c** predicted erosion rates with the modified model

SSC due to liquefaction is not excluded, reproducing effect of the traditional erosion model is poor.

(3) ***Prediction Effect of Modified Model on the SSC Influenced by Liquefaction***

The SSC caused by liquefaction was also separated from the measured total SSC curves with Eq. 8.16 (Fig. 8.16a). It can be found that the SSC caused by liquefaction only exists during the wavy periods. Under calm conditions, sediment resuspension is mainly dominated by shear erosion.

Similarly, time series of field-derived liquefaction erosion rates (Fig. 8.16b) were deduced from corresponding SSC curves (Fig. 8.16a). It is apparent that liquefaction erosion rate increases significantly during wave events, and the absolute value is generally higher than that of the shear erosion rate. Liquefaction erosion rates were furtherly predicted (Fig. 8.16c) with the modified erosion model constructed in Chap. 7 by inputting the field-measured hydrodynamic data. Surprisingly, we found quite a good reproducing effect both in the aspects of peak value and the variation trend.

8.4 Prediction of Erosion Mass and Source with the Modified Model

After the Yellow River changed its channel to Qingshuigou channel in 1976, the sediment supply of Chengdao sea was cut off, and gradually eroded and retreated toward the shore under the action of marine dynamics in the past 42 years, representing the abandoned lobe of the Yellow River delta enters an important historical stage of shoreline erosion. It is of great significance to explore the sediment erosion and resuspension in this stage and its transport laws for the prediction of beach and delta evolution, and the rational development and utilization of coastal resources.

Energetic winds and waves are important causes for sediment movement and topographic changes in the coastal zone. Sediment transport and topographic changes caused by storms are more significant than that accumulated over months or even years under normal sea conditions, so extreme storm events are considered as more important causes for marine geological disasters and offshore engineering facility accidents (Yang et al. 1993). However, it is very difficult to directly record the occurrence process of seabed erosion during extreme sea conditions through in-site observation. Therefore, theoretical prediction of erosion and resuspension in different sea conditions are important and often used by coastal engineers and scholars to help solve practical problems.

Based on the long-term measured hydrodynamic parameters and historical statistics of wave parameters in the study area, in this chapter, the modified erosion model was used to calculate and analyze the effects of seabed liquefaction on sediment erosion and resuspension, the ratio of each source and its impact on the delta evolution in the study area under different sea conditions were analyzed. The results of this section can provide scientific guidance for the engineering practice in the study area, also promoting the development of prediction models for sediment transport and shoreline evolution.

8.4.1 Erosion Mass and Source in a Normal Winter (e.g., December)

Early surveys have indicated that during calm weather, suspended sediment concentration in the Chengdao sea area is ~0.02–0.05 g/l, and in the bottom layer can reach

0.1 g/l or higher. With wind and waves, it can reach 0.25 g/l on average, and in the bottom layer can reach 1 g/l (Yang et al. 1993). High concentration regions often appear in water depths of 8–10 m, which corresponds to the onshore distribution of wave energy and slope of underwater beach. During stormy periods, suspended sediment concentrations will increase sharply, and be well mixed between the upper and bottom layers, to form SSCs of ~1–2 g/l on average or even higher in the bottom layer. Early during the Cooperation Survey between China–USA–Canada in 1987, suspended sediment concentration in the shallow water layer were observed to reach 30 g/l after the storm, showing the feature of underwater gravity flow (Wright et al. 1988).

Although the sediment supply from the Yellow River to the Chengdao sea area had been cut off for many years, the concentration here is still significantly high. This phenomenon indicates that the suspended sediment here is mainly from local seabed erosion and sediment resuspension. This corresponds to the intense seabed erosion found in this area (Liu 2006). The wave breaking zone in the Chengdao sea area is located ~7 m water depth, and the suspended sediment concentration is also the highest in 5–10 m water depth. The dividing point for erosion and sedimentation area on the underwater slope was near 10–12 m water depth, which means that shallower regions are under erosion, and deeper regions are experiencing siltation (Yang et al. 1993). Now, 31 years have passed, the dividing point should have been further moved toward the shore. Liu (2006) detected the temporal and spatial evolution of the abandoned underwater slope profile in the northern Chengdao sea area with geophysical methods. Results show that serious erosion is concentrated in the breaking zone of 5–10 m water depth (Fig. 8.17). It can be inferred that the majority of resuspended sediments in the Chengdao sea area comes from the underwater slope in the breaking zone, where wave energy is the most energetic, and the slope is relatively steep. Therefore, in this section, we mainly discuss the source of eroded sediments in the breaking zone, so the calculated depth is chosen to be 7.5 m on average, which is just consistent with the water depth where our in situ observations carried out (see Sect. 8.3).

After identifying the computation depth range, hydrodynamic parameters were further analyzed according to the in situ observation data in Sect. 8.3. Wave frequency distribution in winter (taking December as an example) was plotted in Fig. 8.18 and summarized in Table 8.5.

The source of resuspended sediments in the study area was predicted using the modified erosion model by inputting different wave conditions. Calculation results are also summarized in Table 8.5. It is clear that the appearance probability of significant wave height ($H_{1/3}$) $\leq$ 1 m ($f_{1/3}$) in December reaches 76%. We know that under the action of such small waves of $H_{1/3} \leq 1$ m, there is no sediment resuspension caused by seabed liquefaction at all. That is, there was no significant liquefaction-induced resuspension in about 23 days of the whole December. During these periods, traditional erosion model is valid. Moreover, we know that under larger $H_{1/3}$ of 1–1.5 m, 31% of the resuspended sediments originated from wave-induced seabed liquefaction. As such periods accounted for 12% of the total December (accumulated time was about 4 days). More importantly, as large winds and waves with $H_{1/3}$ larger than

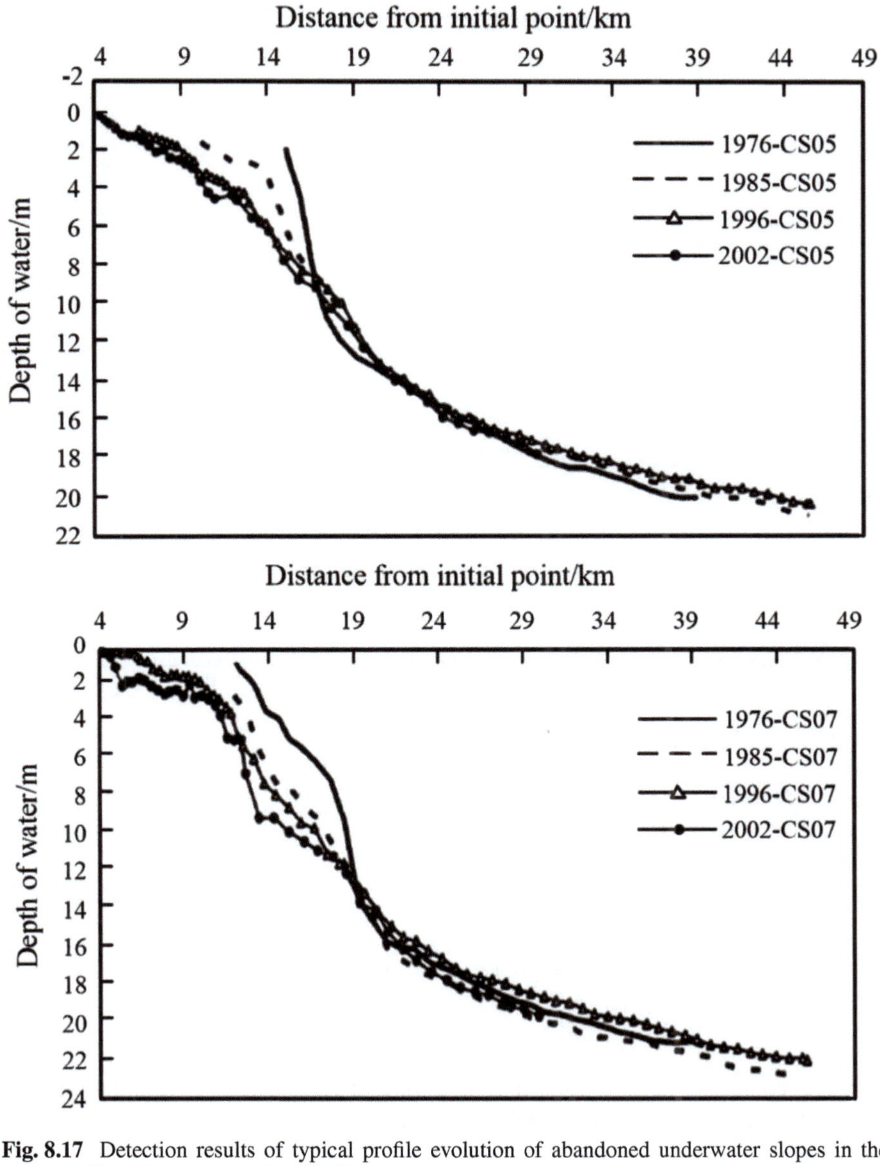

Fig. 8.17 Detection results of typical profile evolution of abandoned underwater slopes in the northern Chengdao sea area (Liu 2006)

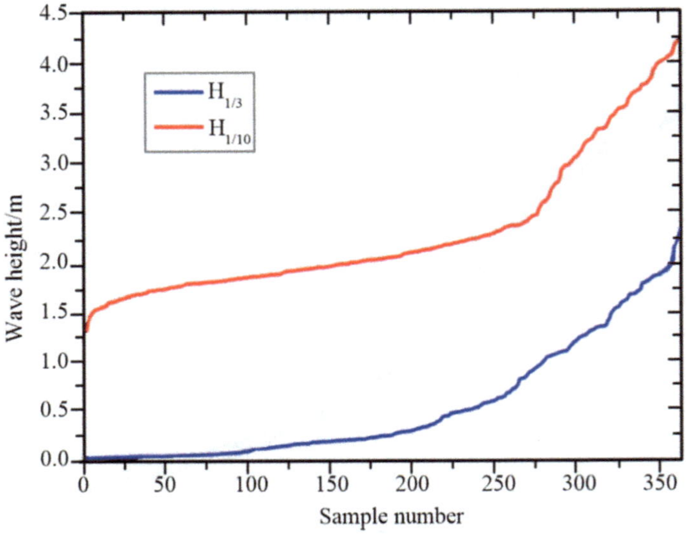

Fig. 8.18 In-situ observed wave height frequency distribution in the Chengdao sea area in December

Table 8.5 Statistics of wave height frequency in Chengdao sea area in December

H	0.0	0.5	1.0	1.5	2.0	2.5	3.0	3.5	4.0	4.5	Sample number
$N_{1/10}$	0	0	4	158	115	20	27	27	19	11	381
$f_{1/10}$			0.01	0.41	0.30	0.05	0.07	0.07	0.05	0.03	
$F_{1/10}$	1.00	1.00	1.00	0.99	0.58	0.28	0.23	0.16	0.09	0.04	381
$N_{1/3}$	231	49	42	35	7						364
$f_{1/3}$	0.63	0.13	0.12	0.10	0.02						
$F_{1/3}$	1.00	0.37	0.25	0.13	0.03						364
C	0	0	31	52	64						

N is the number of occurrence, f is the frequency of occurrence, and C is the calculated contribution of liquefaction to resuspension

1.5 m account for 12% of the whole month, that is, during the remaining 4 days in December, the contribution of liquefaction to resuspension can reach 52–64% (Fig. 8.19).

8.4.2 Erosion Mass and Source in a Normal Year

In this section, we further collected the typical wave parameters in different months of a normal year in the Chengdao sea area. As shown in Table 8.6, the maximum monthly mean wave height $(\bar{H} > 0.7\,\mathrm{m})$ in the Chengdao sea area appears in winter

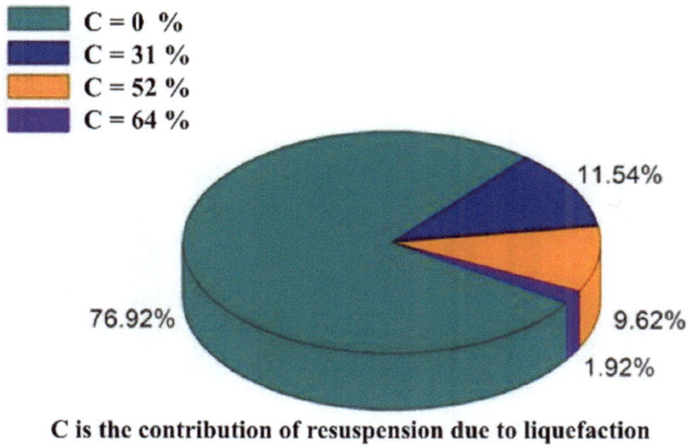

Fig. 8.19 Contribution of liquefaction to resuspension in different sea conditions in the Chengdao sea area (color) and its cumulative contribution frequency in December (pie chart)

Table 8.6 Statistics of monthly wave parameters in the Chengdao sea area (Yang et al. 1993)

Month	1–2	3	4	5	6	7	8	9	10	11	12
H_{max}	*	3.9	3.1	3.2	3.4	2.0	2.2	2.6	4.2	3.6	3.7
$H_{1/10}$	*	2.8	2.7	2.7	2.8	1.5	1.6	1.6	3.6	3.1	2.7
$\bar{H}$	*	0.64	0.41	0.38	0.39	0.45	0.44	0.44	0.62	0.72	0.74
$\bar{T}$	*	8.3	8.9	6.7	7.5	4.8	5.9	5.6	7.6	7.7	6.8

1–2 were in the glacial period during the statistical year, thus wave parameters is missing. The ice beginning time in this area is November 28th at the earliest, and December 23rd at the latest. The ice ending time is on February 17th at the earliest and March 30th at the latest. The longest ice period is 109 days, and the shortest ice period is only 67 days

(November–December). Wave is the strongest in December. That means the contribution ratio of liquefaction to resuspension in December which was calculated in the previous section represents the highest level in a whole normal year. On the other hand, the minimum averaged wave height $\bar{H} < 0.4$ m appears in May and June. Maximum wave height $H_{max} < 3.0$ m appeared in July, August, and September, and $H_{max} > 4.0$ m appeared in October. With respect to the $H_{1/10}$ parameters, waves were the smallest in July and the largest in November.

According to the $H_{1/10}$, wave height parameters of the whole year in Table 8.6, the maximum contribution ratio of each month was also calculated. It was found that liquefaction resuspension intensity was the weakest in July, August, and September, and the liquefaction resuspension intensity was strong in October, November, and December (Fig. 8.20).

It should be noted that the contribution of liquefaction here estimated is the maximum value of the month, it is just a transient contribution, as to the cumulative amount of resuspension due to liquefaction, it further depends on the duration of

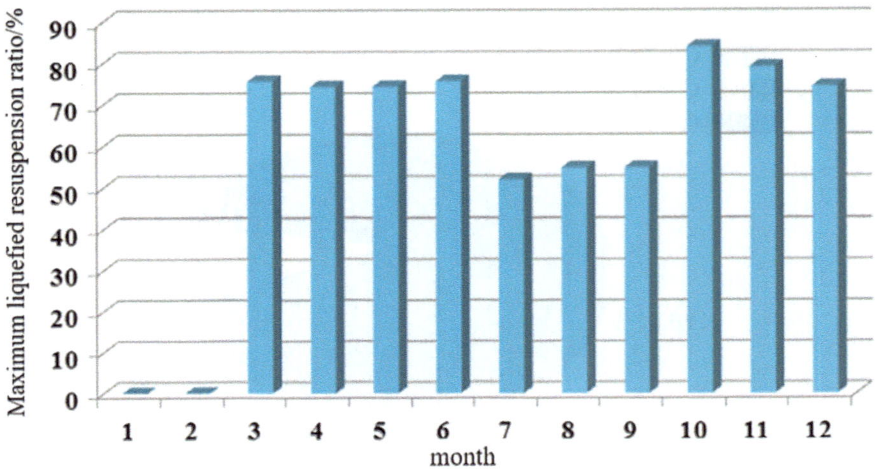

Fig. 8.20 Maximum contribution of liquefaction to resuspension in different months of the year

Table 8.7 Statistics of wave height frequency throughout the year

H/m	0.0	0.5	1.0	1.5	2.0	2.5	3.0	3.5
f	0.600	0.260	0.090	0.030	0.014	0.006	0.000	0.000
F	1.000	0.400	0.140	0.050	0.020	0.006	0.000	0.000

For example, the occurrence probability of wave height $H_{1/10} > 2.5$ m is 0.6%. *f* is sectional frequency and F is the cumulative frequency

corresponding sea condition, which requires the statistics of the frequency of different wave parameters in a year.

Therefore, the frequency of different $H_{1/10}$ in a normal year is shown in Table 8.7. Cumulative time of wave height $H_{1/10} > 2.5$ m accounts for 0.6% of a whole year, $H_{1/10} > 2.0$ m accounts for 2%, $H_{1/10} < 2.0$ m accounts for 98%, $H_{1/10} < 1.0$ m for 86%, and $H_{1/10} < 0.5$ m for 60%.

Based on the annual wave statistics, the sources of eroded sediments in the study area was estimated (Fig. 8.21). There is no liquefaction resuspension in the period of wave height less than 1 m, that is, during 86% time of one year (cumulative 314 days), erosion process can be predicted with the traditional erosion model effectively. However, the contribution of liquefaction to resuspension can reach 31% in 33 days (9% of one year); liquefaction resuspension contributes 52% in 11 days (3% of one year); liquefaction resuspension contributes up to 64% in 5 days (1.4% of one year); in 2 days (0.6% of one year) liquefaction contributes 72% of the total resuspension. To summarize, during 14% of a normal year (51 days), traditional shear erosion model becomes invalid, and the erosion and resuspension process needs to consider the effect of wave liquefaction. The liquefaction erosion model proposed in this book can effectively improve the reproducing effect of erosion process in this circumstance.

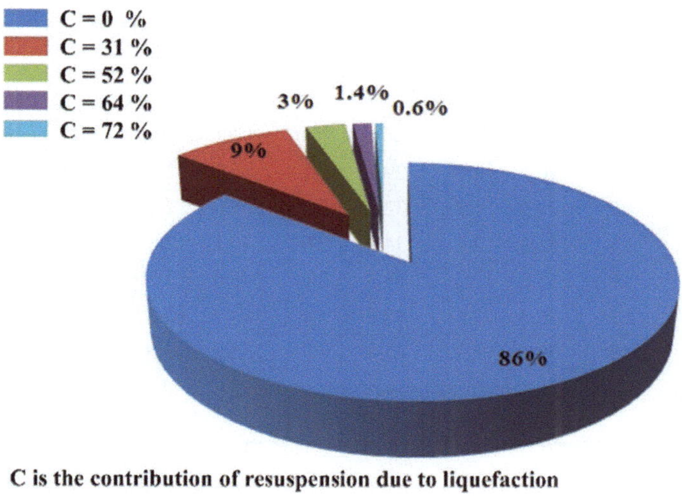

Fig. 8.21 Contribution of liquefaction to resuspension in different sea conditions in Chengdao sea area (color) and its cumulative contribution frequency throughout the year (pie chart)

8.4.3 Erosion Mass and Source Under Different Wind Conditions

Although the calculation results in the previous section show that the period with significant seabed liquefaction contribution is not the normal condition, the amount of erosion caused in a short period of extreme condition is also worthy of further investigation. In order to ensure the safety of marine structures and healthy evolution of shoreline, coastal engineers and scholars often use theoretical calculations to predict the occurrence and intensity of geological disasters in different wind scales or different wave reoccurrence periods. In this section, a similar approach is employed to calculate the proportion resuspended sediments in the study area under 6, 4, 8, 9, and 10 gales and 10, 25, 50 years wave reoccurrence periods. Its impact on the long-term evolution of the delta beach was also analyzed.

Wave parameters under different gales in the Chengdao sea area is shown in Table 8.8. 6 gale refers to the wind with speed larger than 10.8 m/s while 8 gale refers to wind with speed larger than 17.2 m/s. Yang et al. (1993) pointed out that 5–10 m water depth in the Chengdao sea area is the main source of eroded sediments, where the suspended sediment concentration is the largest. Therefore, the calculation in this section is for the water depth range of 5–10 m.

Calculation results show that the contribution of liquefaction to resuspension in the breaking zone of the Chengdao sea area in windy weather with 6 gales or larger can exceed 50%, and the contribution gradually increases toward the shore (Fig. 8.22). Wave energy dissipation in the breaking zone is the strongest, and thus wave-induced seabed liquefaction will add a large amount of "liquefaction resuspension" to the

Table 8.8 Wave parameters caused by different wind levels in the Chengdao sea area

Water depth	Wind level									
	6 gale		7 gale		8 gale		9 gale		10 gale	
	H/m	L/m	H/m	L/m	H/m	L/m	H/m	L/m	H/m	L/m
5	1.5	26.1	1.7	29.7	2.1	32.1	2.3	35.3	2.5	38.1
6	1.7	29.1	1.9	32.5	2.4	35.0	2.6	37.7	2.9	40.5
7	1.8	31.2	2.1	34.7	2.6	40.1	2.8	43.0	3.3	46.0
8	2.0	35.0	2.3	39.3	2.9	45.6	3.2	48.6	3.8	52.5
9	2.2	36.3	2.5	42.6	3.1	49.3	3.5	52.5	3.9	55.4
10	2.3	39.1	2.7	45.4	3.3	55.2	3.6	60.33	4.2	62.1

H is the wave height and L is the wavelength

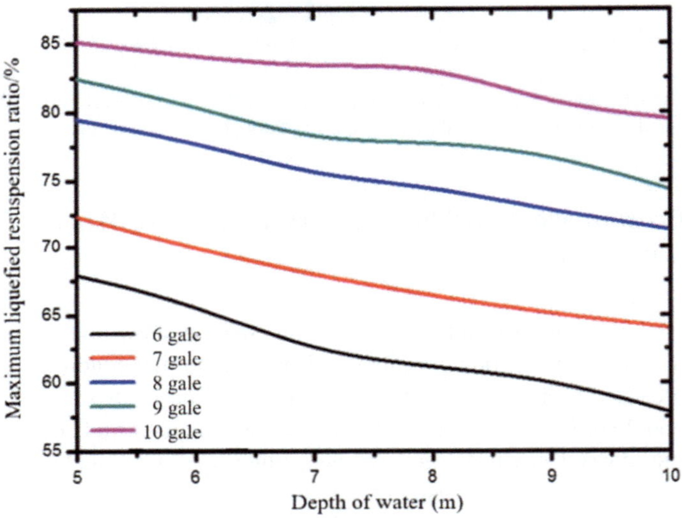

Fig. 8.22 Maximum contribution of liquefaction to resuspension in the wave breaking zone of Chengdao sea area under different wind levels

commonly-known "shear erosion". This supports a previous opinion that collapse pits (Prior et al. 1989) and disturbed strata have close relationship with the wave-induced seabed liquefaction.

It also needs to be noted that the contribution of liquefaction to resuspension calculated here is the maximum estimation, the specific contribution can be different under different gales. To estimate the erosion mass from different sources, it is also necessary to count the cumulative frequency of different gales in a certain period of time. There are two statistical methods for estimating the frequency of wind scale: one is counting windy days per month or year. Table 8.9 shows the number of days with gales larger than 6 and 8 in each month in the Chengdao sea area (Yang et al. 1993). The other is to count the cumulative appearance frequency of gales (Table 8.10).

This frequency is based on the statistics of hourly wind data, which can be used to calculate the cumulative duration of continuous action of different wind levels in a year.

It can be seen from Table 8.9 that there are 6 gale or more in each month of the year. Since the days with the appearance of 6 gales or more is called a windy day. Windy day occurred in November and April for 14.5 and 14.2 days respectively. In March, July and September, the occurrence frequency of strong winds were the lowest, only 8.6, 8.0 and 9.3 days, respectively, and the number of strong wind days in other months was more than 10.

However, the number of windy days can only roughly summarize the occurrence of strong winds. To calculate the duration of the action of different wind levels, it is necessary to further calculate the appearance frequencies of high wind (Table 8.10).

Statistical results show that the occurrence frequency of 6 gale or higher in November, January, and February is relatively higher to 14.6%, 11.4%, and 12.0%, respectively, in June, July, and August the frequencies are lower to 2.6%, 0.6%, and 0.6%, respectively. It indicates that the strong wind in autumn and winter has a long blowing duration, and the cumulative wind period in summer is shorter. The frequency of high winds above 6 gale is 7.11%, i.e., about 26 days; the frequency of high winds above 8 gale is 0.36%, i.e., the cumulative time is about 1.3 days (included in the 26 days above).

To estimate the total erosion and resuspension mass in the Chengdao sea under different gales, it is necessary to measure the corresponding suspended sediment concentration under different gales. The observed wind speed results in Sect. 8.3 have shown that there were roughly six large resuspension events during the observation period, and the corresponding SSC is about 3 g/l. Considering the background SSC is about 1 g/l, the resuspended SSC is estimated as about 2 g/l. This is consistent with the report of Yang et al. (1993): suspended sediment concentration will increase significantly under the wind and wave sea conditions, reaching 1–2 g/l on average, and the strong wind in winter will make the concentration of the entire water body well mixed. It is worth noting that 8 gale also occurred in this observation, but the corresponding suspended sediment concentration did not increase too much. The reason is possibly that the flocculation sedimentation and resuspension of fine sediments reached a dynamic equilibrium. Based on the background information above, in this section we estimated the mass of eroded and resuspended (T_n) sediments in the Chengdao sea under calm sea conditions and wavy sea conditions:

$$T_n = S\int_0^{t_n}\int_0^{d} E_n dh dt \tag{8.24}$$

where S is the area of Chendao sea; t_n ($n \leq 2$) is the duration of each sea condition, where t_1 is the duration of the wavy sea conditions (over 6 level wind) in a year, and t_2 is the duration of the calm sea conditions. According to the statistics above, the frequency of windy sea conditions in the Chengdao sea is 7.11% (i.e., $t_1 = 26$ days, $t_2 = 339$ days); $d = 7.5$ m is the averaged water depth in the wave breaking zone (main source area of resuspension); E_n is the erosion rate that is calculated from the

Table 8.9 Number of 6 and 8 gales in the study area (Yang et al. 1993)

Month	1	2	3	4	5	6	7	8	9	10	11	12	Summation
6 gale	10.8	12.4	8.6	14.2	12.3	10.8	8.0	12.3	9.3	12.5	14.5	11.8	137.5
8 gale	2.1	1.8	0.8	1.6	1.0	0.7	0.7	1.0	1.0	2.3	2.0	0.7	15.7

Table 8.10 Frequencies of 6 and 8 gales in the study area (Yang et al. 1993)

Month	1	2	3	4	5	6	7	8	9	10	11	12	Ave.
6 gale	11.38	11.96	4.2	7.07	6	2.61	0.57	0.57	5.66	9.87	14.58	7.44	7.11
8 gale	1.83	0.47	0.18	0.26	0.26	0.12	0	0.29	0.23	0.49	0.73	0.34	0.36

Ave. is the averaged frequency of occurrence (%)

Table 8.11 Total amount of yearly resuspended sediments in the Chengdao sea area

E_r	Frequency(%)	t(d)	t(s)	Water depth(m)	S(m^2)	T_n (kg)	T_n (t)
1.75 E-04	7.47	26	2.24 E-06	7.5	3 E-08	1.21 E-11	1.21 E-8
1.00 E-05	92.53	339	2.93 E-07	7.5	3 E-08	0.88 E-11	0.88 E-8
						2.09 E-11	2.09 E-8

The Chengdao sea area accounts for about 1/8 of the modern Yellow River underwater delta (~300 km^2)

in situ observation data, here is 1.75 E-04 on average during high winds and 1.10 E-05 kg m^{-2} s^{-1} on average during calm sea conditions.

The total mass of eroded or resuspended sediments (TSM) in the Chengdao sea area can be estimated by the Eq. 8.25. Calculation parameters and results are summarized in Table 8.11:

$$TSM = T_1 + T_2 = S \cdot t_n \sum_{1}^{2} E_n \tag{8.25}$$

Theoretical calculation results show that the annual sediment erosion mass in the Chengdao sea is about 209 million tons. In order to more intuitively understand this erosion status, in this section, we further compared it with the total sediment discharge from the Yellow River into the sea over the recent years. As shown in Table 8.12, when the Yellow River flows into the sea in the study area from 1967 to 1976, the averaged sediment discharge of the Yellow River is 1.026 billion tons/year, which is larger than the annual erosion mass of the Chengdao sea (about 209 million tons). That is, during that period, the amount of sediment into the study area was larger than that of the eroded one, so the delta moved outward to the sea to form a new delta lobe, which is consistent with the historical facts. However, after the diversion of the Yellow River in 1976, the source of sediments into the study area was cut off, and therefore the fact is that the delta lobe was exposed to strong erosion since then, which is also consistent with the historical facts.

Therefore, to maintain the stability of the beach in the study area in a natural way, at least 200 million tons of sediments are needed to be transported to the Chengdao sea area each year. However, in recent years, due to the industrial and agricultural

Table 8.12 Annual sediment transport capacity of the Yellow River at Lijin Station (100 million tons)

Year	1967	1968	1969	1970	1971	1972	1973	1974	1975	1976
TSM	20.86	13.20	5.81	10.87	9.19	4.07	11.97	5.03	12.61	8.98
Year	1977	1978	1979	1980	1981	1982	1983	1984	1985	1986
TSM	9.48	10.23	7.32	3.07	11.51	5.42	10.24	9.33	7.56	1.69
Year	1987	1988	1989	1990	1991	1992	1993	1994	1995	1996
TSM	0.96	5.98	4.69	2.48	4.72	4.20	7.07	5.69	4.38	0.96
Year	1997	1998	1999	2000	2001	2002	2003	2004	2005	2006
TSM	0.16	3.76	1.87	0.22	0.20	0.54	3.77	2.70	1.91	1.49
Year	2007	2008	2009	2010	2011	2012	2013	2014	2015	2016
TSM	1.47	0.77	0.56	1.67	0.93	1.83	1.73	0.30	0.31	0.11

Statistics from the Yellow River Conservancy Commission of the Ministry of Water Resources, Yellow River Sediment Bulletin http://www.yellowriver.gov.cn/

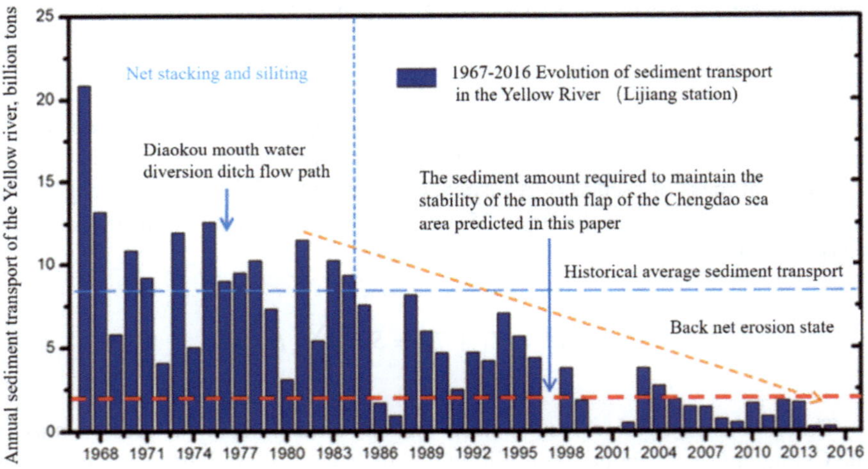

Fig. 8.23 Interannual evolution of sediment discharge at the Yellow River Lijin Station since 1967

water use, construction of dams and the improvement of soil and water conservation in the middle Yellow River area, sediment discharge has been gradually reduced (Yang et al. 2006). The annual sediment discharge after 2014 is only 0.1–0.3 billion tons (Fig. 8.23). Whether the sharply reduced sediment discharge could maintain the siltation status of the present estuary is a matter of great concern.

We believe that the sediment supply required for other lobes in the Yellow River Delta to maintain shoreline stability can also be estimated by the model established in this book, but it is necessary to combine in situ observation data from each lobe. Because the hydrodynamics, the sediment types, consolidation state and the topog-

Table 8.13 Wave heights in the case of 10, 25, and 50 years recurrence period (Yang et al. 1993)

D		3	4	5	6	7	8	9	10	14
$H_{1/3}$	10	1.8	2.3	3.0	3.6	4.1	3.9	4.0	4.1	4.3
	25	1.8	2.3	3.0	3.6	4.1	4.1	4.2	4.3	4.7
	50	1.8	2.3	3.0	3.6	4.1	4.6	4.5	4.6	4.9
H/D		0.6	0.575	0.6	0.6	0.586				

During the observation period, the water depth (D) is 5–7.5 m and the average water depth is 6.326 m

raphy of the underwater delta are different in each lobe, so the mass of eroded and resuspended sediment also should be different.

8.4.4 *Erosion Mass and Source Under Different Wave Recurrence Periods*

Finally, we tried to predict the liquefaction resuspension of sediments in the study area under different wave recurrence period of 10, 25, and 50 years (equivalent to 10 gales) and the contribution ratio with the modified erosion model constructed in this book (Table 8.13).

Calculation results show that the liquefaction resuspension ratio can reach 84%, 86%, and 90%, respectively, under the wave recurrence period of 10 years, 25 years, and 50 years. That is, when extreme storm wave loads act on the seabed in the study area, almost all the resuspended sediments are derived from seabed liquefaction, which completely beyond the prediction ability of classic theory which based on the understanding of traditional shear erosion. If the storm waves with recurrence period of 10, 25, and 50 years continued to act for 24 h, about 12 million tons of sediments will be liquefied and resuspended, which is equivalent to the total amount of sediment discharge from the Yellow River in 2016 (Table 8.12). According to the observation results in the study area: about 5% of the local resuspended sediments would settle in situ, while the remaining 95% would be transported to other places, so the revised mass of liquefaction erosion was about 0.11 billion tons. That is, if one extreme storm wave event lasts for one day, it could erode the total sediment discharge from the Yellow River in 2016.

The modified erosion model constructed in this book strongly supported a long-time hanging point of view: sediment transport caused by a storm event and the topography change it caused, could be much more intense than that induced over months or even years under the normal sea condition, and therefore tend to be the major cause of coastal hazards and engineering facilities disasters.

8.5 Summary

This chapter is the summarization and application of the research knowledge of the whole book. First, a series of controlled erosion experiments were conducted in a newly developed annular flume to test the erodibility of Yellow River silts that are in various fluidization degrees. The influence of seabed fluidization on sediment erodibility was thus successfully parameterized into erosion coefficient of the unified linear erosion formulation, and a modified erosion model which considers the influence of wave-induced seabed fluidization was constructed. Second, the advantages of the liquefaction erosion model are verified by comparing the calculated results of the model with the in situ measured data. Finally, the modified liquefaction erosion model was employed to predict the amount and source of the eroded sediments in the Chengdao sea area under different wave conditions, wind magnitudes and wave reoccurrence periods. It is suggested that an extreme storm event in the study area lasts for 24 h, it can cause 0.11–00.12 billion tons of sediment to be liquefied and resuspended, which should be equivalent to the total amount of sediment from the Yellow River into the sea in 2016. Theoretical analysis and calculation results in this book strongly support a viewpoint that has been difficult to be proved directly through field observations: sediment transport caused by a storm is much more drastic than the accumulated changes of months or even years under normal sea conditions, and therefore could be the major cause of coastal hazards and engineering facilities disasters.

References

Amos CL, Daborn GR, Christian HA et al (1992) In situ erosion measurements on fine-grained sediments from the Bay of Fundy. Mar Geol 108(2):175–196

Baldock TE, Holmes P (1999) Seepage effects on sediment transport by waves and currents. Coast Eng 1998:3601–3614

Bedford KW, Lee J (1994) Near-bottom sediment response to combined wave–current conditions, Mobile Bay, Gulf of Mexico. J Geophys Res Oceans 99:16161–16177

Cao D, Chiew YM (2014) Suction effects on sediment transport in closed-conduit flows. J Hydraul Eng 140(5):04014008

Cao D, Chiew YM, Yang SQ (2015) Injection effects on sediment transport in closed-conduit flows. Acta Geophys 64(1):125–148

Carstens T, Brebner A, Kamphuis JW (1976) Seabed mobility under vertical pressure gradients. In: Proceedings of behaviour of offshore structures (Boss' 76), pp 423–438

Clukey EC, Kulhawy FH, Liu PLF et al (1985) The impact of wave loads and pore-water pressure generation on initiation of sediment transport. Geo-Mar Lett 5(3):177–183

Dawson TH (1983) Offshore structural engineering. Prentice Hall Inc., Old Tappan, NJ

Dade WB, Nowell ARM, Jumars PA (1992) Predicting erosion resistance of muds. Mar Geol 105(1–4):285–297

Drake DE, Cacchione DA (1986) Field observations of bed shear stress and sediment resuspension on continental shelves. Alaska California Continent Shelf Res 6(3):415–429

Dyer KR (1980) Velocity profiles over a rippled bed and the threshold of movement of sand. Estuar Coast Mar Sci 10(2):181–199

Foda MA, Huang C M (2001) Fluidization model for cross-shore sediment transport by water waves. Ocean Wave Measurem Anal 1356–1365

Kuo YY, Chiu YF (1994) Transfer function between the wave height and wave pressure for progressive waves. Coast Eng 23(1–2):81–93

Lambrechts J, Humphrey C, McKinna L et al (2010) Importance of wave-induced bed liquefaction in the fine sediment budget of Cleveland Bay, Great Barrier Reef. Estuar Coast Shelf Sci 89(2):154–162

Le Hir P, Monbet Y, Orvain F (2007) Sediment erodability in sediment transport modelling: Can we account for biota effects? Cont Shelf Res 27(8):1116–1142

Li G, Zhuang K, Wei H (2000) Sedimentation in the Yellow River delta. Part III. Seabed erosion and diapirism in the abandoned subaqueous delta lobe. Marine Geol 168(1):129–144

Liu S (2006) Variations in the water and sediment fluxes entering the sea and their relation to the evolution of the Yellow River Delta. East China Normal University

Lu Y, Chiew YM, Cheng NS (2008) Review of seepage effects on turbulent open-channel flow and sediment entrainment. J Hydraul Res 46(4):476–488

Maa JPY, Lee CH (1997) Variation of the resuspension coefficients in the lower Chesapeake Bay. J Coast Res Special Issue 25:63–74

Mei CC, Fan SJ, Jin KR (1997) Resuspension and transport of fine sediments by waves. J Geophys Res Oceans 102(C7):15807–15821

Neumeier U, Ferrarin C, Amos CL et al (2008) Sedtrans 05: an improved sediment-transport model for continental shelves and coastal waters with a new algorithm for cohesive sediments. Comput Geosci 34(10):1223–1242

Nielsen P (1997) Coastal groundwater dynamics. Proc Coast Dynam'97, Plymouth. ASCE, pp 546–555

Nielsen P, Robert S, Møller-Christiansen B et al (2001) Infiltration effects on sediment mobility under waves. Coast Eng 42(2):105–114

Paphitis D, Collins MB (2005) Sediment resuspension events within the (microtidal) coastal waters of Thermaikos Gulf, northern Greece. Cont Shelf Res 25(19):2350–2365

Parchure TM, Mehta AJ (1985) Erosion of soft cohesive sediment deposits. J Hydraul Eng 111(10):1308–1326

Partheniades E (1962) A study of erosion and deposition of cohesive soils in salt water. University of California, Berkeley

Piedra-Cueva I, Mory M (2001) Erosion of a deposited layer of cohesive sediment. Coastal Estuarine Fine Sediment Proc 3:41–52

Prior DB, Suhayda JN, Lu NZ et al (1989) Storm wave reactivation of a submarine land slide. Nature 341(6237):47–50

Prior DB, Yang ZS, Bornhold BD et al (1986) Active slope failure, sediment collapse, and silt flows on the modern subaqueous Huanghe (Yellow River) delta. Geo-Mar Lett 6(2):85–95

Qiao L, Zhong Y, Wang N et al (2016) Seasonal transportation and deposition of the suspended sediments in the Bohai Sea and Yellow Sea and the related mechanisms. Ocean Dyn 66(5):751–766

Rodriguez HN, Mehta AJ (2000) Longshore transport of fine-grained sediment. Cont Shelf Res 20:1419–1432

Saffer DM, Silver EA, Fisher AT et al (2000) Inferred pore pressures at the Costa Rica subduction zone: implications for dewatering processes. Earth Planet Sci Lett 177(3):193–207

Sanford LP, Maa JPY (2001) A unified erosion formulation for fine sediments. Mar Geol 179(1–2):9–23

Soulsby RL (1983) The bottom boundary layer of shelf seas. 35:189–266 Elsevier

Stone M, Krishnappan BG, Emelko MB (2008) The effect of bed age and shear stress on the particle morphology of eroded cohesive river sediment in an annular flume. Water Res 42(15):4179–4187

Sumer B M, Fredsøe J (2002) The mechanics of scour in the marine environment. World Scientif 552

Thompson CE, Couceiro F, Fones GR et al (2011) In situ flume measurements of resuspension in the North Sea. Estuar Coast Shelf Sci 94(1):77–88

Tsai CH, Huang MC, Young FJ (2005) On the recovery of surface wave by pressure transfer function. Ocean Eng 32:1247–1259
van Rijn LC (1989) Sediment transport by waves and currents. Delft Hydraul
Wang H, Liu H, Zhang M (2014) Pore pressure response of seabed in standing waves and its mechanism. Coast Eng 91:213–219
Wang YH (2003) The intertidal erosion rate of cohesive sediment: a case study from Long Island sound. Estuar Coast Shelf Sci 56:891–896
Wolanski E, Spagnol S (2003) Dynamics of the turbidity maximum in king sound, tropical Western Australia. Estuar Coast Shelf Sci 56:877–890
Wright LD, Wiseman WJ, Bornhold BD et al (1988) Marine dispersal and deposition of Yellow River silts by gravity-driven underflows. Nature 332(6165):629–632
Yang Z, Wang T et al (1993) Marine environment of the Chengdao oil field exploration and exploitation. Ocean University of Qingdao Press, Qingdao
Yang ZS, Wang HJ, Saito Y et al (2006) Dam impacts on the Changjiang (Yangtze) River sediment discharge to the sea: the past 55 years and after the three gorges dam. Water Resour Res 42(4)
Zhang M, Yu G, La Rovere A et al (2017) Erodibility of fluidized cohesive sediments in unidirectional open flows. Ocean Eng 130:523–530

MIX
Papier aus verantwortungsvollen Quellen
Paper from responsible sources
FSC® C105338

If you have any concerns about our products,
you can contact us on
ProductSafety@springernature.com

In case Publisher is established outside the EU,
the EU authorized representative is:
Springer Nature Customer Service Center GmbH
Europaplatz 3, 69115 Heidelberg, Germany

Printed by Libri Plureos GmbH
in Hamburg, Germany